Contemporary Models for Pro-Poor Housing in Urban Sub-Saharan Africa

Lovemore Chipungu ·
Hangwelani Hope Magidimisha-Chipungu ·
Joseph Kamuzhanje
Editors

Contemporary Models for Pro-Poor Housing in Urban Sub-Saharan Africa

Springer

Editors
Lovemore Chipungu
University of KwaZulu-Natal
Durban, South Africa

Hangwelani Hope Magidimisha-Chipungu
University of KwaZulu-Natal
Durban, South Africa

Joseph Kamuzhanje
Cooper Zimbabwe
Harare, Zimbabwe

ISBN 978-3-032-05488-3 ISBN 978-3-032-05489-0 (eBook)
https://doi.org/10.1007/978-3-032-05489-0

This Springer imprint is published by the registered company Springer Nature Switzerland AG
The registered company address is: Gewerbestrasse 11, 6330 Cham, Switzerland

Foreword

Housing is a basic necessity for the survival of all living organisms as it provides the life support systems required for the liveability of human settlements. It is an important element of the urban landscape that determines the wellbeing of human beings. The importance of housing is underscored by Article 25 (1) of the United Nations Declaration of Human Right, which recognizes housing as a basic right that must be provided for all citizens to ensure good life and security. The lack, or inadequacy of housing is associated with a myriad of societal ills such as homelessness, health burdens, socio-economic deprivations, poverty, exposure to violence, and despair.

Despite the importance of housing as a basic necessity, housing for the urban poor is both a universal and perennial problem especially in the developing countries. According to the 2024 Report of UN-Habitat, 2.8 billion people around the world are affected by the housing problem. Many African countries south of the Sahara are experiencing housing crisis as a result of both quantitative and qualitative shortfall in housing. The precarious housing situation in these African countries has been compounded by high and unabated rate of urbanization. Available evidence shows that Africa has the fastest growing population in the world. The population, which was estimated to be 1.2 billion people in 2015, is now predicted to hit the 2.5 billion mark by 2050. This high rate of urbanization and rapid population growth have severe housing implications in the Sub-Saharan Africa. As more people moved into the cities, demand for housing is on the increase and the available units become grossly inadequate for the teeming population. Hence, in 2024, the UN-Habitat estimated that Africa south of Sahara currently needs over 50 million housing units to accommodate its growing urban population.

The housing problem in Africa is worse for the urban poor who have no foothold in the city. The search and ultimate acquisition of housing among the poor is not born out of choice, but out of need. Hence this lack of choice has robbed them of any opportunity to choose where to stay other than in slums and informal settlements, which are easily accessible from financial and locational perspectives. The housing option available for the urban poor is characterized by low quality buildings, inadequate infrastructure and squalid environment. Hence, the narrative of housing for the urban poor in Sub-Saharan Africa features a chronology of informality, illegality, poor housing condition, lack of basic services, unemployment, poverty, crime, and social stigmatization.

Urban authorities across Africa have instituted and implemented a wide range of housing policies and programmes to address the housing problem of the urban poor with varying degrees of success. The reality on the ground shows that governments are investing a lot to address the plight of the urban poor. However, a huge housing gap still exist as approximately 85% of African urban dwellers lack secured access to formal finances and do not have the financial capacity to develop their own homes. As it were, there is a dare need to address the ramifications of housing problem in Africa. The Africa Urban Forum held in September, 2024, recognized the need for affordable housing in Africa, and the forum is committed to knowledge development and sharing activities on affordable and resilient housing delivery. This commitment makes the publication of the book titled: *Contemporary Models for Pro-Poor Housing in Urban Sub-Saharan Africa* timely and apt.

This book is of great importance and relevance to the current nature and diversity of housing issues in Sub-Saharan Africa. The papers addressed pro-poor housing issues across many African countries. The various chapters address contemporary topics on pro-poor housing solutions designed to provide adequate, affordable, and safe housing for the poor by promoting low-cost housing, improving access to finance, basic infrastructure and services, and ensuring inclusivity and accessibility. The papers on housing land and construction examined issues of low-cost and affordable housing scheme provision through diverse models and housing options applied in different Sub-Saharan African countries within the context of land administration and low-income housing provision. Other relevant subjects addressed in the book include the important roles of social movements, cooperatives and self-help groups in expanding housing opportunities for the urban poor; appropriate/alternative housing finance system and inclusivity; as well as gender based pro-poor housing provision.

The value of all the chapters in this book lie in their robust descriptive and empirical approaches. The research problems were adequately investigated, research findings were generated and relevant solutions were recommended for policy action and implementation. The outcomes of the papers in this book provide rich experiences from diverse backgrounds in different countries. These experiences can therefore act as best practices for improving pro-poor housing policies and programmes across Africa. I sincerely commend the resourcefulness of the contributors to this book and the efforts of Prof. Lovemore Chipungu and members of his editorial team for putting this masterpiece together. I therefore recommend that the book be given wide publicity and circulation for the benefit of mankind.

Prof. A. M. Junaid
Professor of Urban and Regional
Planning (Housing Option)
Federal University of Technology
Minna
Minna, Nigeria

Acknowledgements

The journey of compiling this book has been a collaborative endeavour; one that could not have been realized by any single individual. It is a testament to the power of collective intellect, dedicated support systems, and unwavering commitment from all stakeholders involved.

Our story begins in 2022, with the "Pan African City Symposium: Yesterday, Today, and Tomorrow" hosted by the School of the Built Environment and Development Studies at the University of KwaZulu-Natal. This symposium was born out of a shared vision among diverse stakeholders eager to examine contemporary urban developments across Africa. Notable contributions from the Institute of Human Settlements Practitioners South Africa (IHSP) and the US-Africa Collaboration were pivotal in ensuring the success of this gathering. It was during this symposium that the seed for this publication was planted—centred on the compelling papers presented within the Housing Track. Although we have incorporated additional works beyond those presented at the symposium, we remain deeply grateful for this platform that fostered rich ideas, fruitful discussions, and innovative insights.

We extend our heartfelt appreciation to the University of KwaZulu-Natal, whose conducive environment and academic excellence provided the foundation for this project. The university serves not only as a world-class academic institution but also as a vital conduit for research, nurturing knowledge that seeks to impact society positively. The significant support from the School of the Built Environment and Development Studies, complemented by essential institutional resources such as ICT infrastructure, has been instrumental in bringing this project to fruition. Our collective track record of research and publication continues to gain recognition globally; an achievement made possible through this esteemed institution.

Special recognition goes to the SARChI-Chair for Inclusive Cities, whose unwavering support, both during the symposium and throughout the book's production, has been invaluable. The financial backing from the National Research Foundation (NRF) facilitated critical human resources for research, writing, and editing, ensuring that this work met the highest standards.

Finally, we are profoundly grateful to the authors whose contributions form the heart of this publication. Their dedication, patience, and commitment to revising their chapters in line with publisher guidelines ensured that the final product reflects quality and excellence. This arduous yet rewarding process would not have been possible without their unwavering support.

To everyone who played a role in this journey, from conception to completion; thank you. Your collective efforts have made this book possible, and we hope it will serve as a meaningful contribution to the discourse on urban development in Africa.

Contents

About the Editors

Lovemore Chipungu holds a Ph.D. in Town and Regional Planning, a Master's Degree in Rural and Urban Planning as well as a Bachelor of Science Degree (Honours) in Rural and Urban Planning. He is an active member of the Zimbabwe Institute of Regional and Urban Planners as well as a Corporate Member of the South African Planning Institute. He is currently an Associate Professor in the School of the Built Environment and Development Studies, at the University of KwaZulu-Natal. Professor Chipungu is both a practitioner and academic. In addition, he is an editorial board member of various journals and the Managing Editor of the Journal of Inclusive Cities and the Built Environment. He has published widely in peer-reviewed journals and books; and his areas of interest are housing policy, land-use planning and urban design. He is currently a co-co-principal investigator on a project on inclusive cities in Southern Africa.

Hangwelani Hope Magidimisha-Chipungu is a professor in town and regional planning at the University of KwaZulu-Natal. Her record of service spans but not limited to the South African Research Chairs Initiative (SARChI) for Inclusive Cities, National Planning Commission (RSA), Alternate Chairperson Scientific for the South African Council for Planners (SACPLAN), Committee Member for the International Society of City and Regional Planners (ISOCARP), City Planning Commission for eThekwini Municipality, advisory committee of Cooperative Governance and Traditional Affairs (COGTA-KZN), member of KwaZulu-Natal Tribunal and up to late a Board member of SACPLAN, Researcher at the Human Sciences Research Council (HSRC) etc.

Joseph Kamuzhanje has over 30 years of experience in the development sector. He worked for 15 years in the Department of Physical Planning in the Ministry of Local Government, where he rose to be the Chief Planning Officer for Matebeleland South Province between 2002 and 2007. He holds a B.Sc. in Rural and Urban Planning, and M.Sc. in Regional Development Planning and Management and he is a Professor of Practice in Regional and Urban Planning. He has worked in Government, civil society (both humanitarian and development), and in the private sector. He has contributed to several books and papers on rural, urban, and regional development. He is currently working with the University of Venda (South Africa), University of KwaZulu Natal (South Africa) and the Federal University of Technology, Minna (Nigeria), on various publications for books and journals.

Chapter 1
An Overview of Housing and the Urban Poor South of the Sahara

Joseph Kamuzhanje

Abstract The provision of urban land and housing for low-income households has been documented at various levels and in various contexts by many researchers and commentators. There is general consensus that there is a housing crisis among low-income groups in urban areas in Less Developed Countries (LDCs) where the housing issue has defied all efforts to bring it to manageable levels. It would seem that these problems are here to stay unless radical measures are adopted to tame them. The United Nations (UN) observes that the housing problems among low-income people are not a new phenomenon. It notes that since the advent of cities, humanity has been living in inadequately serviced and overcrowded urban societies. It is in light of these continued challenges that this research study contributes to ongoing debates by focusing on the development of urban land and housing delivery systems in contemporary cities south of the Sahara.

Keywords Pro-poor housing · Affordability and access · Sustainability · Urbanisation

1.1 Introduction

Access to affordable and adequate housing is a pressing issue across sub-Saharan Africa, where rapid urbanisation, poverty, and inadequate infrastructure have contributed to a growing housing crisis. Millions of people in the region live in informal settlements or substandard housing, lacking basic services such as clean water, sanitation, and secure tenure. Pro-poor housing, which prioritises the needs of low-income households, is increasingly recognised as a critical solution to this challenge. This book explores the concept and practice of pro-poor housing across sub-Saharan Africa, examining the strategies, policies, and innovations that are transforming housing for the poorest segments of the population.

J. Kamuzhanje (✉)
Cooper Zimbabwe (1992) Private Limited, Msasa, Harare, Zimbabwe
e-mail: josephkamuzhanje1@gmail.com

L. Chipungu et al. (eds.), *Contemporary Models for Pro-Poor Housing in Urban Sub-Saharan Africa*, https://doi.org/10.1007/978-3-032-05489-0_1

The housing crisis in sub-Saharan Africa is compounded by a range of factors, including population growth, limited government resources, and market failures that make housing unaffordable for the majority. According to the United Nations, the region's urban population is projected to double by 2050, with 60% of urban dwellers currently living in informal settlements (UN-Habitat, 2020). The scale of the problem demands urgent attention and creative solutions. Pro-poor housing initiatives, which focus on affordability, community participation, and sustainable development, offer a pathway towards improving living conditions for millions of people while addressing broader goals of social equity and economic development.

Pro-poor housing is essential for addressing housing inequalities and promoting social inclusion, yet it faces significant challenges due to inadequate policy frameworks and limited financial resources (Mutisya & Yarime, 2011). Governments across sub-Saharan Africa have struggled to meet the housing demand, with many policies favouring middle and upper-income populations, leaving the poorest segments underserved (Marx et al., 2013). Furthermore, market failures and the lack of affordable housing finance options compound the problem, making it nearly impossible for low-income families to access decent housing (Buckley & Kalarickal, 2005).

Sustainable Development Goal 11 seeks to "make cities and human settlements inclusive, safe, resilient and sustainable". However, globally, more than 1 billion people live in slums. This is 12.5% of the world population. Of these people, and out of Africa's population of 1.2 billion, 218 million people live is squalor conditions.

Just to illustrate the point, East and Southern Africa, and West and Central Africa have a housing backlog of close to 50 million units. An overview of the housing backlog in selected African countries as shown below (see Table 1.1) substantiates the challenges African countries south of the Sahara are experiencing.

While the crisis will increase in its scope and magnitude going forward due to poorly performing micro- and macro-economic situations, the lack of mortgages and other reasonable housing finance products and the graduation of the youths into adulthood, and therefore needing their own houses, work has been going on across Southern Africa.

This book draws on case studies from across the region, highlighting innovative housing models, from slum upgrading projects in Kenya to community-led housing initiatives in South Africa. It also critically examines the role of governments, non-governmental organisations (NGOs), and private sector actors in promoting housing solutions that are accessible to the poor. By analysing both successful initiatives and ongoing challenges, this book aims to provide a comprehensive understanding of the state of pro-poor housing in sub-Saharan Africa and to offer practical insights for policymakers, urban planners, and development practitioners. Innovative approaches, such as slum upgrading and community-driven housing initiatives, have shown promise in several countries. In Kenya, slum upgrading projects have improved living conditions in informal settlements, offering a model for other nations in the region (Huchzermeyer, 2011). South Africa's state-subsidised housing program is another example, having delivered millions of homes to low-income families since the end of apartheid (Tissington, 2011). However, these initiatives are often constrained by

Table 1.1 Comparison of population and housing backlog in some African countries

Country	Total population (millions)	Housing backlog (millions)
Angola	35	1.9
Kenya	51	2
South Africa	46	2.3
Zambia	20	1.5
Zimbabwe	15	1.25
Rwanda	14	0.9
Uganda	47	1.6
Nigeria	228	17
Ghana	32	1.7
Cameroon	27	1.2
Togo	8	0.25
Mali	24	0.4
Cote d'Voire	32	0.6

Source Author's Construct (September 2024)

limited funding and land tenure issues, which complicate efforts to scale up pro-poor housing solutions (Napier, 2005).

Through this exploration, the book emphasises the importance of an inclusive, people-centred approach to housing development, one that not only addresses the immediate need for shelter but also integrates housing into the broader framework of sustainable development, economic empowerment, and social justice (Gilbert, 2016). As sub-Saharan Africa continues to urbanise, the need for scalable, affordable housing solutions that cater to the most vulnerable populations becomes increasingly urgent. Access to affordable and adequate housing is a critical issue across sub-Saharan Africa, where rapid urbanisation, widespread poverty, and infrastructural deficits exacerbate housing shortages. In fact, over 60% of the urban population in the region currently resides in informal settlements, lacking basic services such as clean water and sanitation (UN-Habitat, 2020). The growing demand for housing, driven by population growth and migration to urban centres, has outstripped supply, leaving millions without proper shelter (Collier & Venables, 2015). This has led to a surge in the interest in pro-poor housing approaches, which aim to provide sustainable, affordable housing solutions for low-income communities (Gilbert, 2016).

The concept of pro-poor housing is not just about providing shelter; it is about creating environments that promote social inclusion, economic stability, and environmental sustainability. It involves a range of strategies, from slum upgrading and affordable housing schemes to community-driven housing initiatives and public–private partnerships that empower low-income populations and improve their quality of life. However, despite its critical importance, pro-poor housing remains underfunded, poorly integrated into broader development strategies, and often neglected in policy frameworks. The intersection of housing and sustainable development is also critical. Without adequate housing, progress towards achieving Sustainable Development Goal 11, which focuses on making cities inclusive, safe, resilient, and sustainable, is severely hindered (Satterthwaite, 2016). Climate change further complicates the situation, as many informal settlements are located in environmentally vulnerable areas, increasing the risk of displacement due to floods, droughts, and other climate-related disasters (Dodman et al., 2017).

This book aims to shed light on the pressing need for pro-poor housing and explore the innovative approaches being adopted by governments, NGOs, and communities to address this issue. Through a series of case studies, policy analysis, and expert insights, it will highlight the successes, challenges, and lessons learned in various pro-poor housing initiatives around the world. In doing so, this book seeks to address these challenges by exploring the various pro-poor housing strategies across sub-Saharan Africa. Through a series of case studies and policy analyses, it further examines the successes and limitations of existing initiatives and proposes actionable solutions for expanding affordable housing in the region. By focusing on the intersection of housing, poverty alleviation, and sustainable development, this book aims to contribute to the ongoing discourse on how to provide dignified, sustainable housing for all in sub-Saharan Africa (Gilbert, 2016; UN-Habitat, 2020).

The chapter authors have put forward compelling arguments on the need to reimagine the ways of tackling the housing crisis, especially as it affects the poor, vulnerable, and marginalised urban communities.

Adamu et al. argue that one of the global urban sustainability hitches of housing for internally displaced persons in the Third World countries had been revealed by unsustain urbanisation, violence insecurities, institutional lapses and environmental changes. The chapter seeks to examine the housing challenges for internally displaced persons using Bonaberi, a cosmopolitan urban area of Douala as a case study. Adamu concludes that there is the need for low-cost and affordable housing schemes for emerging internally displaced households, services optimisation, extension, investment and employment in additions to capacity building, empowerment, local participation, advocacies, and social inclusion for better housing standards. He also advocates for effective policy implementation, technology, coordination and collaboration which could enhance the overall living standards of IDPs.

Yankson puts forward the notion of contrapuntal urbanism which refers to how individual factors work together to create a coherent (but not necessarily unitary) urban system which may be in a state of flux. With specific reference to urban housing for the poor, various factors interact to configure the efficiency of its delivery. The situation is prevalent within the Southern African region in postcolonial societies such as South Africa and Namibia. This chapter seeks to unpack the implications of interconnected factors for housing delivery for the poor in the informal settlements of Namibia. The chapter proposes practical solutions and policy recommendations for the provision of affordable housing in informal settlements.

Nyadzani and Chipungu argue that South Africa is one of the most unequal societies in the world, and the government has a pivotal role in equalising the imbalance and unjust socio-economic conditions exposed to the marginalised and disadvantaged communities. The establishment of the social housing programme promised a change in the state of housing provision and an outlook towards spatial transformation for urban centres still staggered by the legacy of apartheid spatial planning and land dispossessions. The spatial transformation envisioned was to be anchored by restructuring zones entrusted with the determination of suitable areas for social housing projects. The chapter analyses the factors impeding the ability of social housing programme in fulfilling a core government mandate of reversing the spatial imbalances in urban areas. The chapter advances the argument that the social housing restructuring zone guidelines must be reconfigured to ensure compulsory alignment municipal spatial planning plans. It further proposes an increase in the development of inclusionary housing to counter the impact of limited suitable land and costly land assembly process while also sharing social responsibility with the private sector.

Chatiza argues that the degree to which the urban poor's struggles have yielded policy and practical changes in Zimbabwe's housing sector is little understood let alone acknowledged adequately. Messaging about their struggles has overwhelmingly emphasised how they have experienced repeated demolitions and evictions. This focus has overshadowed their efforts towards assisting the Zimbabwean state to tame its penchant to steer elite ordering of urban spaces. Through dialoguing with and exposing state officials to alternative approaches inside and outside Zimbabwe, urban social movements have aided considerable urban transformation through mutual state–society capacitation. These successes have had limited reach resulting in relevant lessons not being available to comparable African urban jurisdictions. His chapter draws on external evaluations of the work of Dialogue on Shelter Trust, Zimbabwe and the Centre for Community Initiatives, Tanzania. Both non-governmental organisations technically support movements of the urban poor in the two countries. The movements are affiliated to Slum Dwellers International through which they access good practices, solidarity for their efforts, and get empowered to act locally. Chatiza concludes that housing social movements have made important advances towards realisation of the right to adequate housing from which other poor urban Africans and those governing them can learn for sustainably and inclusively scaling low-income housing.

Ogunsanya puts forward the argument that the affordable housing crisis is a pressing issue facing many communities around the world. The high cost of housing, particularly in urban areas, has left a significant portion of the population unable to afford suitable living arrangements. Traditional models of affordable housing for low-income families have faced numerous challenges, including high operating costs, inadequate infrastructure, stigmatisation, and a lack of flexibility to meet the changing needs of residents. Smart affordable housing refers to residential developments that incorporate technology and design elements to reduce the cost of living for residents. Ogusanya proposes covering new and innovative technical advances in the building industry in production of affordable housing systems made up of case studies predominantly from Africa and various countries around the world. The intention is to contribute to the knowledge within the housing sector about how alternate models of smart technology can be implemented to improve the quality of life for low-income families, reduced housing costs and maintenance, enhanced security and privacy and provide better housing flexibility and adaptability to changing needs of its users.

Mangara tackles quite a pertinent issue in housing provisions. He takes a deep dive into the challenges facing prospective home owners as they struggle to access reasonable finance products in South Africa. The housing finance system in South Africa has mainly been inadequate in addressing the country's housing challenges. A shortage of affordable housing, high levels of debt and financial stress among homeowners, and a lack of access to housing for many characterise the housing finance system. These challenges are a legacy of South Africa's historical inequalities, economic constraints, and a lack of political will to provide low-income housing, especially for the urban poor. Mangara concludes that the current housing finance system is inadequate with providing accessible, affordable housing for all South Africans. The government's housing subsidies and grants have often been insufficient or poorly targeted, leading to a lack of meaningful impact on the overall housing situation. Therefore, this chapter urges the South African and other governments to improve the effectiveness, efficiency, and accessibility of the existing housing finance systems, create and devise innovative housing finance mechanisms and instruments, and promote equal and affordable access to housing finance for all people.

Lovemore Chipungu and Hangwelani Magidimisha-Chipungu focus on housing finance. They argue that finance is the most critical factor which affects housing production. It cuts across the housing production process, i.e. from the time a piece of land is acquired to the time individual housing units are sold. It is also a key factor in the refurbishment of old housing stock. Unfortunately, evidence from around the world shows that financial institutions are more willing to give housing loans to high and middle income than to the low-income groups. The argument behind such a scenario is that low-income households have limited capacity to service their loans. In fact, they are viewed as a liability and financial institutions do not want to risk their resources. This chapter revisits the issue of housing finance by analysing it through the window of low-income housing that emerged in post-2000 in Harare, Zimbabwe.

Myeni argues that the creation of sustainable low-income human settlements is a challenging task in the South African housing provision practice. This is because sustainable human settlements can only be achieved if the principle of integrated development informs most human settlements projects especially in the post-apartheid South Africa. This may include the use of space in relation to the provision of basic service, strategic location of settlements in relations to access and availability of employment opportunities, the consideration of social and environmental aspects of the settlements and the provision of sport and recreational facilities. The non-provision of basic services has led to several challenges especially for you youth. His chapter emphasises the provision of sport and recreational facilities as forming part of basic needs for any community that intends to pursue the goal of sustainable development and argues that the provision of sporting facilities must be thought out as early as when settlements are conceptualised so that such facilities form part of project implementation.

Darazo and Kawu make a case for public housing in order to address the pertinent housing issues especially in urban areas. Urban residents, particularly government employees, are known to place high value on living in public housing and always welcome any process that can help achieve this goal. In pursuing this objective, Bauchi State Government, in North-eastern Nigeria, have been embarking on public housing provision for over four decades through different strategies. From the beginning, the mass housing programmes have had modifications that inculcated participatory project execution, ownership transfer agreements, and payment arrangements; however, in spite spanning through different shades of leaderships and democratic eras, products of the scheme still lack the expected public acceptance. The authors propose that detailed assessments of the local social and environmental peculiarities are conducted before any wholesale adoption of a particular strategy.

Medayese et al. focuses on the challenges of infrastructure provision and delivery which have been a major point of discourse across the global planning space. According to the authors, African cities have also grappled with these infrastructure challenges which ranges from transport infrastructure, healthcare, and housing infrastructure, among others. Countries like Nigeria in West Africa currently has a housing deficit of about 28 million housing units with respect to the over 200 million population. This housing infrastructure challenge is however more system in terms of the causative issues than physical. Most of the statutory agencies of government that are saddled with the responsibility of providing housing and housing infrastructure have oftentimes proposed mass housing programmes which are public private sector driven but which eventually fail as a result of failures on either the part of government to meet its own obligations or sharp practices on the part of the private sector to maximise profit. The chapter identified poor coordination, absence of plans and poor political will on the part of government as the major factors limiting infrastructure provision, while, unearthing the challenge of sharp practices, poor monitoring and implementation and resistance to agreed memorandums of understanding (MOU) as the related issues and concerns for the private sector operators in meeting their obligations of planning infrastructure provision and especially housing.

Lehobo et al. seem to suggest that housing problems can be reduced with more innovative approaches. His argument is that while traditionally housing has been provided through zoning from high to low density, it is possible to be inclusionary and adopt a mixed approach. Housing is undoubtedly a huge problem for the urban poor, particularly in developing countries. Sub-Saharan Africa bears the brunt of this housing challenge. This chapter proposes inclusionary housing and mixed-use development to include poor people in housing and development and respond to the need to redress spatial inequalities and imbalances.

For Chipungu and Magidimisha-Chipungu, the fast-track land reform programme in Zimbabwe (in 2000) provided a platform for popular movement of people to access land in and around cities legally and illegally. This mass movement, analysed from an urban perspective, signified a determined occurrence by the general populace to shrug off decades of denial to the city and its associated benefits. The chapter analyses the acquisition of land and the eventual emergence of housing on such land. These houses, while accepted by beneficiaries, were scorned and criticised by others who saw the informalisation of urban space under the guise of social justice. At the centre of these dynamics were tenure issues which, despite the government's determination to protect beneficiaries in the newly acquired space, still left a cloud of uncertainty. This was not surprising given that successive Zimbabwean governments have failed to embrace popular modes of housing production in the past. Informal housing schemes have always been treated with apathy and indifference—hence whenever they mushroomed, they were ruthlessly destroyed. What could have changed the scenario in 2000? In response to this question, this paper analysed emerging housing schemes in Harare and noted that the tenure status of beneficiaries was varied with "chance" and "determination" being the most determinant among most of them. In a bid to unravel the structural determinants of these developments, the study's main focus was in the period 2000–2009. The author employed both qualitative and quantitative research methods using mapping, interviews and observations in order to collect data. The selected housing schemes of Nehanda Housing Scheme provided cross-cutting perspectives that enabled the extrapolation of key findings at both city and national levels. The findings emanating from this study show that the FTLRP provided a platform for innovativeness among key stakeholders and was fuelled by the unpronounced national urban vision and the prowess of the political leadership—thus guarantying low-income households' the right to the city despite prevailing challenges of infrastructure.

Adamu also argues that access to adequate housing is a fundamental human right of every citizen. Yet, this remains unrealised in the lives of over 300 million women in sub-Saharan Africa characterised by severe patriarchal cultural practice, unequal rights in legislative and policy frameworks and underlying systemic and structural gender inequalities norms due to acute poverty, violence, disabilities and migration status. Illustrating with the Buea, a cosmopolitan city in Cameroon, this chapter examines the challenges faced by women in acquiring adequate housing and the gaps in housing policies implementation practices. The study adopted an explanatory and exploratory design. Adamu calls for a rethink of housing policy and services through the promotion of equality, dignity, finances and worth for all while accommodating women's difference and participation in decision-making to ensure housing equality in sub-Saharan cities.

According to Le Roux, post-apartheid South Africa has seen a significant rise of urbanisation. According to official census figures a total of 53.7% of the South African population were living in urban areas in 1996 and increased to 62.9% by 2011. One of the consequences of this rapid urbanisation is the escalating demand for housing in urban areas. In response to the impact of government initiatives and the inadequacy of the housing market to provide sufficient affordable and sustainable housing, backyard dwellings manifest itself as an innovative response to the housing backlog. On a national scale the occurrence of backyard dwellings as a proportion of total housing has risen from 4.1% in 2001 to 4.9% in 2011. This represents an increase in absolute numbers from 459,528 to 712,956 (Stats SA 2001; Stats SA 2011), and according to Lemanski (2009) is increasing at a faster rate than informal settlements. In the City of Cape Town backyard dwellings has risen from 4.3% of total housing stock in 2001 to 7% in 2011, and in absolute numbers increased from 32,747 to 74,957. The occurrence of backyard dwellings at this scale has a definite impact on urban structure and poses a number of urban management challenges. The paper investigates the spatial trends and the socio-economic characteristics of backyard households in light of its possible impact on urban form and management within the City of Cape Town.

In summary, it must be admitted that urbanisation is rapidly increasing in sub-Sahara Africa, and this puts pressure on adequate and effective urban planning. The rapid growth of cities results in degradation of natural habitats, changes in species composition, cities micro-climate, energy flows, subsequently urban, unemployment, urban congestion, change in living conditions and an increase in living conditions (utilities and basic services). This book analyses shifts and continuities in prevailing housing issues, tracing their emergence from within the democratic movements including labour and community or civic organisation, and the private sector. Most intervention measures seem to be ignoring the needs of the poorest sector in society. Further, the inadequately integrated location of subsidised development for the poorest remains unchanged. The perception of local government, whether facilitatory or as an implementer, seems to be struggling to address local realities due to dynamic factors that underscore the housing question.

References

Buckley, R. M., & Kalarickal, J. (2005). *Housing policy in developing countries: Conjectures and refutations.* World Bank Publications.

Collier, P., & Venables, A. J. (2015). *Housing and urbanization in Africa: Unleashing a formal market process.* Oxford University Press.

Dodman, D., Leck, H., Rusca, M., & Colenbrander, S. (2017). African urbanization and urbanism: Implications for risk accumulation and reduction. *International Journal of Disaster Risk Reduction, 26*, 7–15.

Gilbert, A. (2016). *Rental housing: Lessons from international experience and policies for emerging markets.* World Bank Publications.

Huchzermeyer, M. (2011). *Cities with 'slums': From informal settlement eradication to a right to the city in Africa.* University of Cape Town Press.

Lemanski, C. (2009). Augmented informality: South Africa's backyard dwellings as a by-product of formal housing policy. *Habitat International, 33*, 472–484.

Marx, C., Stoker, T., & Suri, D. (2013). The geography of informal work in developing countries: Implications for the global South. *Urban Studies, 50*(4), 745–759.

Mutisya, E., & Yarime, M. (2011). Understanding the grassroots dynamics of slums in Nairobi: The dilemma of Kibera informal settlements. *International Transaction Journal of Engineering, Management, & Applied Sciences & Technologies, 2*(2), 197–213.

Napier, M. (2005). Core housing, enablement and urban poverty: The consolidation paths of households living in two South African settlements. *Urban Forum, 16*(4), 317–336.

Satterthwaite, D. (2016). Missing the millennium development goal targets for water and sanitation in urban areas. *Environment and Urbanization, 28*(1), 99–118.

Tissington, K. (2011). *A resource guide to housing in South Africa 1994–2010: Legislation, policy, programmes, and practice.* Socio-Economic Rights Institute of South Africa.

UN-Habitat. (2020). *The state of African cities 2020: Africa's urbanization dynamics.* UN-Habitat.

Joseph Kamuzhanje has over 30 years of experience in the development sector. He worked for 15 years in the Department of Physical Planning in the Ministry of Local Government, where he rose to be the Chief Planning Officer for Matebeleland South Province between 2002 and 2007. He holds a B.Sc. in Rural and Urban Planning, and M.Sc. in Regional Development Planning and Management and he is a Professor of Practice in Regional and Urban Planning. He has worked in Government, civil society (both humanitarian and development), and in the private sector. He has contributed to several books and papers on rural, urban, and regional development. He is currently working with the University of Venda (South Africa), University of KwaZulu Natal (South Africa) and the Federal University of Technology, Minna (Nigeria), on various publications for books and journals.

Part I
Embracing Inequality in the Production Process

What emerges clear from housing among the poor is a narrative of inequality that manifests itself spatially, economically, socially and architecturally. While this seems to be a label that the urban poor cannot shrug off, it also emerges as a survival strategy which sustains their settlements. On one hand the poor urban households emerge as the most vulnerable as seen through the window of unrelenting displacement, yet on the other hand, they suffer discrimination as seen through gender inequality. Yet in the mist of these elements of inequality, emerges resilience which keeps them going to thus enabling them to create acceptable homes for their households. One attribute which stands out clearly among the urban poor is that they are vulnerable. This attribute sticks out as a label that hinders them from a number of opportunities among which is the fact that their voice is muffled and no one takes them seriously. Municipal officials do not seem to recognise them since they do not own property (such as land) on which they build their houses. Hence within the urban space, they seem not to exist on municipal maps—thus making it easy to move or displace them. It is this lack of recognition which makes it easy to overlook or simply displace them. The situation is even made more complex when inequality further manifests itself along gender lines—with women emerging as the least recognised in the housing production system. Thus, inequality paralyses the whole housing production process since it denies the urban poor access to opportunities with women bearing the worst brunt.

Chapter 2
Housing Facilities Challenges for Internally Displaced Households in Bonaberi-Douala, Cameroon

Baba Adamu, Balgah Sounders Nguh, Nformi Beatrice Maluh, Humphrey Ndl Ngala, and Toumba Oumarou

Abstract One of the global urban sustainability hitches of housing for internally displaced persons in the Third World countries had been revealed by unsustain urbanization, violence insecurities, institutional lapses and environmental changes. This paper seeks to examine the housing facilities challenges for internally displaced persons using Bonaberi, a cosmopolitan urban area of Douala. A sample of 281 households was used to represent the entire population of Bonaberi. Stratified random sampling technique was used to map out the appropriate location of internally displaced households. Results were analysed using Statistical Package for Social Science Standard (SPSS version 23.0) and presented in the form of tables, figures and plates. Results revealed diverse demographic and socio-economic profile amongst internally displaced households in Bonaberi. Despite their diverse housing features, conditions and services patterns, unsustain urbanization, urban planning deficiencies and coordination continue to prevail resulting to acute overcrowding, limited accessibility and services deliveries, pollution and disease proliferations. Although, diverse stakeholders' alternative housing facilities strategies exist for emerging internally displaced persons, there were ineffective due to weak policies, language barriers, stigmatization among others. Therefore, there is the need for low-cost and affordable housing schemes for emerging internally displaced households, services optimization, extension, investment and employment in additions to capacity building, empowerment, local participation, advocacies and social inclusion for better housing

B. Adamu (✉) · B. S. Nguh · T. Oumarou
Department of Geography, Faculty of Social and Management Sciences, University of Buea, Buea, Cameroon
e-mail: baba.adamu@ubuea.cm

N. B. Maluh
Department of Town Planning, National Advanced School of Public Works, Yaoundé, Yaoundé, Cameroon

H. N. Ngala
Department of Geography, Higher Teachers Training College, University of Yaoundé, Yaoundé, Cameroon

L. Chipungu et al. (eds.), *Contemporary Models for Pro-Poor Housing in Urban Sub-Saharan Africa*, https://doi.org/10.1007/978-3-032-05489-0_2

standards. Also, ensuring effective policy implementations, technology, coordination and collaboration could enhance the overall living standards of IDPs and host communities in Bonaberi and other Third World Cities.

Keywords Housing facilities · Internally displacement · Challenges · Livelihood enhancements

2.1 Introduction

Accounting for more than 70% of the land use in most cities, housing determines urban form and densities, providing employment and contributes to growth (UN HABITAT, 2016). Housing is one of the most basic human needs indicators for living standards, economic development and inclusiveness in cities (EC, 2005; The World Bank, 2015). Its needs in cities have never been so important in the past than today due to intense conflicts, poverty, climate shocks, natural disasters, terrorism, institutional and political changes (Gilbert, 2013; IDMC, 2013, 2019). Hence, cities are complex environments, and displacement to, from and within them is highly dynamic and complex (World Bank, 2015). One of the biggest issues of cities is to provide adequate and affordable housing to displaced people even in the developed world (UN Habitat, 2010; IOM, 2015). Conversely for developing countries, displaced persons are living, among whom, and within whose local jurisdictions are critical factors in their ability to support themselves, their families with basic social services and also contribute to local development (Nunez-Ferrera et al., 2020).

Today, approximately 40 million people fleeing insecurity (Cantor & Cardona-Fox, 2021; UNHCR, 2017) and violence have been living in protracted situation of displacement in city slums and squatter settlements (Stein & Lord, 2011). An additional 24 million are driven by disasters and draughts while an estimated 15 million people are displaced by development projects (UN, 2018) every year. More than half of all displaced persons are under the age of 18 (UNHCR, 2017). In Latin America for instance, over 83% of the population live in cities (Warn & Adamo, 2014) while 70% of people who had been internally displaced by conflict and violence in Iraq were living in cities (Florczyk et al., 2019). This phenomenon is acute in Brazil and Venezuela with considerable internal displacement over the past half Century with Colombia currently suffering from one of the world's largest internally displaced crises in the world (IOM, 2015). Similarly, episodes and chorionic forms of violence-induced displacement and housing in Central American and Mexico cities are provoked by state and gang violence (Muggah, 2017). Africa is home to more than a Third of the World IDPs with over 18 million in sub-Saharan Africa (SSA) alone in 2019 just from 5 million in the past decades mostly living in poverty and unemployment in cities (UNHCR, 2019). Hence, like many developing cities housing shortage in SSA is causing an upsurge in price and demand of housing units, social services, sewage and insecurity resulting to increased socioeconomic and health risks.

Available literature revealed that, irrespective of the reasons for displacement, rapid urban growth, urban sprawl and poor city planning always leaves a negative socio-economic footprint on millions of people through slums development, increase pressure, high demand for housing, high land value, social facilities and housing, crimes, traffic, pollution and food shortage (Michael et al., 2014). Currently, UN-Habitat (2019) noted that over 55% of the world's population live in urban areas, a proportion that is expected to increase to 68% by 2050. Africa urban growth rate in this context is almost 11 times greater than that of Europe and as such negative of conflicts, disasters and spatial expansion of urban settlement through the invasion and reclassification of rural areas continue to prevails (UN HABITAT, 2016). Thus, adding to the increasing pressure on natural resources and services as over 80% of displaced person now live in cities. Davies (2012) noted that in 2010, out of the 14.7 million displaced persons protected and assisted by UNHCR, an estimated 52% lives outside the camp and half of the 54 countries monitored by IDMC (2019) have few or no camps thus amplifying pressure on host communities service deliveries in slums (Alan, 2000; UN, 2021).

Even though, stakeholders, municipal authorities and governments agencies have recently scaled up short-term delivery aids including housing facilities to IDPs through humanitarian social development actions to reduce household vulnerabilities, long-term solutions to housing and social services continue to remains elusive for most Third World Cities (IDMC, 2018; Kimengsi et al., 2015) due to acute financial cost, insecurity limited facilities, high demands, investment and coordination (OCHA, 2021). Currently, UNDP (2022) noted an unprecedented decline in housing indicators in SSA cities due to limited planning, droughts, insecurity and conflicts. There was the need for the full implementation the UN (2030) especially goal 11 on making cities and human settlements inclusive, safe, resilient and sustainable. The goal addresses rapid pace of urbanization and it associated challenges like slums, pollution and inadequate infrastructure. Despite this, Durosaro and Ajiboye (2011) have noted acute IDPs sufferings from emotional stress, abuses, threats and assaults from host community members in their quest for inclusive, reliable and accessible housing. In many cases, the urgency, suitability and loss of displaced persons actions means housing situations are featured by acute over crowdedness, spacing, social cost, accessibility, employment, inadequate housing and deprivations from sharing already meagre resources (Alekseyenko et al., 2021). Similarly, high cost of connection, duration of humanitarian assistants, scarce developmental actions and bottlenecks in administrative procedures and implementation of city strategies, stigmatization and discrimination, and public participation continue to compound these efforts (Muggah, 2017; UNHCR, 2017). In this respect, involving urban IDPs and host communities in broader urban development processes by stakeholders, community engagement and inclusivity can help identify those priorities in housing and social service delivery, and housing development that can contribute most towards achieving solutions to urban displacement and sustainable urban development.

In 2019, more than 54% of Cameroonian live in cities where in decades, residents have enjoyed relative socio-economic and political stability compared to other African cities in the sub region (UN-HABITAT, 2019). This is however changing due to negative implications of rapid urbanization characterized by environmental and political changes. Although, internal moments have enhanced the urbanization process at destination areas of the Coastal Lowlands of Cameroon, Douala is now recently concerned with the phenomenon due to natural disasters and conflicts with concomitant impacts on housing facilities, limited suitable living environment, social services and livelihoods enhancement opportunities. Displaced persons are barely assisted by their host communities (MINEPAT, 2022). Significantly, unsustained urbanization continues to crammed IDPs into makeshift housing with very limited hygiene and sanitation conditions and waste mismanagement very prejudicial for their health and wellbeing. MINEPAT (2022) further noted that IDPs accounts for 51.9 and 53.6% of Douala and Yaoundé the two cities of millions in Cameroon. In Douala, high influx of displaced persons are mostly concentrated in Bonaberi (38,692 IDPs), with the hope of finding better socio-economic opportunities. In particular, income-generating activities amongst most of the youths who are displaced persons often come down to informal sector business or prostitution including girls. Despite stakeholders' attempt towards containing housing facilities challenges, residents in Bonaberi continue to report increased pressure on housing services, the precariousness of which may already be remarkable long before the arrival of displaced persons with households having to cope with acute services and facilities due to the failures of urban planning to meet the need of all people. The response seems deficient in views of the housing availability, crime, congestion and basic services.

Therefore, the need to increased investments, innovations, change in behaviour, good leadership and overall collective solutions to reduce housing facilities vulnerabilities for internally displaced families and households are becoming more eminent than ever before. This point was more emphasized as opined by the UN Under Secretary for Humanitarian Affairs "no one behind". The Secretary-General rightly identified addressing displacement was a critical ingredient towards achieving the SDGs 11 (UN, 2030) on "inclusive, safe, resilient and sustainable cities" while reducing the suffering of millions of people who are without access to housing or basic social services. In Cameroon, the UNHCR (2019) in support of the Presidential Humanitarian Peace Plan (HPC, 2023), noted that the adoption of inclusive urban planning policies for internally displaced persons and their host communities through the provision of access to adequate, safe and affordable housing and basic social services and upgrade slums at all levels of policies formulations and implementations was necessary to reduce housing vulnerabilities among the displaced and their host communities in urban areas. Despite the success, this initiative is still hampered by weak government policies, effective participation and collaboration efforts amongst stakeholders in Bonaberi. Thus, as targets towards achieving Cameroon National Development Strategy (NDS, 2030) on city sustainability, inclusion and reduce vulnerability and risks for marginalised urban communities including internally displaced families and households, this study sets to examined the housing facilities challenges for displaced persons in order to proffer better housing alternatives that is inclusive, affordable and safe, with effecient and improved accessibility, quality wellbeing and environmental protection in Bonaberi of Douala, Cameroon.

2.2 Study Area

Bonaberi is situated in the South east banks of the Wouri Estuary in the Douala IV Municipality of the Littoral Region of Cameroon. It is located on Latitude 4′ 03″N of the Equator and 9′ 41″E Longitude of Greenwich Meridian with an estimated population of 250,626 inhabitants (National Population Census, 2005). It is characterized by two main seasons: a rainy season (May to October), with an average total rainfall exceeding 3000 mm, and the dry season (November to April) with a yearly average temperature of 26 °C. It is found on the part of Douala basin experiencing severe flooding with large expand of informal settlement units sometime swamped beyond recognition (Asangwe, 2006).

2.3 Research Methods

This study made use of stratified random sampling technique to map out IDPs locations in Bonaberi. A total of seven neighbourhoods were randomly selected: namely, Ndobo, Mabanda, Grand Hanger, Sodiko and Boanmikano. Bikoko and Ngwele. A total of 281 questionnaires were administered to the male and female IDP household heads between the ages of 25 and 65 years old who have been displaced or fled to Bonaberi from diverse regions due to peaceful nature and cosmopolitan characteristics. This was done taken into considerations the population of the study sites. The study purposively focuses on people who were displaced due to natural and human challenges like violent conflict, natural disaster, development projects and others from the ten regions of Cameroon including Divisions of origins. The target population for the survey were IDPs residing in dwellings (independent, apartments, flats, informal settlements or slums) for over five years. Simple sampling technique was used to select IDPs households head who were picked for response of the questionnaires which were open and close-ended. This was used to collect the primary data for this study. In addition, photographs and field survey, empirical data were also vital for the study. This somehow helped in the review of literature and the discussion of results. Ethical considerations were obtained from the Douala City Council (DCC) and concern of the IDPs were sought before questionnaire administration. Data were analysed using the Statistical Package for Social Science (SPSS Version 23.0) and presented using tables, figures and plates.

2.4 Results and Discussions

2.4.1 Demographic and Socio-economic Profile of IDPs Households

Generally, results revealed that the proportion of female IDPs was higher 54.1% (152) than that of the male 45.9% (129) thus aligning somehow with the demographic distribution of the Littoral Region. In terms of age group of IDPs, significant majorities were young adults between the ages of 18–37 and 28–37 as represented by 38.1% of the respondents respectively. This was followed by respondents who were between the ages of 38–47 with a proportion of 15.5%, those above the age 48 whom represented only 7.1% while those with >18 represented just about 11.7% of the respondents. The age–sex structure of Bonaberi is somewhat modified by emergence internal displacement. Thus, with the growing proportion of youth including children, the government is faced with the challenge of meeting the needs of the youthful population. The search for better livelihood opportunities therefore explains why single, female and youth IDPs move into Bonaberi for hawking, house help engagements in cooking, cleaning, washing and providing general home care like laundry and gardening and babysitting, hair dressing, prostitution, home teaching of children, barbing and hair dressing saloon, educational, commercial and other informal activities. All these requires a lot of strength, time and distance to be covered to work or IDPs attending school. Displaced persons who are currently unemployed or underemployed also search for better opportunities in industrial setup within Bonaberi even though without adequate required skills. Based on marital status, a majority (39.1%) of IDPs were single, followed by 27.8% who were married, 19.9% were divorce and only 13.2% were widower(s).

In view of displaced persons' educational status, significant majority (27.5%) of the respondents had basic school certificates, followed by 24.6% who had secondary school certificates, 16% had high school education certificates while only 15.7% had University Degrees. Those with informal education were represented by 13.9% of the respondents while post-secondary or vocational education represented only 2.5% of the respondents. Based on the number of IDSs per room in Bonaberi, results revealed that a significant majority (46.3%) were between 7–9 per room, followed by those between 4–6 with a proportion of 26.4%, 18.8% for those between 1–3 and the least 8.6% for IDPs above 9 people living in a single room. These results signified a growing rate of overcrowding among the IDPs in Bonaberi in addition to congestion in other public spaces. For instance, the education structures hosting IDPs struggled to include the additional students with an average pupil per classroom ratio of up to 200 children and limited infrastructure. This is in line with the finding of Maluh et al. (2019) where 66.1% of population in the Bamenda City have room density of 4–6 persons indicating overcrowding. Household income determined the financial capacities of displaced persons monthly expenditure on social services and payment of rents for those who are not landlords. IDPs who earn <50,000 FCFA were the highest with a significant proportion (39.1%) of the respondents, followed by those who earn between 50,001–100,000 FCFA with a proportion of 31.7%, then those earn between 100,001–150,000 FCFA with a proportion of 22.4% of the respondents while respondents with income levels between 150,001–200,000 FCFA and >200,000 FCFA were the least with a proportion of only 4.6% and 2.1% of the respondents respectively. The proportion of IDPs input to the above income range per month vary significantly with the highest unrivalled proportion of about 84.4% for those who contributed <20% of their monthly income towards household needs.

Based on the levels of satisfaction with these aforementioned monthly IDPs contributions (see Table 2.1), a significant majority (48.8%) of respondents were not satisfied, 31.3% of the respondents were satisfied while a minority of the respondent (19.9%) were neutral. The reasons given ranges from inconsistencies in rent payment resulting to conflict with landlords, domestic violence and stress which lead to exposure and vulnerabilities of many IDPs. A significant majority of IDPs who live with host families are taken in free of charge or in exchange for very little financial contributions towards the upkeep of family needs. In terms of residential status, a significant majority (48.4%) were IDPs, followed by immigrant as represented by 35.6% and the least indigents with a proportion of only 16% of the respondents. Conversely, in terms of IDPs' occupational status, a significant majority (34.9%) were students, followed by 22.8% who were traders, 20.3% were semi-skilled workers, 11.7% were skilled workers while just 10.3% of the sampled respondents were farmers. Moreover, results revealed a significant majority (38.4%) of IDPs lives between 7 and 9 persons in the house, followed by 29.9% for those who live between 4–6 persons, 18.5% for those who live more than 9 persons in the house while the least (13.2%) for who live between 1 and 3 persons in house (see Table 2.1).

In terms of household members that are IDPs, a significant majority (41.6%) were between 1–3 persons followed by those who were between 4–6 persons with a proportion of 36.8%, those who were between 7 and 9 with a proportion of 12% and the least those who were 9+ with a proportion of only 8% of the respondents. Based on the duration of IDPs in the study area, a significant majority (35.2%) live more than 5 years, followed by those who live between 2 and 3 years with a proportion of 28.1%, then those who live between 4 and 5 years with a proportion of 24.9% and the least those who live less than 1 year with a proportion of only 11.7% of the respondents. Results from Table 2.2 presented a statistically significant association between the number of persons in the house and the number of IDPs. This means that, the proliferations of IDPs in Bonaberi have swilled the number of household members thus adding pressure on housing services and facilities due to high room densities and overcrowding.

2.4.1.1 Trends of IDPs Families

The rate of IDPs into Bonaberi was categorized as slow (63%), rapid (25%) and constant (28%) across the different months and years between 2016 and 2022 (Table 2.3). As for those who indicated the rapid, a significant majority (89%) noted push factors such as violence insecurity, impacts of developmental projects

Table 2.1 Demographic and socio-economic characteristics of IDPs households

Variable	Category	Frequency (n)	Percent (%)
Age group (Years)	<18	33	11.7
	18–27	107	38.1
	28–37	78	27.8
	38–47	43	15.3
	48–57	20	7.1
Sex	Female	152	54.1
	Male	129	45.9
Education level	Informal education	39	13.9
	Basic	77	27.4
	Secondary	69	24.6
	High school	45	16.0
	Post-secondary/ vocational	7	2.5
	University	44	15.7
Marital status	Single	101	39.1
	Married	78	27.8
	Divorced	37	13.2
	Widowed/widower	56	19.9
Residential status	Immigrant	100	35.6
	Indigent	45	16.0
	IDPs	136	48.4
Monthly income levels (FCFA)	<50,000	110	39.1
	51,000–1,000,000	89	31.7
	100,001–1,500,000	63	22.4
	1,500,001–200,000	13	4.6
	200,000- 250,000	6	2.1
IDPs contributions to household monthly income(%)	<20	237	84.4
	21–40	23	8.3
	41–60	11	3.9
	61–80	5	1.8
	80–100	5	1.8
Occupation	Farming	29	10.3
	Trading	64	22.8
	Semi-skilled workers	57	20.3
	Skilled workers	33	11.7
	Students	96	34.9

(continued)

Table 2.1 (continued)

Variable	Category	Frequency (n)	Percent (%)
Estimated number of persons in the house	1–3	37	13.2
	4–6	84	29.9
	7–9	108	38.4
	>9	52	18.5
Estimated number of household members that are IDPs	1–3	166	41.6
	4–6	64	36.8
	7–9	32	12
	9+	19	8
Duration of stay in Bonaberi (years)	<1	33	11.7
	2–3	79	28.1
	4–5	70	24.9
	>5	99	35.2
Satisfaction levels of IDPs contributions to household monthly income ranges	Satisfied	31.3	31.3
	Dissatisfied	48.8	48.8
	Neutral	19.9	19.9
Number of IDPs per room	7–9	130	46.2
	4–6	53	26.4
	1–3	74	18.8
	9+	24	8.6

Sources Fieldwork (2022)

Table 2.2 Association between number of persons in the house and number of IDPs

Spearman's rho	Number of persons in the house	Correlation coefficient	1.000
		Sig. (2-tailed)	–
		N	281
	Number of household members that are IDPs	Correlation coefficient	0.292[a]
		Sig. (2-tailed)	0.000
		N	281

[a] Correlation is significant at the 0.01 level (2-tailed)
Sources Fieldwork (2022)

and environmental changes to be the triggers. The population size of Douala has been increasing over the years due to the above factors resulting to housing facilities challenges for the IDPs who live in slums and sub-standard houses. For instance, the population of Douala has increased tremendously from 458,426 inhabitants in the 1976 to 809,852 inhabitants in 1987 given a growth rate of 5.3% and 1.931977 inhabitants in 2005 given a population growth rate of 4.7% and to more than 2 million inhabitants (NIS, 2005). The population is estimated to increase to 4 million inhabitants by the year 2025 and expected to reach 5, 965,650 inhabitants by the year 2035. Therefore, the city has witnessed a tremendous growth in its population over the years from 458,426 in 1976 to 809,852 in 1987 to 1,931 977 in 2005 given a population increase of 1,122,125 inhabitants between 1987 and 2005. Bonaberi has a population of about 250,626 inhabitants with an average growth rate of 4.85% (Cameroon – Third General Population and Housing Census, 2005). It receives about 50,000 immigrants every year predominantly the youths in search of educational opportunities, employment, safety nets and security from within and outside Cameroon. Hence, the number of IDPs has incessantly amplified from 2016 to 2022 due to violence crises, development projects and natural disasters in Cameroon (see Table 2.3).

From Table 2.3, Bonaberi is facing exponential demographic growth due to displaced persons, and this is likely to keep on extending in the future if no urban strategic plans are ellaborated and established. The Multi-Sector Needs Assessments (MSNAs) in August and Septembers 2021 noted an increase the number of IDPs from the North and South West in the Littoral Region has from 7000 to 8000 due to violent conflicts. Among those displaced were unaccompanied women and separated children thus adding to the figure of IDPs received in Bonaberi. According to IOM, in 2019, internal displacement continues to be very frequent and as of August same year, IOM estimates the number of IDPs in the Borno, Adamawa, Yobe region at 2,018,513 (of which at least 1.32 million IDPs were in Born State), showing a slight but definite upward trend.

Table 2.3 Evolution of IPDs in Bonaberi between 2016 and 2022

Month/Years	2016	2017	2018	2019	2020	2021	2022
Jan	123	2432	3943	4321	1443	3211	1003
Feb	45	2654	2901	3791	1143	1782	1062
Mar	2341	2821	1002	2753	1345	2653	1013
Apr	1321	1454	2111	2001	561	2301	345
May	943	1221	2543	1132	7543	1132	4543
June	2111	2943	1211	2291	2101	1161	2101
July	800	1002	1218	1593	1216	1004	902
Aug	1021	1234	1400	800	1243	650	1043
Sept	1894	1132	2314	901	321	901	321
Oct	4532	2913	3001	3211	1432	2111	1013
Nov	321	1232	1421	3650	2321	1250	521
Dec	345	900	792	1321	2342	1021	1140
Total	15,797	21,938	23,857	27,765	23,011	19,177	15,007

Sources OCHA MSRA (2021) and IDCM (2022)
Source Author (2022)

Significantly, armed conflict in the Far North Region and the socio-political (anglophone crisis) have led to a steady rise of displaced families and households into Bonaberi since 2014. For instance, in 2017 over 15,000 IDPs move out of the Far North Region and move into Bonaberi. The number of people affected by the crises in the North and South West Regions have dramatically increased in the second half of 2018 due to increasing violence making it one of the fastest growing displacement crises in Africa in 2018. It has since continued to grow displacing over 1.1 million people as of October 2020 with major destination Bonaberi. The trends are also increasing due to the influx of IDPs from Central African Republic (CAR) coming through the East, Adamawa and the North Regions of Cameroon with most of their final destination urban centres such as Bonaberi. This makes Bonaberi one of the fastest developing centres in Cameroon. Upon arrival displaced persons settle in slums, putting great pressure on the already saturated basic services including housing. Mabanda neighbourhood consists of about 24% of IDPs of the entire city of Douala. Results further revealed that the influx of IDPs is constantly growing, and continued violence crisis have by the end of 2019 alone, generated over 970,000 IDPs, some 240,000 of whom were newly displaced including those from Bonaberi. IDPs of various ethnic origins and regions were attracted by educational facilities, cheap land, job opportunities and the present of relatives. The Ministries of Basic and Secondary Educations also noted that almost 15,000 displaced children are attending basic education in schools in the Littoral region from the North West and South West Regions and most schools have either relocated or have open their branches in Bonaberi for easy accessibility. This trend will still continue in Bonaberi due to diverse opportunities, proximity and attractions. These corroborates the views of Fombe and Balgah (2012); Nunez-Ferrera et al. (2020) who pointed that rapid urbanization process in developing cities have offered diverse opportunities for work, access to healthcare, education and a place of hope and aspiration. Contrary to pull factors of displacement, influx IDPs in Bonaberi were linked to arrival experiences to natural disasters, violent crises and poverty. Regardless, urban sprawl remains a critical issue for neighbourhoods like Sodiko and Mabanda which are chaotically growing and characterized by insufficient housing and social services. Significantly, the period 2021 and 2022 witness a drastic decrease in the number of IDPs due to normalcy in areas of origin which ease the return of many IDPs.

2.4.1.2 Ethnicity and Flows of IDPs Families

The ethnicity of the IDPs in Bonaberi showed that 87% were none native and only 8.9% were native of the region from the Littoral Region (Mongo and Sanaga Maritime Divisions). There were attracted into Bonaberi due to floods, limited social amenities and land dispute around the airport area between the urban development authorities and the local communities which result to destruction of home and livelihoods. The 87% constituted IDPs from all regions of Cameroon. The South West represented 34.9% of the IDPs into Bonaberi specifically from Momo, Ndian, Manyu and Fako Divisions and the North West 33.1% specifically from Bui, Meme, Donga

Mantung, Ngokitunjah, Boyo, Menchum due to violence insecurity, natural disaster and land disputes. As of now one of the critical challenges facing the region is Fulani-headmens crisis and kidnappings for ransom. There were no significant differences in proportion between the former and the later. Nonetheless, the proportion of the IDPs from the South West dominates due to its closed proximity to Bonaberi in the Littoral Region. Most migrants or IDPs choose to move over short than long distances and are likely attracted by opportunities. The West (10.3%) of IDPs into Bonaberi were specifically from Menoua, Nde, Mifi and Noun Divisions due to floods and landslides, inter-communal disputes and traditional practices such as witchcraft. IDPs from the Far North (5%) specifically from Logone et Chari, Diamare, Mayo Kane Mayo Savo and Dany Divisions were displaced due to the Boko Haram Sec atrocities, draught, famine, floods, inter-communal disputes, among others. This region has experienced one of the most recurrent droughts (2009–2011) and floods (2010, 2012, 2014, 2019 and 2020) in the last few decades resulting to severe environmental degradation and loss of shelter, crops and fields. Significantly, the resurgent of inter-community violence around the Lake Chad basin between farmers and headers have led to forced displacement of people down South into Bonaberi. In an article entitled "*the shifting frontier of displacement in Latin America*", Muggah (2017) noted that over 7 million IDPs of which 6 million were from Colombia alone. IDPs from the North (4%) and the Adamawa (3%) specially from Faro, Mayo Ray, Mayo Loti and Benue Divisions were displaced due to high levels of theft and terror, natural disasters and inter-communal disputes. Conversely, the IDPs from the East (7%) were displaced into Bonaberi specifically from Lom et Djerim and Ocean Division due to war in the Central African Republic (CAR) which has been going on since 2010 until date. IDPs from the South (4%) and the Centre regions (4%) were displaced into Bonaberi specifically from Nfoundi, Mbeme et Kim and Haute Sanaga Division due to natural disasters and the impacts of developmental projects. Overall, there were overall high representation of IDPs into Bonaberi from the various regions and divisions of Cameroon. The arrival of IDPs in Bonaberi has been followed by an increase in housing and social services demands with severe risks for IDPs. This rapid demographic growth manifested through internal displacement has generated acute deprivation, poverty and inequalities.

2.4.1.3 Drivers of IDPs Concentration in Bonaberi

The factors for the displacement of respondents into Bonaberi were both anthropogenic and human in nature. They range from violence crises or conflicts, natural disasters, development projects or terrorism (see Table 2.4).

As the most significant factor of displacement of IDPs, respondents vary in their responses. Although, natural forces were found to contribute to enormous loss of lives and properties, the contribution of human forces makes the impacts very severe and devastating. From Table 2.4, respondents noted violence crises in the North West and South West Regions, the East Region, Adamawa and the Far North Regions of Cameroon. A significant majority of displacement were noted by respondents to be

Table 2.4 Causes of IDPs concentration in Bonaberi Neighbourhood

Reasons	Category	Frequency	Percent (%)
Development projects	Road, plantation, HEP generation and dam projects	38	13.4
Natural disasters/ environmental change	Drought, floods, famine and landslide, deforestation etc.	66	23.4
Violent crises	Communal, land grabs, ethnic, socio-economic and political	168	59.7
Terrorism	Theft, harassment, kidnappings for ransom etc.	10	3.5
	Total	281	100

Source Fieldwork (2022)

due to natural disasters like landslides, draughts and floods which affected lives and livelihoods opportunities. Also, the impact of development projects embarked upon by stakeholders and government at various levels have led to the repossession of land and valuable livelihood opportunities of people thereby contributing to displacement of IDPs. The Humanitarian Response Report (HPC) (2023) of the President of the Republic (MINEPAT, 2022) noted that the most people affected by violence crises especially in the East, Adamawa, North West and South West Regions are youths, children and women. They also noted that, the youths were highly at risk of unemployment, under employment, radicalization and arbitrary arrest warrants due to their displacement into Bonaberi for safety and opportunities.

In the Far North, the insurgency of Boko Haram compounded by chronic vulnerability and the going impact of climate change have causes over 1.2 million people in need of humanitarian assistants since 2014. OCHA (2022) noted that the region is the second most affected by Boko Haram insurgencies. In the Lake Chad region alone, over 52,700 persons were internally displaced Southwards with destination urban centres like Bonaberi. A study by Ginnetti et al. (2017) noted major causes of displacement including natural disasters (floods, volcanic eruptions, landslides and earthquakes); environmental issues (desertification, land degradation and global warming) and human-induced causes (accidents and radioactivity) in Asia and Latin America. A recent report by UNHCR (2019) noted Russian annexation of Crimea have recently resulted over 1,432,290 IDPs in need of public support as of December 2019 (UN, 2021). IDMC (2019), Researching International Displacement (2022) noted over 2.7 million IDPs in urban centres of Nigeria and Somalia due to violent insecurities and extended draughts with acute housing issues. Consequently, displaced persons would prefer to live in Bonaberi due to a greater availability of housing, job offers, cultural diversity and social connections.

2.4.2 Housing Characteristics, Conditions and Services for IDPs Households in Bonaberi

The housing conditions, characteristics and service provisions in Bonaberi were clear indications of IDPs wellbeing and standard of living. These were determined by their levels of household income, disposable income, expenditure, employment, impacts of humanitarian assistance or government affordable low-cost housing. Significantly, results from the study revealed the interrelationship between housing and basic social services like water supply, electricity, food, waste management, sanitation, accessibility and social security available for IDPs in Bonaberi.

2.4.2.1 Nature of Housing for IDPs

With regards to the characteristics and conditions of housing facilities for IPDs, results from the study classified houses in Bonaberi using four classifications namely, standards, sub-standards, semi standards and slums settlements based on UN-Habitat (2019) classifications for settlement types (see Table 2.5). Hence, majority of IDPs faced significant complexity of accessing adequate and efficient housing facilities and axiliary services. IDPs were often house in precarious settlements of dilapidated and deplorable conditions due to unsustainable income which makes it difficult for them to live in standard neighbourhoods therefore exposing them to various type of risks.

Table 2.5 revealed a significant majority (43.7%) and (31.7%) of IDPs lives in slums and sub-settlements unlike in semi standard and standards settlements usually reserved for high-income earners (see Plate 2.1). Low-land values in Bonaberi attracts most unemployed IDPs who buy plots to build homes of their own in these settlements which are liable to floods and forced evictions by the urban planning authorities of Douala. In Mabanda neighbourhood which is flood prone settlements with seasonal inundation, IDPs housing are likely to attracts health and socioeconomic issues.

Many reasons accrued for IDPs choices or decisions to leave in slums and sub-standards settlements including low land values (44.6%), businesses opportunities

Table 2.5 Nature of housing occupied by IDPs in Bonaberi

Nature of settlement	Frequency	Percent %
Standard	29	10.3
Semi standard	40	14.3
Sub-standard	89	31.7
Slum	123	43.7
Total	281	100

Source Fieldwork (2022)

Plate 2.1 Evidence of housing conditions and structures for IDPs in slums and sub-standard settlements of Bonaberi. NB: **a** overview of housing structure with dilapidated roof tops **b** a house constructed with cement inhabited by IDPs **c** a house constructed with Wood inhabited by IDPs. *Sources* Fieldwork (2022)

provided by markets and proximity to job (28.3%) respectively, present of relative (19.9%) and friend influence (7.3%). Business opportunities were strong site determinants in Bonaberi including proximity to job sites. It is normal for IDPs to choose these high accessibility areas despite the risk of eviction by urban planners. Mabongunje et al. (1978), Belfinger et al. (2019) identified residential areas based on the level of income of households and concluded that there are limited prospects for residents because the high-income dwellers get the first choice of houses and social services while the emerging low-income displaced families are obliged to seek poor structural quality of housing with acute services.

Significantly despite the dilapidating nature of housing for IDPs, a majority of IDPs who are not land lords in Bonaberi do not own their homes and therefore pays high rents for single rooms or apartments for their families. For instance, about 52% of respondent pays <15,000 FCFA, followed by 29.9% who pays between 15,000–25,000 FCFA while only 9.9% pays between 25,000–50,000 FCFA. Linda a 17 years Fruit vendor was noted saying "*am stranded with a six-month old child—living in a house without a roof because, I am unable to pay six-month rents demanded by my landlord who decided to take off the roof of his house so I could look for another house*". According to IOM (2015), only 11% of IDPs in Ukraine now live in their own homes, about 60% have been renting apartments for more than a year since the conflict stared, and 17% live with relatives or host families (Alekseyenko et al., 2021). Based on the types of building used, results revealed that a significant majority (63.3%) of IDPs lives in single rooms, 20% live in duplexes, 10% have multiple apartments while only 2% of IDPs lives in a storey building. Additionally, the highest housing material used by IDPs in Bonaberi was zinc (100%) due to it availability and less cost, followed by cement (93%), woods materials 56%, titles (20%) and uncemented floor (19%). Housing deteriorations characterized by wall and floor cracks, leaking roofs, broken windows continue to be a norm for displaced persons in neighbourhood like Mabanda due to limited renovations, late payment of rent or rent evasion. The use of temporal building materials (cardboard, old wood, plastic films), their non-compliance with planning regulations, and inadequately maintenance results to the growth of slums occupied by IDPs. This is in line with the findings of Fombe (2005), Ujuku and Mengue (2017), Fombe and Abotoko (2014) who revealed that more than 25% of houses in Douala and Buea are sub-standard, dilapidated and built with temporal or non-durable materials.

2.4.2.2 Housing Facilities for IDPs

Due to high congestion, the Bonaberi slums also revealed elements of deprivations and low accessibility, efficiency, availability, affordability, high costs and unreliability of social services delivery (see Table 2.6). Alan (2000) expressed this and noted that, residential areas for low-income earners in metropolitan peripheral areas of many SSA cities are generally known to have limited or no access to services, poor sanitation and are mostly informally developed settlements.

Table 2.6 revealed a significant majority (41.6%) of IDPs were strongly satisfied with the water supply as they sourced water from unimproved sources of wells, streams and boreholes with significant health complication due to contamination, 28.5% were satisfied due to access to centralized system offered by Cameroon Water Utility Company (CAMWATER), 12.5% were dissatisfied because they depend on vendors or trek long distances to source for water, 9.3% were neutral and only 8.2% were very dissatisfied. Overall, a significant majority (70.4%) of IDPs were dissatisfied with water supply in Bonaberi as against only 20.6% who were very satisfied

Table 2.6 Levels of satisfaction with housing facilities/services for IDPs

Housing facilities	Levels of satisfactions				
	Strongly satisfied	Satisfied	Dissatisfied	Strongly dissatisfied	Indifferent
Water supply	41.6% (117)	28.5% (80)	8.2% (23)	12.5% (35)	9.3% (26)
Electricity	22.1% (62)	47.7% (134)	12.5% (35)	8.9% (25)	8.9% (25)
Waste management	17.4% (49)	20.3% (57)	27.8% (78)	24.5% (68)	10% (29)
Sewage/ Sanitation	23.5% (63)	16.0% (41.5)	19.2% (19.2)	31.7% (409)	9.6% (29)
Accessibility	20.6% (58)	12.5% (35)	22.8% (64)	28.5% (80)	15.7% (44)
Security	21% (59)	22.8% (64)	24.6% (69)	23.1% (65)	8.5% (24)
MRA	25.0% (421)	30.0% (404)	18.6% (213)	16.1% (271)	10.4% (176)

Source Fieldwork (2022)

and satisfied. The growing number of IDPs and the inability of CAMWATER to provide adequate potable water supply production, distribution extension programs or rationing to slums and sub-standard settlements characterised by mismanagement have forced many IDPs to adopt alternative but doubtful sources resulting to high level of contaminated water sources (see Plate 2.2), such as boreholes (48%), wells (24%) and rivers/streams (17%) and rainfed water (12%). IDPs are uplarged to consume and use these water sources for their daily activities. Human health incidence of cholera, dysentery and typhoid are becoming very common among IDPs especially children and women. This findings confirmed the findings of Baba and Humphrey (2017) in Buea and Ako et al. (2009) in Bonaberi who revealed high rates of unimproved water sources, unreliability and accessibility. While young, displaced girls have to trek long distance to fetch water, queue at water sources, many of their families also purchase water at high costs thus negatively affecting productivity and exposure to severe health risks (Baba et al., 2020).

Based on the level of satisfaction with electricity supply situations for IDPs, a significant majority (47.7%) were dissatisfied, 22.1%, were very dissatisfied, 12.5% were very satisfied, and 8.9% were either neutral or satisfied. Overall, IDPs in Bonaberi were highly dissatisfied and dissatisfied with the electricity supply situations due to inaccessibility, unreliability, high cost, congestion, low voltage and poor rationing by the Cameroon Electricity Company (ENEO) with a proportion of 69.8% as against only 21.4% who were highly satisfied or satisfied with the company services to the IDPs and the host communities. These have slowed down the rate of socio-economic activities and damage household food supplies items kept in regurgitators. It has also intensified the rate of crime waves due to constant black out at night. This situation is more acute in settlements like Ngwelle and Ndobo where over 80% of IDPs were dissatisfied due to cables dilapidation, abandonment or mismanagement, high cost per unit consumption and illegal connection. IDPs only options in these settlements are to protect the exposed cable from touch by kids. Limited awareness on the dangers of cable exposure could result to fatal accident and death from shocks.

Plate 2.2 Evidence of alternative water sources and their location for IDPs. NB. Well water sources in Ndobo (**a**) and toilets located very close to well water sources in Sodiko (**b**). *Sources* Fieldwork (2022)

Based on the IDPs' level of satisfactions with waste disposal, a significant proportion (27%) and (20.6%) were very satisfied and satisfied with the Waste Management Company (HYSSACAM) and DCC waste management department in terms of area coverage for waste collection, personnel employed and accessibility. However, 20.3% of IDPs were dissatisfied due to delayed payment of officials and subsequent abandonment of waste mount and garbage littered on streets, public places, drains, gutters and poor ethics of some displaced residents despite the placement of waste collection cans or noticed board (see Plate 2.3). Social services were limited for IDPs due to massive slum development, poverty and migration unlike natural increase with severe risks. For instance, inaccessible of neighbourhoods to ensure any effective and efficient waste management coverage for IDPs in Ndobo and Sodiko neighbourhoods result to high incidences of dissatisfied (17.4%) and indifferent (13.9%) in respondents' views. The inadequate waste disposal in Bonaberi have thus increased the amount of waste thereby increasing pest and insect causing Malaria, Typhoid and Cholera incidences. This confirms the finding of Fombe (2005), Igwe et al. (2017) who noted correlation between poor housing conditions and human health pathologies of residents due to poor waste disposal, poverty and unemployment.

In terms of the IDPs' level of satisfactions with sewage and sanitation facilities, a significant proportion (31.7%) were dissatisfied, 23.5% were strongly satisfied, 19.2% were very satisfied, 16% were dissatisfied while only 9.6% were indifferent. Although, sanitation and sewage related services such as drainage and water facilities are strong determinant of human health adaptations to diseases in Bonaberi, sanitation and toilet services were conspicuously limited for many IDPs and their low-income host in settlements like Ndobo, Sodiko and Mabanda due to high cost, distance, low plain and accessibility (see Plate 2.4). Many IDPs practice open defecation especially in nearby streams and gutters, queue in toilet or connect their toilet sewage drains directly and into gutters. The aftermath is usually delay at work, unhygienic conditions, limited privacy and pollution. During the rainy season toilets are left open for runoff to flush sewage thus raising the rate of infections among IDPs.

Field observations from the study noted that, IDPs dwellings in Bonaberi slums were constructed very close to each other, in some cases >2 m away from each other. Some of the dwellings in Mabanda and Grand Hangar neighbourhoods for instance constitutes public path ways as most of the houses lack accessible roads

Plate 2.3 Poor wastes disposal by IDPs in Sodiko (**a**) Ndobo (**b**) Neighbourhoods. *Sources* Fieldwork (2022)

Plate 2.4 Poor sanitation situations for IDPs in Ndobo (**a**) Sodiko (**b**) and Mabanda (**c**) Neighbourhoods. *Source* Fieldwork (2022)

while others are constructed within marginal lands which are prone to floods and subsidence. While, a significant majority (28.5%) of IDPs were strongly satisfied with neighbourhood accessibility, 22.8% were very satisfied, 20.6% were very dissatisfied, 15.7% and only 12.5% were indifferent and dissatisfied due to limited evacuation of emergency health situation intense flooding of streets and buildings due to poor land use and non-respect of town planning norms (see Plate 2.5). This confirms the finding of IOM (2019; Bagui, 2009) who observed emerging IDPs residing in risky zones and makeshift houses below transition lines or along recently constructed highways faced diverse socioeconomic and health risks.

Housing challenges, large flows of displaced people into Bonaberi have constricted local economy and society, as pressure increases on housing and basic services continue to mount. It has led to the proliferation of slums, since demand for low-income housing outstrips supply. IDPs of informal settlements find themselves vulnerable to repeated displacement, through eviction, natural hazards or social unrest due to acute poverty, insecurity and unemployment as significant majority (24.6%) of respondents were dissatisfied, 23.1% were strongly dissatisfied, 22.8% were strongly satisfied, 21% were satisfied and only 8.5% were neutral. Unfortunately, due to high rate of joblessness and integration of social services, the rate of theft and crimes have intensified even though, the security situations across neighbourhoods did not varies significantly. IOM (2019), UN (2018) noted that crime and pollution are the most challenging issues of cities migrants given rise

Plate 2.5 Housing accessibility situations characterized by narrow and floods streets with poor drainage systems in Sodiko (**a**), Ndobo (**b**) and Sodiko (**c**). *Source* Fieldwork (2022)

to violence. Thus, high concentration of IDPs means high insecurity and illnesses in Bonaberi where urbanization changes have reduced households resilient and adaptation strategies.

2.4.2.3 Housing Facilities Challenges for IDPS in Bonaberi

Cities have been the destinations for most of the world's displaced persons, given their degree of socio-economic, cultural and educational importance. Taking the stocks of the challenges of housing facilities, impacts and opportunities are critical for sustainable urban development. Hence, the emerging internal displacement patterns in Bonaberi have led to diverse levels of satisfaction for IDPs ranging from increased overcrowding, social tension, limited privacy, pollution, reduce savings among others (see Table 2.7).

Table 2.7 revealed that due to limited accommodation, affordability and high costs of housing, a significant majority (51.6%) of respondents strongly agreed with overcrowding as the main housing challenge faced by IDPs, followed by 18.5% for those who agreed, 14.2% for those who disagreed and 11% for those who strongly disagree and only 4% for those who were indifferent. Over-crowding indeed is a very common housing problem among IDPs in Bonaberi due forced displacement. This have been a leading cause of environmental problems such as air and water pollution for many IDPs. As a result, the living condition among IDPs are deteriorating while the prices of houses are pushed up to excessively high just because the supplies do not satisfy the demand forcing many IDPs to crammed into rooms, houses and public spaces. Beside these, there are other ills and diseases related to congestion such as traffic jam, overcrowding in schools and hospitals especially with the upsurge of COVID 19 in late 2019 and early 2020. This often course delays and loss of valuable hours that could have been used for other benefits.

In terms of social tensions in Bonaberi, a significant majority (49.5%) of IDPs agreed with as the main housing challenges, followed by 20.6% disagrees, 13.2% strongly disagree, 8.9% agree, while 7.8% were indifferent. Overall, 58.4% of the respondents strongly agreed and agreed that social tension was a major factor affecting IDPs. This is very important as most IDPs have to cope with conflicting situations due to limited trust, stress, violence and suspicions. Social tension associated with xenophobia and discrimination and to violence in neighbourhoods, work place and schools have been reported. In Bonaberi, due to its cosmopolitan nature, this could easily lead to family separation and the destruction of community support which are very essential for the prevention of protection risks in the absent of preexisting social protection services among the IDPs. As noted by the Geneva Convention on Migrants and Cities, many cities in Africa experience xenophobia and violence behaviour due to difference in clans and tribe among people (IOM, 2015). A significant majority (53.1%) of respondents agreed to limited privacy as the main challenges faced by IDPs.

A significant majority (26.7%) agreed to noise pollution from bikes, cars and industries are a very common problem faced by IDPs in Bonaberi, followed by

Table 2.7 Housing challenges for IDPs in Bonaberi

Housing facilities challenges for IDPs	Stretched				
	Strongly agree	Agree	Disagree	Strongly disagree	Indifferent
Overcrowding	51.6% (145)	18.5% (52)	14.2% (40)	11.0% (31)	4.6% (13)
Social tension	8.9% (25)	49.5% (139)	20.6% (58)	13.2% (37)	7.8% (22)
Lack of privacy	30.6 (86)	22.8% (64)	21.0% (59)	14.2% (40)	11.4% (32)
Noise	24.6% (69)	26.7% (75)	18.1% (51)	23.1% (65)	7.5% (21)
Reduces saving	21.4% (60)	28.8% (81)	20.6% (58)	16.0% (45)	13.2% (37)
Indoor pollution	25.6% (72)	33.8% (93)	14.6% (46)	19.2% (54)	6.4% (18)
Increase social costs	32.0% (90)	26.3% (74)	21.4% (60)	16.0% (45)	4.3% (12)
Insecurity/ Crime	22.8% (64)	39.1% (110)	11.7% (33)	19.6% (55)	6.8% (19)
Leakage of facilities	22.8% (64)	34.5% (97)	7.1% (20)	21.7% (61)	13.9% (39)
Diseases/Pest transmission	25.3% (71)	29.5% (83)	21.4% (60)	16.0% (45)	7.8% (22)
Psychological stress/trauma	32.4% (91)	30.6% (86)	15.3% (43)	14.2% (40)	7.5% (21)
MRA	24.7% (764)	30.9% (956)	21.2% (655)	15.5% (478)	7.7% (237)

Source Fieldwork (2022)

24.6% who strongly agreed, 18.1% disagreed and only 23.1% strongly disagreed. In terms of reduction in saving capacity a significant majority (28.8%) of IDPs were optimistic due to high costs, unemployment and low-income levels which affects their purchasing power. This was followed by 21.4% who strongly agreed, 20.6% agreed, 16.0% strongly disagreed and only 13.2% were neutral. In terms of indoor pollution, the quality of housing for IDPs was generally deplorable with limited toilet systems, potable water and electricity supplies and high room densities. The proliferation of cockroaches, rats and mouse have resulted to disease outbreaks. Hence, a significant majority (33.8%) of IDPs agreed indoor pollution as the main housing facilities challenge to IDPs, followed by 25.6% for strongly agreed, 14.6% for disagreed, 19.2% for strongly disagreed and only 6.4% for those who were neutral.

Overall, indoor pollution was a significant housing facilities problem for IDPs in Bonaberi with a proportion of 49.8%. The continuous influx of IDPs into Bonaberi has added to the pre-existing difficulties of accessing basic services which leads to a shortage of educational and economic opportunities due to high social cost. Hence, a significant majority (58.3%) of IDPs agreed and strongly agreed to it as a pertinent housing problem for IDPs resulting to low standard of living. Decreasing economic resources and destruction of the social safety net also have left many IDPs to opt for negative behaviours and crimes, including survival sex, drugs and alcoholism resulting to arbitrary arrests, physical and sexual violence.

The rate of crime was significantly high with a proportion of 39.1% of IDPs, followed by 22.8% for strongly agreed unlike 11% for disagreed, 19.6% disagreed and only 6.8% were neutral. A significant majority (34.5%) of IDPs agreed to leakages of facilities as the main challenges faced by IDPs especially during intense floods. This sometime leave most of the belonging wet. This was followed by 22.8% for strongly agreed, 7.1% disagreed, 21.7% strongly disagreed and only 13.9% were neutral. A significant majority (29.5%) of respondents agreed to the proliferation of pests and diseases as the main housing facilities problem for IDPs, followed by 25.3% for strongly agreed, 21.4% for disagree strongly, 16.1% for disagreed and 7.8% for neutral. Usually, psychological trauma and stress are pertinent issues among the IDPs due to intense competition, abused and harassment from other individuals. Results noted that a significant majority (64.6%) of the respondents experienced stress and trauma respectively while other IDPs parents noted secondary schools refusing to grant kids' admission due to non-possession of birth certificates. This could lead to bribery and corruption resulting to unnecessary rise in citizens social costs. Also, limited finance and unemployment means IDPs, and their families go hungry for days, do not have enough access to safe drinking water and do not have a roof over their heads.

2.4.2.4 Stakeholders Adaptation Strategies to Housing Facilities Challenges for IDPs

If the chronic underfunding of the humanitarian response and government and civil society assistants are not addressed through effective coordination and collaboration, hundreds of IDPs in Bonaberi just like their counterpart in many developing cities could continue to be left without vital humanitarian assistance and protection against risks. Hence, integration amongsts stakeholders could be a key to solving the housing facilities quandaries of IDPs in Bonaberi. In this study these were assessed through the adoption of infrastructural development, education and sensitization, food rationing, capacity building, free medical care and women empowerment strategies by stakeholders (see Table 2.8).

Table 2.8 revealed that infrastructural development was the highest adaptation strategy by IDPs stakeholders in settlements like Bikoko (86.1%), followed by Ngwele (77.5%), Sodiko (73.8%) and Grand Hanger (70.5%) unlike Mabanda (68.5%). Also, the expansion of social services and infrastructures was highly perceived adaptation strategy in Ngwele neighbourhoods (42.5%), followed by Grand Hangar (40.9%) unlike in Mabanda (25.9%). The proportion of scholarship scheme

Table 2.8 Perceived stakeholders' adaptation strategies to housing facilities challenges for IDPs across Neighbourhoods

Adaptation strategies	Stats	Neighbourhood						
		Ndobo	Grand hangar	Mabanda	Sodiko	Bikoko	Ngwele	Total
Infrastructural development	n	28	31	37	31	31	31	189
	%	58.3	70.5	68.5	73.8	86.1	77.5	
Expansion of social services	n	19	18	14	13	13	17	94
	%	39.6	40.9	25.9	31.0	36.1	42.5	
Scholarship schemes	n	26	22	19	18	14	12	111
	%	54.2	50.0	35.2	42.9	38.9	30.0	
Women empowerment	n	24	19	23	21	16	14	117
	%	50.0	43.2	42.6	50.0	44.4	35.0	
Food rationing	n	24	25	29	23	18	22	141
	%	50.0	56.8	53.7	54.8	50.0	55.0	
Capacity building	n	22	22	28	18	19	17	126
	%	45.8	50.0	51.9	42.9	52.8	42.5	
Education and sensitization	n	27	26	27	22	18	24	144
	%	56.3	59.1	50.0	52.4	50.0	60.0	
Free medical care	n	21	24	22	19	19	17	122
	%	43.8	54.5	40.7	45.2	52.8	42.5	
Counselling services	n	15	18	23	16	10	18	100
	%	31.3	40.9	42.6	38.1	27.8	45.0	
Total		48	44	54	42	36	40	264

Source Fieldwork (2022)

was significantly higher in Ndobo (54.2%) and Grand Hanger (50%) unlike in Ngwele (30%). Women empowerment and food rationing were perceived almost similarly across settlements while capacity building was highest in Grand Hangar (52.8%), Mabanda (51.9%) and Ngwele (50%) unlike in Ngwele (42.5%) where displaced persons in collaboration between humanitarian agencies and municipal authorities provide free handicraft, industrial and agricultural training to young, displaced persons with little financial support as start-up capital after three-months training and capacity building. Education and sensitization and medical care provision for displaced families did not vary significantly across settlements in Bonaberi as many humanitarian agencies provide rapid health support for displaced families including referral services for those individuals in critical health conditions. Aware of the situations of schooling among the IDPs, stakeholders in partnership with some humanitarian agencies have retrieved over 5000 loss documents for IDPs whose children are now able to set in for public exams or avoid harassment from law enforcement officials. This initiative helped to overcome harassment and exploitation from some school authorities in Bonaberi. Counselling services for displaced families and individuals were offered more in Ngwele (45%) unlike in Ndobo (31.3%) as noted by the respondents.

These findings relate to the results of UN (2021) who noted humanitarian and government efforts at ensuring adequate financial support, investment, knowledge, provision of vital documents, local participation and capacity building to manage growing IDPs emergency issues through transparency, efficiency, collaboration, partnership and coordination among actors. IDMC (2019) also noted that after settling at their place of destination many IDPs starts a new phase of livelihoods through empowering and capacity building in driving, metal works, carpentry, motor repairs, tailoring, carpet weaving which they have already built skill in with the help of financial assistance and support from humanitarian advocacy groups mentioned above. This was evident among many IDPs in Bonaberi. A UN Planning Framework for the Construction of UNHCR (2017) in Iraq considered investment and the stimulation of urban economy to create livelihood opportunities and jobs for IDPs with emphasis on low-cost transport, housing and social services (IOM, 2015). IDPs more especially the youths between the ages of 18 to 38 years have also engaged in non-formal activities as source of income from financial assistance in Bonaberi.

2.4.2.5 Barriers to Stakeholders' Adaptation to Housing Facilities Strategies for IDPS

Despite the multitude efforts made by actors in the last few years to regenerate the housing conditions of IDPs, there are still limited humanitarian agencies and government financial support, accountability, coordination, access to services, facilities and collaboration to the plight of housing challenges faced by IDPs in Bonaberi. Significantly results from the study noted high rate of poverty and inequality, limited climate adaptation and language barriers as the main hindrances to IDPs adaptation strategies by stakeholders (see Fig. 2.1).

From Table 2.8, results revealed that a significant majority (81.8%) of IDPs indicated that limited and high cost of land, social services and infrastructure as the major barriers to IDPs adaptations, followed by limited control and weak policies and coordination of projects and programmes with a proportion of 43.1% of respondents, language barriers with a proportion of 38.4%, limited funding and finances with a proportion of 33.4%, difficulties to cope with climatic related hazards with a proportion of 23.8%, conflict, inaccessibility and biased distributions of IDPs items and aids with a proportion of 17.4%, limited participation and collaboration with a proportion of 12.8%, stigmatization, bullying or harassments with a proportion of 9.3%, unemployment and urban poverty with a proportion of 6%, inadequate social inclusion and integration with a proportion of 4.3%, corruption and favouritism with a proportion of 3.2%, limited empowerment with a proportion of 3.2% and the least limited statistics/data (including gender, age and disability status and housing situations) with a proportion of 1.1% of the respondents. Since Bonaberi is very dynamic, knowing the precise location of displaced person is very difficult at any particular

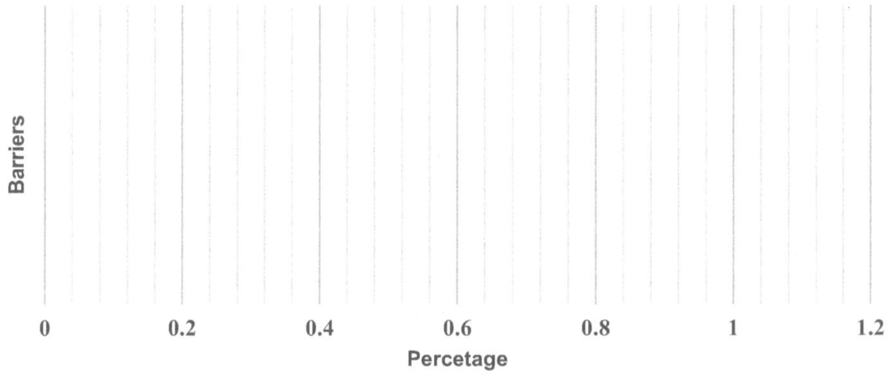

Fig. 2.1 Barriers to stakeholders' adaptation to housing facilities strategies for IDPs. *Source* Fieldwork (2022)

time coupled with many most displaced persons anonymity seek due to potential threat to their security. IDPs' financial resources, access to education and sensitization and social facilities in Bonaberi to rebuild their livelihood were limited. These was noted by Ellen et al. (2009) as the main adaptation challenges to displaced persons in Northern Uganda. Hence, the need for government and support groups to provide livelihood support, increase access to services, security, access to land, data, training and capacity building. These are particular challenges for humanitarian first responders trying to provide adequate protection and assistance to women, aged, disabled persons and children deprived of social services and educational opportunities thus generating high illiteracy, poverty and underemployment. Poor coordination by government agencies also hampered the work of many humanitarian agencies' expansion and distribution of aids due to corruption and limited transparency. Significantly, their work has been reduced to certain areas and threatening the implementation of programs intended to assist IDPs in additions to, limited government response plan (rehabilitation, subsidies and financial assistants). Widoyoko (2007) reveal corruption in public housing programme resulted to limited housing facilities and services for IDPs with rising informal housing cost in Indonesia. Hence, building municipal capacity and community dialogue can improve protection, promote of inclusive climate actions and encouraging urban planning that take account of both displaced and host community needs.

2.5 Conclusions and Policy Implications

This study examined the housing facilities challenges for IDPs in Bonaberi with key results revealing significant demographic and socio-economic profile in terms of income range, occupation, sex and level of education among others. Results noted the increasing rates of IDPs in the last few years characterized by push and pull factors such as violence conflicts, socio-economic environmental causes, urban growth and economic development. The majority destination of IDPs into Bonaberi were from north and south-west regions due to high incidence of insecurity. Although there was diverse housing facilities characteristics, conditions and service provisions for IDPs and the host communities in Bonaberi, the study revealed limited level of satisfaction. These have resulted to diverse challenges such as overcrowding, limited privacy, noise, limited social services, diseases and psychological stress with profound effects on health and wellbeing of IDPs. Despite the multitude adaptation strategies by stakeholders through infrastructural development, optimization and expansion of social services, scholarship opportunities and women empowerment, there were limited due to limited coordination, social development control, participation, collaboration, language barriers and climate change.

Therefore, ending forced displacement and identifying solutions to housing facilities challenges for IDPs in Bonaberi should remain the ultimate priorities goals/targets of stakeholders, government agencies and municipal authorities who aimed at finding long term solutions to internal displacement and sustainable development. Significantly, effective planning and partnership for resource mobilization, coordination and control by state agencies could ensure land allocation, urban regeneration and densification, resettlement plans, dialogue, reconciliation and tolerance towards solving housing challenges for IDPs in Bonaberi. Also, adequate housing infrastructures and social services provision could ensure environmental quality while ensuring overall sustainable livelihoods through collaboration, spatial planning, creation of public spaces and environmental protection zones and optimization.

In addition, access to education, documentation, communication, language centres, counselling experts and field data (including gender, age, disability status, income and access to services and capacities) using recent technologies like mobile phone data, satellite imagery analysis and community mapping, could help to overcome these challenges and analyses could provide actors with skills and capacities needed to contribute towards local economic and social integration. Moreover, increase saving, cash assistance and the development of loans association could avoid IDPs slipping into acute poverty while increasing access to education, shelter, enhance livelihoods, food needs. Humanitarian advocacies, training, capacity building, scholarship schemes, counselling and psychosocial support crisis could enhance educational quality and wellbeing of IDPs. Ultimately, inclusive urban planning measures could reduce IDPs vulnerabilities and adaptations.

References

Ako, A., Elambo, N., & Eneke, T. (2009). Water quality and occurrence of water-borne diseases in the Douala V District, Cameroon. *Water Science and Technology*, 23–25.

Alan, G. (2000). Housing in third world cities: The critical issues. *Geographic Association, 185*(2), 43–45. https://www.jstor.orgstable/40573408

Alekseyenko, L., Tulai, O., Petrushenko, Y., Kuznietsov, N., & Derkash, J. (2021). Affordable housing for internally displaced persons: The priorities for investment and development in

Ukraine. *Journal of Investment Management and Financial Innovations,* 101–15. https://doi.org/10.21511/imfi.18(1).2021.09
Asangwe, K. (2006). *The Douala coastal lagoon complex: Cameroon environmental issues.* St. Cloud State University.
Baba, A., & Humphrey, N. N. (2017). Changing trends in water sources and related pathologies in small to medium size African Cities. *Geo Journal, 83*(4), 8855–896. https://link.springer.com
Baba, A., Ndi, H., & Balgah, S. (2020). Evaluation of current issues in water supply systems and implications for sustainable urban water management in Limbe, Cameroon. *Current Journal of Applied Science and Technology, 39*(45), 52–74.
Baqui, A. (2009). Community-based validation of assessment of new borne illnesses by trained community health workers in Sylhet District of Bangladesh. *Journal of Tropical Medicine and International Health, 14*(12), 1448–1456.
Bilfinger, C., Eiztze, J., Kaizer, M., & Weck, W. (2019). *Conception and implementation of agenda 2030 basic laws of sustainability.* Retrieved August 10, 2022, from https://agenda2030basiclaws
Cameroon - Third General Census of Population and Housing. (2005). IPUMS Subset, Bureau Central des Recensements et des Études de Population, Minnesota Population Center Report generated on: August 27, 2019. https://microdata.worldbank.org/index.php
Cantor, D., & Cardona-Fox, G. (2021). UN high-level panel on internally displacement: Can a new development-based approach brings solutions to this global crisis? *IDRP, 13*(11), 32–38. https://doi.org/10.1080/02680939.2013.790079
Cheserek, G. J., & Opata, G. P. (2011). Environmental and housing problems of low-income household in Eldoret Municipality, Kenya. *Journal of Emerging Trends in Economics and Management Sciences, 2*(4), 1–5.
Davies, A. (2012). *IDPs in host families and host communities: Assistance for hosting arrangements.* Division of International Protection Operational Support. https://www.calpnetwork.org/wp-content/uploads/2015/10/4fe8732c2.pdf
Durosaro, I. A., & Ajiboye, S. K. (2011). Problems and coping strategies of internally displaced adolescents in Jos Metropolis, Nigeria. *International Journal of Humanities and Social Science, 1*(20), 256–262.
Ellen, M., Celia, P., & Acidri, J. (2009). Livelihoods in crisis: A longitudinal study in Pader, Uganda Evidence for Development. https://www.researchgate.net/publication/
European Commission. (2005). *Housing finance systems for countries in transition.*
Florczyk, A. J., Corbane, C., Ehrlich, D., Freire, S., Kemper, T., Maffenini, L., …, GHSL Data Package. (2019). EUR 29788 EN, Publications Office of the European Union, Luxembourg. https://doi.org/10.2760/290498. JRC 117104. https://human-settlement.emergency.copernicus.eu/documents/GHSL_Data_Package_2019.pdf?t=1478q532234372
Fombe, L. (2005). *Substandard housing and slums development in Douala: An urban development perspective.* Thesis submitted to the Department of Geography of Buea University.
Fombe, L. F., & Agbortoko, M. (2014). Housing standards, household health and disease nexus in the Buea Municipality. *Journal of Sustainable Development, 7*(4), 34–35.
Fombe, L. F., & Balgah, S. N. (2012). *The urbanization process in Cameroon: Processes, patterns and implications.* Nova Science Publishers, Inc. Retrieved August 10, 2022.
Gilbert, A. (2013). Free housing for the poor: An effective way to address poverty. *Habitat International, 4*(1), 253–261.
Ginnetti, P., Mario, V. M., & Valenrie, W. (2017). Estimating the loss of economic production caused by internal displacement due to disaster. *International Journal of Disaster Risk Sciences, 9*(1), 18–19. https://doi.org/10.1007/dss.13753-018-0190-9.28
Henilane, I. (2016). Housing concept and analysis of housing classification. *Baltic Journal of Real Estate Economic and Construction Management, 4*(1), 168–179.
Humanitarian Programme Cycle. (2023). *Humanitarian Response Plan Cameroon.* file:///C:/Users/user/Downloads/CMR_HRP_2023_v5_20230405.pdf
IDMC. (2013). *Nigeria: Increasing violence continues to cause internal displacement: A profile of the internal displacement situation.* IDMC, NRC.

IDMC. (2018a). The ripple effect: Economic impacts of internal displacement.

IDMC. (2018b). *City of challenge and opportunity: Employment and livelihoods for internally displaced people in Maiduguri, Borno State.* Internal Displacement Monitoring Centre. Available at: https://api.internal-displacement.org/sites/default/files/publications/documents/20180209-idmc-nigeria-case-study_0.pdf#page=6.07

IDMC. (2019a). Unveiling the cost of internal displacement. Thematic Series. The ripple effect: Economic impacts of internal displacement.

IDMC. (2019b). Global Report on Internal Displacement 2019. Geneva: IDMC. Available at: https://api.internal-displacement.org/sites/default/files/publications/documents/2019-IDMC-GRID.pdf

IDMC. (2022). Children and youth in internal displacement Internal displacements by conflict and disasters in 2021. https://www.scribd.com/document/575097471/IDMC-Grid-2022-Lr

Igwe, P. U., Okeke, C. A., Onwurah, K. O., Nwafor, D. C., & Umeh, C. N. (2017). A review of housing problems. *International Journal of Environment, Agriculture and Biotechnology, 2*(6), 453–458. https://doi.org/10.22161/ijeab/2.6.40

IOM. (2015). *Overview World Migration Report 2015. Migrants and Cities: New Partnerships to Manage Mobility.* https://www.iom.int/sites/g/files/tmzbdl486/files/country/docs/syria/IOM-World-Migration-Report-2015-Overview.pdf

IOM. (2019). Global report 2019. Operations and emergencies global highlights 2019. https://www.iom.int/sites/g/files/tmzbdl486/files/documents/IOM-Global-Report-2019-Operations-and-Emergencies.pdf

Kimengsi, J. N., Akhere, S. G., & Ngala, L. E. (2015). Urban housing improvement and conflict prevention framework for Cameroonian cities: Lessons from Kumba, Cameroon. *JGRP, 8*(7), 181–188. https://doi.org/10.5897jprp.2015.0504

Mabongunje, A., Hardoy, J., & Misra., R. (1978) *Shelter provision in developing countries: The influence of standards and criteria.*

Maluh, B., Balgah, S., & Kimengsi, J. (2019). Housing situation in the Bamenda urban space: Challenges and the way forward. *Journal of Geography, Environment and Earth Science International, 24*(4), 1–18.

Michael, A. O., Mary, A., & Matthew, T. O. (2014). Crisis induced internal displacement: The implication on real estate investment. *ILESD, 5*(4), 20–14. www.iiste.org

MINEPAT. (2022). *National human development report 2022. What human capital for the structural transformation of the Cameroonian economy?* https://minepat.gov.cm/wp-content/uploads/2020/07/NHDR-2022_Final-1.pdf

Muggah, R. (2017a). Chapter 19: The shifting frontiers of displacement in Latin America. In P. Bourbeau (Ed.), Handbook on Migration and Security (pp. 395–410). Edward Elgar Publishing. https://doi.org/10.4337/9781785360497.00030. https://www.elgaronline.com/collection/Social_and_Political_Science_2017

Muggah, R. (2017b). World Economic Forum. *Safe havens: Why cities are crucial to the global refugee crisis.* Available at: https://www.weforum.org/agenda/2017/06/safe-havens-why-cities-are-crucial-to-the-global-refugee-crisis/

NDS30. (2020–2030). *National development strategy for structural transformation and inclusive development.* Emerging Cameroon Ministry of Economic Planning and Regional Development. Development Partners and UNDP. www.minepat.gov.cm

NIS. (2005). *Household health survey.* https://www.statistics-cameroon.org/downloads/pets/2/Rapport_principal_Education_anglais.pdf

Nunez-Ferrera, I., Aubrey, D., Earle L., & Loose, S. (2020). *IDPs in towns and cities—Working with the realities of internal displacement in an urban world.* IIED, JIPS and UN-Habitat. www.urbancrises.org

OCHA. (2021). *Global humanitarian overview: News and press release.* NHCR Works.

OCHA. (2022). *Cameroon situation report on internal displacements.*

Researching Internal Displacement. (2022). *Working Paper working paper No. 23: How urban are IDPs? Between forcibly evicted urban poor and conflict IDPs Rebecca.* Enobong Roberts Technische Universität Berlin. Retrieved August 10, 2022.

Stein, M. I. A., & Lord, J. E. (2011). Enabling refugee and IDPs law and policy: Implications of the U.N. convention on the rights of persons with disabilities. *Arizona Journal of International & Comparative Law, 28*(2), 213–234.

Ujuku, T., & Mengue, A. (2017). *Urban growth and front development on risk zones GIS application for mapping of impacts on Yaoundé North Western Highland of Cameroon. Current Urban Studies.* University of Yaoundé I.

UN. (2018). *Global perspective human stories UN launches global plan to strengthen protection of IDPs.* https://news.un.org/en/news/topic/migrants-and-refugees

UN. (2021). *Housing, land and property issues in the context of international displacement: Report of the special rapporteur on the human rights of IDPs.* 47 Human Right Council Session.

UN Habitat. (2010). Urban planning and infrastructure in migration contexts in Douala Cameroon: Special issue for better future. *1*(1), 32–35.

UN HABITAT. (2016). *Using the framework of the EU Global Human Settlement: Layer.* https://ec.europa.eu/jrc/en/global-human-settlement-layer

UN-HABITAT. (2019). *The future of Asian and Pacific Cities: Transformative pathways towards sustainable urban development.*

UNDP. (2022). *Africa sustainable development report 2022. Building back better from the coronavirus disease (Covid-19) while advancing.* The full Implementation of the 2030 Agenda for Sustainable Development. https://www.undp.org/sites/g/files/zskgke326/files/2023-06/asdr_2022-en-full_report-final.pdf

UNHCR. (2017). *Global Trends Report: Forced Displacement in 2017.* https://www.unhcr.org/africa/media/unhcr-global-trends-2017

UNHCR. (2019a). *2019 Global Refugee Forum Background Note for the First Preparatory Meeting.* https://www.unhcr.org/sites/default/files/legacy-pdf/5c98a70d7.pdf

UNHCR. (2019b). *Cameroon situation: Responding to the needs of IDPs and Cameroonian refugees in Nigeria.* https://reporting.unhcr.org/sites/default/files/UNHCR%20Cameroon%202019%20Supplementary%20Appeal%20%28March%202019%29.pdf

UNICEF & World Bank. (2015). *Progress on sanitation and drinking water: Update and MDGs assessment* (p. 50). WHO. https://files.unicef.org

Warn, E., & Adamo, S. B. (2014). World meteorological organization, "The impact of climate change: Migration and cities in South America". *Bulletin, 63*(2). Available at: https://public.wmo.int/en/resources/bulletin/impact-ofclimate-change-migration-and-cities-south-america

Widoyoko, D. (2007). *Tackling corruption to improve housing services in Indonesia: Partnering to combat corruption series.* World Bank.

World Bank. (2015). *Stock taking of housing sector in sub–Saharan Africa summary report: The International Bank for Reconstruction and Development/World Bank 181-h street.* NW Washington DC. www.worldbank.org

Baba Adamu is currently a Senior Lecturer of Geography at the University of Buea, Cameroon, where he also earned his Ph.D. in Human Geography. He is an active member of African Geographic Society (AGS). His research expertise spans urbanization and development, urban planning policy, urban water management/governance and urban social justice. He is the co-author of changing trends in water sources and related pathologies in small to medium size African cities (Springer, 2018).

Balgah Sounders Nguh is a Professor in Urban and Development Studies with particular interest in Remote Sensing and Its applications to land use dynamics, natural resources and resource management/development. He has published extensively in these areas in learned scientific journals and is a co-author of two text books and has made significant contributions in Development studies. He is a consultant in these areas and in planning issues.

Nformi Beatrice Maluh is currently a Lecturer in the Department of Town Planning at the National Advanced School of Public Works. She holds a Ph.D. in Geography and Planning in 2021 and a master's degree in Geography in 2016 all from the University of Buea, Cameroon. Her area of research interests focuses on land use planning, urban development and planning, human geography and natural resource management. She has a wealth of publications in high impact journals on issues like land use change and planning, land use conflicts, housing development, urban green spaces and assessments of housing regulations. She was a part time lecturer in the Department of Geography at the University of Buea from 2017 to 2022 and Landmark Metropolitan University from 2022 to 2024.

Humphrey Ndl Ngala is Cultural Counsellor at the High Commission for the Republic of Cameroon in London. He is also professor of Health Geography at the University of Yaounde I, Cameroon where he has supervised many postgraduate students. He holds a M.Sc. degree in diplomacy, foreign policy and statecraft from Loughborough University. He has also undertaken postgraduate training in ethics, leadership and governance as well as in multilevel government at King's College London.

Toumba Oumarou holds a Ph.D. in Physical Geography, environmental studies and Regional Planning, a Master's Degree in Geography and Land Planning as well as a Licence in Geography. He is an active member of the World Geomorphology Society, the Cameroon Geomorphology Society, the African Geography Society, the African Regional and Urban Planning as well as a Member of the Geomatics Laboratory of the University of Ngaoundere. He is currently the Head of Department of Geography and an Associate Professor at the University of Buea. Professor Toumba is both a practitioner and academician. In addition, he is a Board Member of Faculty of Social and Management Sciences, of Academic Planning Committee and of the Senate of the University of Buea. He has published widely in peer-reviewed journals and books; and his areas of interest are Geomorphology/Geo-risks, Natural resource management, land planning and sustainable development. He is currently a co-principal investigator on a project on smart growth cities in Cameroon, Non-Governmental Organisation and sustainable development participation.

Chapter 3
Housing Delivery for the Poor in Informal Settlements: Contrapuntal Urbanism in the Postcolonial Namibian Context

Eric Yankson

Abstract The notion of contrapuntal urbanism refers to how individual factors work together to create a coherent (but not necessarily unitary) urban system which may be in a state of flux. With specific reference to urban housing for the poor, a plethora of factors may interact in various ways to configure the efficiency of its delivery. The situation is prevalent within the Southern African region in former colonial and apartheid societies such as South Africa and Namibia. This proposed chapter seeks to assess the implications of interconnected factors for urban housing delivery for the poor in the informal settlements of Namibia. The choice of Namibia has been informed by the relatively less focus on this country in the extant academic literature on urban housing delivery. The largely qualitative study relies on document and discourse analyses, as well as interviews to make its arguments. The research dissects cases of both successful and less successful models of housing delivery for the poor in Namibia's informal settlements. For successful models of housing delivery, factors such as housing financing and cooperative production systems interact to ensure an increase in supply. And for less successful models of housing delivery for the poor, postcolonial dynamics such as land use regulations and rural–urban migration are important interacting factors. The chapter concludes by proposing a postcolonial relational schema to ensure improved urban housing delivery for the poor in these settlements. The model prioritises interventions such as private non-commercial, cooperative and public sector housing production systems, as well as reforms in institutional and policy frameworks.

Keywords Housing · Informal settlements · Poverty · Contrapuntal urbanism · Postcolonialism

E. Yankson (✉)
Department of Architecture, Planning and Construction, Namibia University of Science and Technology, Windhoek, Namibia
e-mail: eyankson@nust.na; ericyankson22@gmail.com

L. Chipungu et al. (eds.), *Contemporary Models for Pro-Poor Housing in Urban Sub-Saharan Africa*, https://doi.org/10.1007/978-3-032-05489-0_3

3.1 Introduction

This chapter applies the concept of contrapuntal urbanism to assess the combined ramifications of housing financing, cooperative production systems, land use regulations and rural–urban migration for housing delivery in Namibia's informal settlements. According to the United Nations Sustainable Development Goals, housing is a basic right which every urban resident is entitled to. This notwithstanding, the supply of this resource has historically failed to keep up with demand in many parts of the Global South. In Southern Africa, the situation has been compounded by the vestiges of colonialism and apartheid which have engendered inequality in the urban landscape. As a result, poor neighbourhoods in particular are particularly placed at a disadvantage when it comes to accessing good quality and affordable housing. In the post-independence era, the emergence of informal settlements has further exacerbated the problem. These settlements spring up partly due to the unaffordability of housing, especially for low- and middle-income residents in more affluent areas. Also, the inability of housing supply to match up with demand creates the problem. Moreover, the migration of citizens from rural to urban areas in search of green pastures accounts for the situation. However, existing urban land use regulations often fail to adequately plan for the increased influx, resulting in the emergence of these settlements (Yankson & Allotey, 2024).

Informal settlements are unplanned residential units which normally emerge at the urban periphery (Soyinka & Siu, 2018, p. 24). These settlements are associated with little or no adherence to building codes, road standards and zoning regulations (Chipungu & Magidimisha, 2021, pp. 5–6). They are thus characterised by poor security of land tenure, low-quality or non-existent facilities and physical infrastructure. Due to their disadvantaged position in relation to formal neighbourhoods, a major outcome associated with the emergence of these settlements is socio-spatial exclusion (Soyinka & Siu, 2018, p. 24). Also, the relatively high numbers of rural–urban migrants implies that informal settlements are associated with high population densities. This is because these settlements tend to be concentrated in places where urbanisation is prominent (Ren et al., 2020, pp. 1–2, 12). Moreover, while government interventions such as slum clearance may be geared towards managing the growth of informal settlements, the inability to address the housing deficits which result in the emergence of these settlements in the first place implies that the results are only short-lived. Thus, they could re-emerge or evolve in different forms which further exacerbate the problem (Magidimisha & Chipungu, 2011, p. 469). Besides, while informal settlements may primarily emanate from poor planning, governmental interventions in the form of accelerated land reform could also result in the emergence of settlements at the urban periphery which do not conform with urban master or structure plans (Chipungu et al., 2022, pp. 45–46).

As already noted, this research is based on a case study of Namibia. The choice of Namibia has been informed by its rapid population growth, fuelled by urbanisation, which has led to the proliferation of informal settlements at its urban periphery. Namibia's population witnessed an increase from 1,409,915 to 2,113,077 between the years 1991 and 2011. Over the same period, there was a nearly three-fold increase in the urban population from 382,280 to 903,434. However, the increase in the rural population was 220,000 between 1991 and 2001. From 2001 to 2011, there was actually a decline in the rural population by 20,000 people (Weber, 2017, p. 16; Weber & Mendelsohn, 2017, p. 15). Between 1991 and 2011, the proportion of Namibia's population residing in urban areas increased from 27 % to 41%; and this is projected to rise to 67% by 2041 (Namibia Statistics Agency, 2015, p. 19). From 1991 and 2011 for instance, the majority of cities and towns in Namibia experienced a growth rate of over 4% per annum (Weber, 2017, p. 16; Weber & Mendelsohn, 2017, p. 15). Regions like Khomas and Erongo which contain major cities such as Windhoek, Walvis Bay and Swakopmund were found to have more than 85% of their populations living in urban areas as of 2011 (Namibia Statistics Agency, 2015, p. 19).

Rapid urbanisation in Namibia has contributed to the proliferation of informal settlements. As of 2018, a total of 995,000 people out of the country's total population of 2.4 million resided in informal settlements (Nakale, 2018). These settlements epitomise the geography of inequality based on class and race which still define the urban landscape in post-independence Namibia. Thus, they are primarily inhabited by low-income black residents and associated with the concentration of poverty (Chiwara & Lombard, 2020, pp. 31–32). Even though Namibia is an upper-middle-income country, its Gini coefficient of 0.59 makes it one of the most unequal countries in the world (World Bank, 2017). Moreover, an estimated 17.4% of people in Namibia were below the country's poverty line in 2016. This poverty line is calculated as having an upper echelon of 520.8 Namibian dollars per month, with the Namibian currency being pegged to the same value as the South African rand (Namibia Statistics Agency, 2019). Besides the income definition of poverty, multidimensional poverty is also evident in terms of poor access to assets such as land and housing necessary for improved quality of life (Chiwara & Lombard, 2020, pp. 34–37). Moreover, there is limited access to public services such as water, sanitation and electricity (Karuaihe & Wandschneider, 2018, pp. 472–477).

While several efforts have been made over the years to provide housing for the poor especially in Namibia's informal settlements, these have been characterised by limited success. For instance, the Namibia Housing Enterprise, established in 1993 to help address the nation's housing shortfall, managed to provide only 17,113 housing units between its inception and mid-2019. From 2016 to 2019, only 896 housing units were delivered. While this entity was established to provide housing for low- and middle-income Namibians, many of the houses constructed thus far have not been affordable to low-income groups due to the primary focus on civil servants (Ndjavera, 2022). Moreover, the mass housing programme intended to provide 10,000 housing units per year between 2013 and 2030 towards the ultimate target of 185,000 houses did not meet its target partly due to lack of affordability. These public–private partnerships suffered from challenges such as poor municipal capacity in specialist financial services, as well as low rigour in loan book management and capital recovery (Ndjavera, 2022).

In Africa, many of the studies on housing delivery in informal settlements appear to focus on major countries such as South Africa and Nigeria. In nations like Namibia, there is a bourgeoning academic literature on urban informality even though these could benefit from nuanced analysis to elevate them to a similar level comparable with research already conducted in other countries. Equally important is the need to analyse the interlinkages among factors which account for the provision of housing in informal settlements. This is important for understanding the intricacies of the phenomenon and how they can be addressed. The rationale for the study thus emanates from the need to examine the context-specific factors which configure urban housing in Namibia's postcolonial era. The research objectives are as follows: to assess the nature and evolution of informal housing and underlying factors in the postcolonial era; to unpack the interrelationships between the factors which configure successful and less successful approaches to urban housing delivery in informal settlements within the postcolonial epoch; and to propose a postcolonial relational schema for improved housing delivery in informal settlements. The chapter starts by reviewing relevant literature. It then dissects the conceptual framework and methodology before discussing findings, recommendations and conclusions.

3.2 Housing, Poverty and Informal Settlements

Urban poverty and market conditions are correlated with the availability of and access to housing units. For instance, in urban areas where there is a right to shelter policy, higher levels of poverty and market rents may result in more pronounced levels of homelessness (Hanratty, 2017, pp. 652–653). Similarly, urban poverty is associated with poor conditions of housing units and municipal infrastructure. For instance, the location of housing units in poor areas affects their property values. This occurs because the unattractiveness of an area may result in declining socio-economic activities leading to the concentration of poverty (Kisiała, & Rącka, 2021, p. 1). The unaffordability of housing and its concomitant challenge of low housing supply have emerged as major issues in contemporary urban planning. These occur in part because housing-related expenses increasingly exceed the rise in income levels. The situation has been exacerbated by rapid urbanisation and increased inequality (Wetzstein, 2017, p. 3159).

The inability to provide housing for the urban poor is a major factor accounting for the emergence of informal settlements. Towards addressing this situation, a participatory planning approach which incorporates inputs from the urban poor is critical (Average, 2019, p. 927). Such a strategy entails the involvement of low-income residents in the design of housing models which meet their needs and preferences. The success of these endeavours however requires critical support from the government towards ensuring the provision of sustainable housing (Average, 2019, p. 927).

Moreover, notwithstanding the numerous challenges associated with housing in informal settlements, factors such as their affordability and the viability of their acquisition make them attractive options for potential residents. This appeal can be further enhanced through improved access to financing for homeownership, security of tenure and upgrading to improve the state of these settlements (Sullivan & Olmedo, 2015, pp. 1050–1051). Ensuring affordability is a major strategy for the provision of sustainable housing. One way of attaining this is through self-help initiatives. This is however limited by poor funding, locational constraints and poor adherence to urban planning regulations. Another strategy entails public–private partnerships in which local governments provide land and the private sector serves as the source of funding (Chirisa & Matamanda, 2016).

Moreover, informal settlements are characterised by challenges such as shortage of housing to meet the needs of their residents. Different policy responses have thus emerged over the years to manage the dynamics associated with these settlements. These include demolition, upgrading and formalisation (Ren, 2018, pp. 79–88). In many instances, there are contesting perspectives on what policy interventions are best placed to address the challenges faced by informal settlements. In terms of upgrading for instance, evidence points to the fact that physical improvements and provision of services such as water and electricity are critical. However, these need to be complemented with social and economic interventions to enhance their effectiveness. In the same vein, the socio-economic strategies are configured by structural factors which determine whether or not their desired impacts would be maximised on the ground (Brown-Luthango et al., 2017, p. 471).

3.3 Housing Financing and Production Systems

Besides the housing-poverty nexus in informal settlements, the financialisation of housing may be evident in terms of private equity investments in the urban landscape. The situation could however heighten uneven development, resulting simultaneously in gentrification and the emergence of poor neighbourhoods. These market dynamics may be reinforced by municipal policies which reduce rental protections, leading to the charging of commercial rates for hitherto affordable housing units (Fields & Uffer, 2016, pp. 1486, 1497–1498). Moreover, a boom in housing credit within the framework of the system could be inherently linked with a rise in housing prices. This may be evident in terms of both household and firm credit especially in countries with high lending or mortgage rates (Cerutti et al., 2017, p. 1).

Alternative sources of financialisation can however be envisaged. These may entail the liberalisation of existing sources of housing finance. Also, the privatisation of public housing schemes and speculative investments by private hedge or equity funders is another strategy (Wijburg & Aalbers, 2017, p. 968). Moreover, notwithstanding the increased financialisation of housing provision, the resurgence of public housing in certain contexts may be a strategy for dealing with market failures in the provision of equitable housing (Chen et al., 2014, p. 534).

The unaffordability of housing has become a topical issue in contemporary urban planning discourse. This situation has partly been caused by rapid urbanisation and its concomitant population boom. Also, people who move to cities for various reasons including employment, education and the general pursuit of better opportunities result in increased demand for housing units, creating price rises (Wetzstein, 2017, pp. 3161–3162). Moreover, increased intra-urban inequity has exacerbated the problem. The development puts heightened financial and social strain especially on low-income residents. Moreover, workers including police officers, nurses and teachers who provide key services are priced out the urban housing market (Wetzstein, 2017, pp. 3161–3162).

Besides the role of financing in the actual provision of housing, funding can also play a vital function in improving the condition of existing housing units. In a deprived area for instance, community-based sources of funding derived from local banks could help to enhance housing conditions as a strategy for improving quality of life and reducing disease burden. These funding sources are critical for low-income residents unable to access opportunities from mainstream financial institutions (Mponzi et al., 2023, p. 1). Towards enhancing participatory planning, government may also devolve some of its functions in the delivery of housing to curb the existing backlog. Such a strategy may be geared at enhancing social inclusion and a more localised approach to housing provision (Patel, 2016, pp. 2738–2739, 2752–2753).

3.4 Land Use Regulations and Rural–Urban Migration

Apart from housing financing and production systems, land use regulations and rural–urban migration are also critical in the provision of affordable housing. To begin with, the value of urban land is correlated with rates of urbanisation. Generally, the value declines as distance decreases from the central business district implying more affordability. This notwithstanding, land use regulations such as land banking, restrictions on floor area ratios, construction land quotas and land supply controls can impact the ability of specific parts of the city to contain urban growth (Han et al., 2020, pp. 1–3). Moreover, zoning regulations can determine which land uses are allowed or proscribed and limit the kind of housing units which can be put up, thus portending negatively for affordability by low-income residents. Similarly, density regulations impact affordability as low-density houses may be more expensive and thus beyond the means of low-income persons (Goytia et al., 2023, pp. 1–3). Also, the legal requirement for developers to provide infrastructure in certain places results in increased land values, thus pricing them out of the range of certain prospective residents. Simultaneously, the illegal nature of many informal settlements implies that there are legal constraints to the provision of infrastructure in these places (Goytia et al., 2023, pp. 1–3). Generally, cities with more stringent requirements for obtaining building permits are associated with higher values for land. This in turn drives up the prices of housing units (Kok et al., 2014, p. 136).

The issue of non-compliance with land use regulations is also an important factor which affects the nature of urban growth. For instance, developers may exceed floor area ratios if the expected benefits surpass the costs. Moreover, developers who enjoy special relationships with political officials may be able to outwit the system and thus incur little or no costs for doing so (Cai et al., 2017, pp. 223–224, 232). For instance, a lax enforcement of land use regulations may result in the poor provision of physical infrastructure and housing. This can lead to a lower-than-expected pace of urban growth (Monkkonen, 2013, pp. 255–256). Moreover, interested actors such as residents may also engage in an inverse form of planning by assuming some responsibilities which local authorities have been unable to fulfil. For instance, a bottom-up process of producing a subdivision plan may result in a formal approval and regularisation by municipal authorities (Chiodelli & Mazzolini, 2019, pp. 165–166, 176–178).

Land use regulations have inherent relationships with rural–urban migration. For instance, perceptions of land tenure insecurity can influence the decision by people to move to cities. This is especially the case in rural areas with poorly developed land markets where residents attach great importance to land titles or certificates (Ma et al., 2014, pp. 1–3, 18–20). Moreover, the economic opportunities available to rural–urban migrants in cities affect their attitudes towards land ownership in rural areas. For instance, their ability to build secure employment and high incomes implies they may abandon land in rural areas as they pursue more prospects in cities (Xie & Jiang, 2016, pp. 262, 266–267).

Rural–urban migration is driven by a number of underlying factors. To begin with, migrants have perceptions about higher income earnings in urban areas. Given the relatively low-income levels in many rural areas, this serves as a strong pull factor (Niva et al., 2019, pp. 4–6; Selod & Shilpi, 2021, pp. 3–5). Moreover, the presence of amenities in urban areas is an important reason. These include communal facilities such as malls, schools, hospitals and recreational facilities (Selod & Shilpi, 2021, pp. 3–5). Rural–urban migration also serves as a form of insurance against risks such as climate change and conflicts. Rural areas sometimes bear the brunt of these disasters, causing residents to move elsewhere in search of better prospects (Rahman et al., 2017, pp. 1–3, 16; Selod & Shilpi, 2021, pp. 3–5). Similarly, this form of migration creates avenues for the diversification of incomes since migrants can explore alternatives beyond what may be available to them in their rural origins (Selod & Shilpi, 2021, pp. 3–5).

3.5 Conceptual Bases and Methodology

This chapter is based on the concepts of contrapuntal urbanism and postcolonialism. The term "contrapuntal" in contemporary discourse largely emanates from the work of Said (1994) who adopted it from music where it refers to melodious sounding parts working together. Said used the term in relation to imperialism to refer to diverse dynamics which work together to configure this endeavour. With specific reference to urban planning, the term contrapuntal may be used to refer to the relational geography associated with cities. Thus, contrapuntal urbanism entails the coalescing of individual factors towards attaining a combined effect within the urban landscape. It embodies spatial-temporality with its changing nature which may not follow a clearly defined pattern. The dynamics entailed may be associated with fragmentation and differences, but these nevertheless work together to produce a certain outcome (O'Callaghan, 2012, pp. 1930–1931: Wyly, 2023, pp. 108–113).

Due to their conceptual complementarities, contrapuntal urbanism has been broached in conjunction with postcolonialism in some of the existing literature. The term postcolonialism generally refers to the period after colonial rule and the immanent socio-spatial and political economic dynamics associated with this (Myers, 2014; Robinson, 2011, 2014). This school of thought argues that cities in the Global South have unique attributes which largely distinguish them from their counterparts in the Global North. These include urban informality, low security of land tenure, underdeveloped infrastructure and poor sanitation (Roy, 2005; Storper & Scott, 2016, p. 1123). Thus, the Eurocentric models normally applied to analyse cities of the developed world are not necessarily applicable to those of developing countries (Myers, 2014; Robinson, 2011, 2014; Grubbauer, 2019, pp. 469–472).

This analysis argues that the dynamics of a postcolonial setting constitute the bases for the manifestation of contrapuntal urbanism with respect to housing delivery in Namibia. Specifically, the nature of housing financing, cooperative production systems, land use regulations and rural–urban migration are configured by the country's postcolonial emergence; and these work together to shape housing in its informal settlements. The study is based on a qualitative approach. Specifically, it utilises document and discourse analyses, as well as interviews with selected planners and policy makers to make its observations. Moreover, it undertakes field observations of housing units in selected informal settlements across Namibia. The documents analysed include land use planning regulations, policies, reports and journal articles. Also, the interview respondents came from the following: Ministry of Urban and Rural Development; Ministry of Agriculture, Water and Land Reform; the City of Windhoek; Namibia Housing Enterprise and Shack Dwellers Federation of Namibia. The respondents were selected based on their respective expertise on housing financing, cooperative production systems, land use regulations, rural–urban migration and housing delivery in Namibia. All the respondents were assured of anonymity and indicated that their responses did not necessarily reflect the perspectives of the organisations they work for. The informal settlements visited were mostly in Windhoek, Walvis Bay and Swakopmund (Fig. 3.1).

3.6 Informal Housing and Underlying Factors

In this section, the chapter addresses the first research objective of assessing the nature and evolution of informal housing and underlying factors in the postcolonial era. The identified factors include housing financing, production systems, urban land use regulations and rural–urban migration. To begin with, the rapid pace of urbanisation in Namibia has been characterised by the emergence of informal housing units and settlements (Table 3.1). For instance, while the houses made from bricks and blocks nearly increased two-fold from a total number of 73,881 in 1991 to 163,793 in 2011,

Fig. 3.1 Housing in informal settlements. *Source* Author (2018)

the number of shacks witnessed an over sevenfold increase from 10,288 to 77,899 over the same duration. As of 1991, brick and block houses comprised as much as 86% of all dwellings in cities and towns, with only 12% of them being shacks. However, by 2011, shacks constituted 32% of all houses in urban areas (Weber, 2017, p. 16; Weber & Mendelsohn, 2017, pp. 16–17).

Moreover, while the estimated number of formal housing units increased from 194,447 in 2015 to 211,894 in 2017, the number of projected informal housing units rose from 114,393 to 138,668 over the same period. The present number of shacks (i.e. 140,000) is over ten times the number recorded in 1991. This number is expected to increase to over 500,000 shacks in about 13 years' time. It was also projected that by 2025, there would be more shacks than formal housing units made from bricks or blocks (Weber, 2017, p. 16; Weber & Mendelsohn, 2017, p. 17–18).

Given the acute housing situation associated with the emergence of shacks, microlending institutions have emerged as an important source of financing for the provision of low-cost housing in informal settlements. For instance, Letshego which is Namibia's largest microlending entity has launched an affordable housing scheme to cater for the needs of those who cannot afford to own housing under existing market conditions due to low incomes and stringent requirements. Moreover, the International Finance Corporation has provided substantial funding worth N$ 820.9 million to Letshego to enable it provide affordable mortgage options to low-income clients (Centre for Affordable Housing Finance in Africa, 2024).

In terms of housing production systems, the National Housing Enterprise, an initiative of the Ministry of Urban and Rural Development, seeks to provide low-cost housing, especially for government employees. Moreover, the Build Together programme, which is a state-funded microlending scheme launched in 1992, aims to provide affordable housing for the low- and ultra-low income who earn less

Table 3.1 Profile summary of informal housing and related factors

Factors	Attributes
Informal housing	• Proliferation of shacks – 1991: 10,288 – 2011: 77,899 – 2015: 114,393 – 2017: 138,668 – 2030 (projected): 500,000
Housing financing	• Microlending initiatives
Housing production systems	• National Housing Enterprise • Build Together Programme – Urban/rural housing loan sub-programme – Social housing sub-programme – Single quarters transformation sub-programme – Informal settlement upgrading sub-programme
Urban land use regulations	• National land policy of 1998 • Communal land reform act of 2002 • Flexible land tenure act of 2012 • Urban structure plans • Town planning schemes • Integrated regional land use plans • Urban and regional planning act of 2018
Rural–urban migration	• Migration for employment, education, income and better quality of life

Source Author from various sources (2024)

than N$3,000 per month (Centre for Affordable Housing Finance in Africa, 2024). This initiative (i.e. the Build Together programme) comprises four sub-programmes focused on urban/rural housing loans, social housing, single quarters transformation and informal settlement upgrading. The urban/rural housing loan programme provides loans to low-income families who cannot access such facilities from banks or building societies. The rationale is to help them either erect new houses or upgrade existing ones. Moreover, the social housing programme provides small local authorities and regional councils with funds to supply welfare housing on a sustainable and subsidised basis to the physically challenged, aged and other disadvantaged groups (Republic of Namibia: Ministry of Urban and Rural Development, n.d.). Also, the single quarters transformation programme seeks to change the single quarters by demolishing old houses and erecting new ones. Similarly, the informal settlement upgrading programme aims to transform informal settlements by providing basic services such as water, electricity and road infrastructure. The key stakeholders involved in the Build Together programme include the Ministry of Urban and Rural Development, local authorities, regional councils, build together committees, community-based organisations and beneficiaries of the scheme (Republic of Namibia: Ministry of Urban and Rural Development, n.d.).

The emergence of informal housing in the postcolonial era may also be partly explained by Namibia's land use regulations. These emanate from the national land policy of 1998, the communal land reform act of 2002, the flexible land tenure act of 2012, urban structure plans, town planning schemes, integrated regional land use plans and urban and regional planning act of 2018. The land policy of 1998 prioritises the needs of previously disadvantaged groups such as women and the poor. Moreover, it underscores the need for tenure security and diverse forms of and rights. Overall, the policy stipulates land as a form of property and promotes equality before the law with respect to this resource. With respect to urban land, the policy promotes land delivery and protects tenure rights for the poor. Moreover, it facilitates informal settlement upgrading and service provision for residents of these areas (Republic of Namibia: Ministry of Lands, Resettlement and Rehabilitation, 1998, pp. 1–10).

Besides the land policy, the communal land reform act of 2002 establishes land boards to oversee the administration of communal land to protect customary land rights (Republic of Namibia, 2002, pp. 5–6). Also, the flexible land tenure act of 2012 provides alternative forms of land tenure (such as a starter title and a land hold title) which can be administered more easily. Additionally, it promotes tenure security for residents of informal settlements and seeks to empower the poor (Republic of Namibia: Ministry of Land Reform, 2016, pp. 4–5).

Moreover, urban structure plans determine the allocation of land uses through the use of tools such as zoning and subdivision. Also, town planning schemes emanate from the town planning ordinance of 1954 which seeks to promote orderly development by imposing regulations that promote the health, safety and general welfare of the population (Weber & Mendelsohn, 2017, p. 40). Integrated regional land use plans aim to proffer suitable land uses for Namibia's regions to promote sustainable development (Republic of Namibia: Ministry of Lands and Resettlement, 2015a, pp. 2–3; 2015b, pp. 5–6). Moreover, the urban and regional planning act of 2018 provides the overarching legal and regulatory ambit for the preparation of urban and regional structure plans, as well as a national spatial development framework (Republic of Namibia, 2018, pp. 11–28).

Besides land use regulations, rural–urban migration is a primary reason accounting for the growth of informal settlements in Namibia. As of 2011, more than 40% of those residing in the two most urbanised regions of Khomas and Erongo were born outside the regions (Namibia Statistics Agency, 2015, p. 23). These migrants are attracted by the prospects for employment and high incomes in urban areas (Pendleton et al., 2014). Similarly, those who migrate to cities for further education are likely to stay after their studies since this is where most of the jobs are found. Coupled with this is the relatively high levels of poverty in rural areas. The situation may be partly attributable to few job opportunities available in sectors such as agriculture. Moreover, the low incomes in this sector especially for subsistence farmers implies that many rural residents rely on social grants and pensions for their income. Thus, young people perceive urban areas as better options for improving their quality of life (Weber & Mendelsohn, 2017, p. 20–21).

3.7 Contrapuntal Urbanism and Housing Delivery in Namibia's Informal Settlements

Having addressed the first research objective, the chapter now unpacks the interrelationships between the factors which configure successful and less successful approaches to urban housing delivery in informal settlements within the postcolonial epoch.

3.7.1 Successful Models of Housing Delivery

To begin with, the chapter adopts contrapuntal urbanism to assess the relationships between housing financing and cooperative production systems, as well as their implications for successful housing delivery for the poor in informal settlements (Fig. 3.2). Housing financing determines the cost of housing. Thus, housing units funded through private commercial ventures tend to be expensive while those created through cooperatives and non-commercial vehicles are more affordable. Housing financing also affects the structural integrity of homes. As a result, housing units in informal settlements associated with limited funding have low quality while those which attract funding are associated with structural improvements. Also, funding determines the kind of services associated with residential housing units. Houses in informal settlements which do not have sufficient funding therefore suffer from poor quality services such as water, sanitation and electricity due to funding challenges. Houses which are well funded however have good services. Ultimately, financing results in strategic investments in housing provision. The strategic aspects emerge from both the cost-effectiveness of the housing provision and its contribution to enhancing quality of life.

Besides housing financing, cooperative production systems define housing supply. This is evident in terms of the roles they play in making up for the shortfalls associated with existing approaches to housing delivery. Moreover, they are pivotal in shaping the typologies of housing units produced. These seek to move away from the dominance of shacks to emphasise decent, but still low-cost housing forms. Also, cooperative produced housing units prioritise the social dimensions of residential units. It may therefore be asserted that these production systems define a sense of community. They are characterised by communal considerations due to the focus on addressing the needs of low-income residents. Housing cooperatives also create a platform for alternative stakeholders to be involved in housing provision. The rationale is to bring on board others such as non-profit and community-based actors who can help to make up for the financial shortfalls associated with existing strategies.

A number of interrelationships may be observed between housing finance and cooperative production systems. To begin with, housing finance or the lack thereof determines whether or not housing cooperatives exist. As may be implied from earlier statements, these cooperatives emerge to make up for the gaps associated with

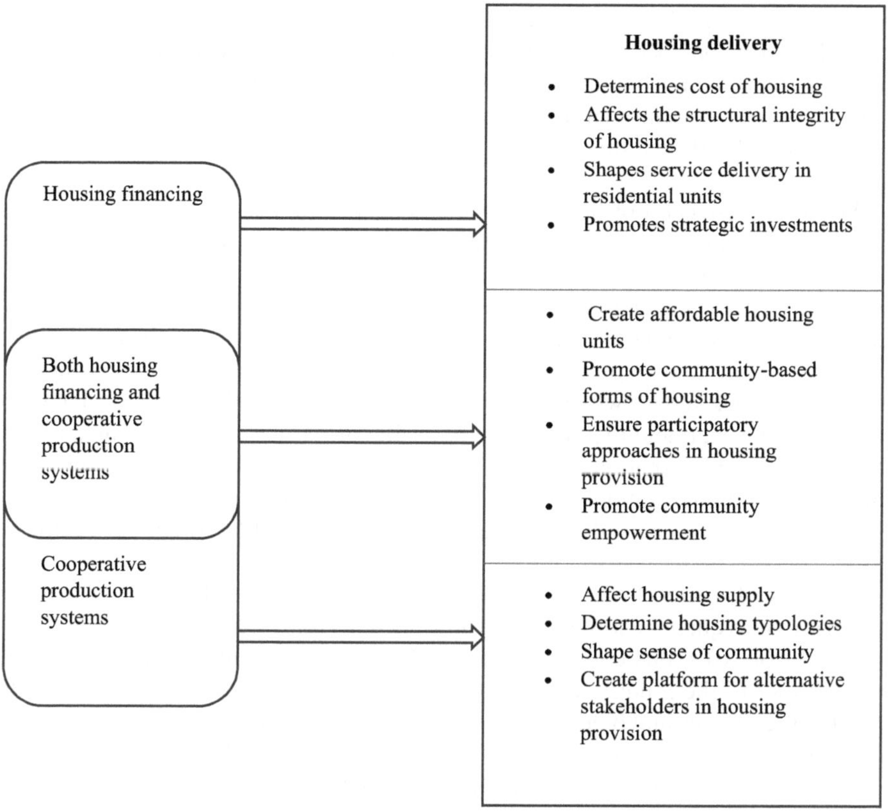

Fig. 3.2 Contrapuntal urban framework for relatively successful housing delivery in informal settlements. *Source* Author (2024)

existing approaches to housing financing. Simultaneously however, notwithstanding their non-profit nature, cooperatives require sufficient financing to undertake their operations. Cooperatives may therefore be regarded as innovative and alternative approaches to mobilise resources for the provision of affordable housing to low-income residents.

The interlinkages between housing finance and cooperative production systems affect housing delivery in various ways. For instance, they define the availability of affordable housing units in informal settlements. This is evident in the funding and cost complementarities they bestow upon each other. Moreover, the two concepts result in the production of community-based forms of housing. These housing approaches prioritise the needs of residents in low-income areas, thus placing less emphasis on market considerations. Similarly, housing finance and cooperative production systems result in participatory approaches to housing provision. Specifically, they bring on board residents, together with community and non-profit organisations to make up for funding shortfalls. Ultimately, housing financing and cooperative production systems therefore serve to promote community empowerment. This occurs through affordability and the provision of decent shelter which improves quality of life.

While the provision of housing for the poor in Namibia's informal settlements has been fraught with many challenges over the years, a number of relatively successful examples can nonetheless be identified. A few years before independence in 1990, a housing cooperative known as Saamstaan (which means "stand together" in Afrikaans) was created in 1987 as a way of empowering the low-income black population to own immovable property for the first time. This was located in Katutura, a black suburb where residents had to depend completely on local authorities under the apartheid regime for the provision of services. It was in tandem with the theme of the International Year of the Homeless. Subsequently, the government through the National Building and Investment Company provided 100% of the loans needed by low-income residents to acquire housing (Habitat International Coalition, 2024).

The Shack Dwellers Federation of Namibia is also a network of 605 savings groups which provides quality and sustainable housing units for residents of informal settlements. As of 2019, this community-driven housing programme comprised 20,400 members across the country. The use of locally produced hollow blocks coupled with the planting of trees and other green concepts help to ensure environmental sustainability in the provision of housing to low-income residents (Centre for Affordable Housing Finance in Africa, 2024). The activities of the Shack Dwellers Federation of Namibia are supported by the Namibia Housing Action Group, a non-governmental organisation which is an umbrella group which facilitates the work of the various savings groups associated with the federation. It was established in 1999 as a trust and separate entity which nonetheless promotes the work of the federation (Shack Dwellers Namibia, 2024).

The various initiatives to provide housing for the poor in Namibia are localised, especially by major municipalities such as Windhoek. For instance, the City of Windhoek has created the Beneficiary Housing Project comprising the Build Together Housing Scheme and Windhoek Housing Scheme. These initiatives are aimed at addressing the housing needs of low-income residents. The Ministry of Urban and Rural Development initially implemented the Build Together Housing Scheme, but this function was devolved to local authorities such as the City of Windhoek in 2000 (Adewunmi et al., 2023, p. 10). The ministry however still provides grants to fund the programme in order to ensure the availability of affordable housing. The beneficiaries of the initiative include individuals and housing groups such as private organisations and the Shack Dwellers' Federation of Namibia. Besides this programme, the Windhoek Housing Scheme provides funding for low-income residents to either construct affordable houses or upgrade their existing housing units. Unlike the Build Together Housing Scheme, funding for the Windhoek Housing Scheme is only provided to those who already own land, but do not have the money to build houses (Adewunmi et al., 2023, p. 10).

Besides the aforementioned initiatives, the Informal Settlement Upgrading Affordable Housing Pilot Project was also launched in 2020 to provide affordable housing units to residents of various informal settlements in Windhoek. These included Havana Proper, Havana Extension 1, Greenwell Matongo, Onyika Number 2, Goreangab Extension 4, Otjomuise Extensions 8 and 9, as well as Freedomland A and B (Rasmeni, 2022). The initiative is a collaborative endeavour comprising the Ministry of Urban and Rural Development, City of Windhoek, Namibia Housing Enterprise and Khomas Regional Council. Between 2020 and 2022, 410 housing units were delivered under this project while 40 were under construction (Rasmeni, 2022).

3.7.2 Less Successful Models of Housing Delivery

Having assessed relatively successful instances of housing delivery, the chapter now looks at constraining factors resulting in less successful approaches to housing provision. Specifically, it applies the concept of contrapuntal urbanism to analyse the interrelationships between urban land use regulations and rural–urban migration, as well as their implications for housing delivery in informal settlements within the postcolonial epoch (Fig. 3.3). To begin with, land use regulations determine the kind of housing units which are allowed in Namibia's urban areas. This is because the regulations affect the value of urban land, which in turn shapes affordability of housing units. Thus, the value of urban land is higher in the central business districts, implying that peripheral areas are more attractive for residents of informal settlements. Coupled with this is the relatively lax enforcement of land use laws in peripheral areas as opposed to more formal neighbourhoods in the city. This makes it easier for shacks to spring up in informal areas than other parts of the city. Moreover, from the historical context, land use regulations promoted spatial segregation based on social dynamics such as race and class. This trend has somewhat continued even after independence further perpetuating the exclusion of low-income residents from formal parts of cities and towns.

Besides land use regulations, rural–urban migration fuels an increased demand for urban housing. This puts pressure on local authorities and the central government to redouble their efforts to attend to the housing needs of residents. However, the inability of these formal actors to supply enough housing results in the emergence of two streams of private sector players who seek to fill the gap. To begin with, some private developers aim to provide affordable housing to meet these needs. Moreover, individual landlords seek to cash in on the housing shortage by renting out their units to migrants. Ultimately, the housing pressures posed by rural–urban migration implies informal and alternative housing typologies such as shacks are more likely to emerge in the urban landscape.

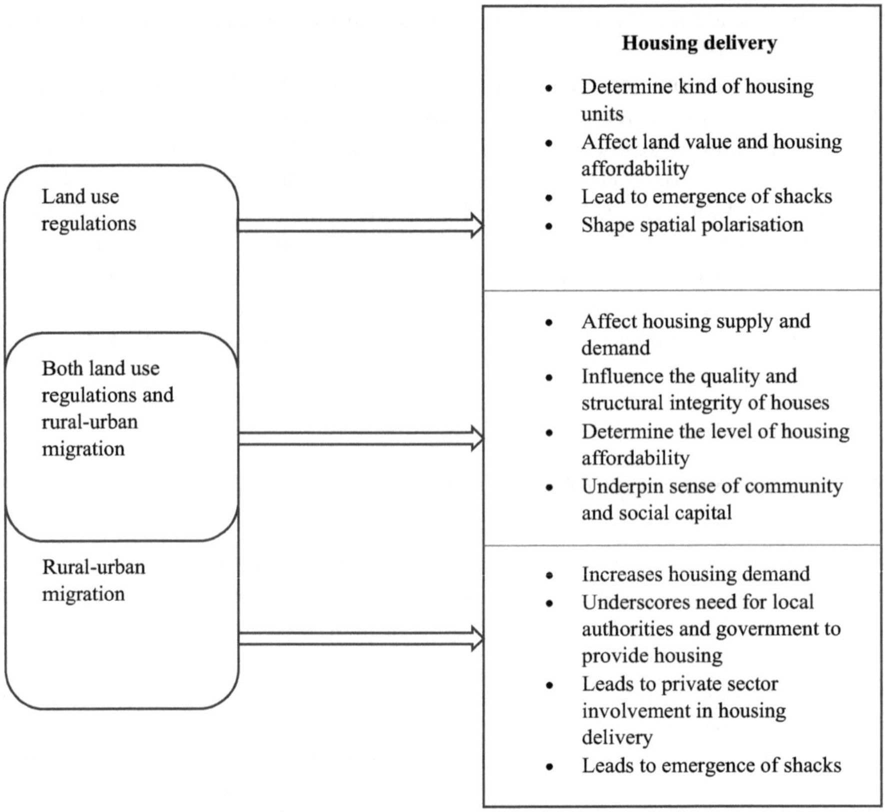

Fig. 3.3 Contrapuntal urban framework for less successful housing delivery in informal settlements. *Source* Author (2024)

A number of inter-relationships may be identified between rural–urban migration and land use regulations. To begin with, rural–urban migration creates the need for speeding up the land delivery process with the aim of enhancing the quality of life by low-income residents in informal settlements. Generally however, Namibia's slow land delivery process with the resultant challenges in land servicing implies less access to land for the provision of housing. Moreover, government financial allocations to the National Housing Enterprise have yielded less returns in terms of the quantum of housing provided as opposed to efforts of the Shack Dwellers Federation of Namibia. This has been worsened by the fact that the actual delivery of housing units through the entity appears to have slowed down from an average of 600 per annum between 1990 and 2022 to 253 between 2003 and 2011 (Sweeney-Bindels, 2011/10, pp. 3–4).

While Namibia's land delivery process is very slow, evidence points to the fact that the pervasive nature of the problem continues to exert strong pressure on municipal authorities to enhance its efficiency albeit with limited success. Moreover, rural–urban migration has led to a temporary form of land use tenure in urban areas. While this is not legal in the strictest sense, it reveals the importance of providing for the needs of a rapidly growing population as a stop-gap measure towards ultimately finding long-term solutions to the challenges they face. Rural–urban migration also creates the need for the allocation of more land for residential purposes especially at the urban periphery. Due to the importance of attending to the housing delivery needs of these migrants, it underscores the urgency of meeting new and future demands for this service.

The interlinkages between land use regulations and migration influence urban housing delivery in a number of ways (Fig. 3.3). To begin with, they configure housing demand and supply. The housing demand is created by the huge influx of migrants eagerly seeking opportunities in cities. For instance, the removal of segregation-based laws after independence made it possible for many citizens to migrate from the north and other parts of the country to major cities such as Windhoek. The issue is exacerbated by natural population increase created by those who have already migrated. In terms of supply, municipal authorities are compelled to make provisions for the housing delivery needs of residents in response to growing demand. Moreover, private homeowners eager to cash in on rising demand also provide their own housing units to target the growing market.

Additionally, land use regulations and migration determine the quality and structural integrity of houses in Namibia's informal settlements. The dynamics create the need for more robust laws which guarantee quality housing units to ensure decent lives for residents. Moreover, they proscribe the use of low-quality building materials to protect the lives of residents. Simultaneously however, the rising population associated with migration means that realistically, residents in informal settlements generally live in low-quality housing. These include backyard housing units which are normally overcrowded. The units, usually made from materials such as wood and scrap metal, also suffer from poor structural integrity.

The interlinkages between land use regulations and migration also affect the affordability of housing units in Namibia's informal settlements. The situation has been exacerbated by the fact that the percentage of government budgetary allocation to the housing sector in Namibia (i.e. 0.3%) is quite low compared to the 2% allocation by neighbouring South Africa. Moreover, unlike countries like the United Kingdom and Philippines which have requirements for a certain minimum number of affordable housing units to be provided by the private sector, no such requirement exists in Namibia (Sweeney-Bindels, 2011/10, pp. 3–4). The situation thus underscores the need to prioritise affordable housing units to satisfy the needs of low-income residents. Municipalities have to embark on strategic interventions for enhancing the delivery of serviced land for purposes of providing housing to meet the needs of low-income residents. Also, property owners need to provide housing units within the income bracket of the target market.

Additionally, the emergence of community networks and social capital in informal settlements is a function of the dynamics of land use regulations and rural–urban migration. These laws promote transience due to the fact they make it illegal to erect permanent structures. The implication is that many of the community networks formed may be temporary in nature. Thus, while migration could result in the emergence of cultural melting pots, the poor security of tenure associated with informal settlements portends negatively for long-term social capital.

3.8 Postcolonial Relational Schema of Housing Delivery in Informal Settlements

Based on observations from the Namibia case study, this chapter proposes a postcolonial relational schema of urban housing delivery for the poor in informal settlements. This model may be applicable in other postcolonial settings besides Namibia, even though some modifications may be required to account for context-specific dynamics. The propounded schema comprises a number of attributes broadly categorised into: spatial and structural planning; and strategic interventions (Table 3.2). Spatial and structural planning comprise the spatiality of informal settlements, propinquity of housing units and housing typologies. Moreover, strategic interventions entail housing financing, housing production systems, urban land use regulations, rural–urban migration, institutional and policy frameworks (Table 3.2). The spatiality of informal settlements is important for understanding the geographical layout of the neighbourhoods within which these residential units are configured. Propinquity of housing units helps to assess the nearness or distantiation of these houses from each other. Also, housing typologies underscore the specific kinds of houses which define informal settlements. Housing financing is important for determining the funding options available for the provision of affordable housing in informal settlements. Besides, housing production systems help to define the options available for the provision of affordable housing to address the existing backlog. Urban land use regulations define the planning regime of cities which affect their allocation of land for informal settlements. Rural–urban migration explains the response of urban areas to population increases which may result in the proliferation of informal settlements. Moreover, institutional and policy frameworks entail organisational approaches and strategic actions by government or municipalities to address housing challenges in informal settlements.

3.8.1 Spatial and Structural Planning

Currently, the spatiality of Namibia informal settlements is primarily characterised by monofunctional and low-income neighbourhoods. This occurs due to vestiges of the apartheid-era form of planning which implies that a high level of spatial segregation is still evident in many urban areas. The implication is that informal settlements are associated with poorly planned neighbourhoods which do not effectively address

Table 3.2 Postcolonial relational schema for housing delivery in informal settlements

Postcolonial attributes	Current situation	Urban futures
Spatial and structural planning		
Spatiality of informal settlements	• Monofunctional neighbourhoods • Low-income neighbourhoods	• Mixed income neighbourhoods • Urban villages
Propinquity of housing units	• Sprawled city model	• Compact city model • Suburban city model
Housing typologies	• Shacks • Backyard housing	• Planned unit developments • Alternative housing typologies
Strategic interventions		
Housing financing	• Market-based home financing options • Emergent cooperative sources of funding	• Provision of credit to low-income persons • Community-based sources of funding
Housing production systems	• Ad-hoc approaches • Governmental and municipal initiatives	• Private non-commercial systems • Cooperative systems • Public sector housing production systems
Urban land use regulations	• Emergent and unfavourable land use laws • Reactive planning	• Zoning for informal settlements • Proactive planning
Rural–urban migration	• Unplanned migration	• Planned migration
Institutional and policy frameworks	• Municipal and government piecemeal initiatives • Private home ownership	• Long-term municipal and government policy interventions • Public–private partnerships

Source Author (2025)

residential housing needs. Towards attaining a more desirable urban future, this chapter proposes mixed income neighbourhoods as a solution to the current urban form. This entails making informal settlements attractive so they attract a variety of residents from different income categories. It will invariably contribute to the quality of housing units as middle- and high-income residents will be able to afford more upscale housing units. Moreover, urban villages are a viable alternative to informal settlements. This implies transforming the settlements into liveable spaces that are accessible and provide housing, amenities, jobs and ecologically friendly neighbourhoods. Specifically, municipalities must invest in creating alternative neighbourhoods that will provide quality residential units and communal facilities which attract new migrants from rural areas.

As regards the propinquity of housing units in informal settlements, the current siting of houses appears to be defined by the sprawled city model. Due to Namibia's abundant land area vis-à-vis its apartheid origins, urban sprawl is a very dominant feature. This is manifested in the relatively wide distances between some housing units, implying a non-optimal usage of land. Moreover, the relative distantiation of informal housing units from formal neighbourhoods lends credence to this assertion. Towards addressing the challenges posed by urban sprawl, future models of urban planning for housing delivery in informal settlements should underscore the compact city model. This entails a more prudent utilisation of land to increase housing densities and cut down on the waste currently associated with land use planning. Admittedly, informal settlements currently have high population densities. However, this is neither commensurate with the density nor quality of the housing units available. Besides the compact city model, informal settlements must also prioritise the suburban city concept. This entails reconceptualising these settlements as a new form of urbanism at the urban periphery with more improved housing conditions to make them more akin to formal housing units.

Currently, the preponderant housing typologies in informal settlements appear to be shacks and backyard housing units. The shacks are sub-par residential housing units built with materials such as wood and scrap metal making them less resistant to hazards and inclement weather. Moreover, as already noted, these housing units are associated with poor quality and low structural integrity. Similarly, many housing units in informal settlements are backyard residential units. These are houses behind the main dwelling with relatively lower rental charges but nonetheless generally sub-par quality. A planned unit development is one possible way in which the challenges associated with current housing units in informal settlements can be addressed. These developments allow municipalities greater flexibility in zoning and land use. The implication is that they can allocate more land for affordable housing units and also require developers to provide physical infrastructure and public amenities which improve the quality of life in informal settlements. Similarly, the rolling out of alternative housing typologies by municipalities would help informal settlements to have better quality of residential life. These alternative housing units would utilise a variety of building materials and also meet the needs of diverse residents.

3.8.2 Strategic Interventions

In terms of strategic interventions, the financing system for the provision of housing is presently characterised by market-based mechanisms. Specifically, private banks and lending groups charge commercial rates on loans granted with the aim of maximising profits. The implication is that many of the housing units available are not affordable to the majority of low-income residents in informal settlements. Coupled with this is the fact that many of the cooperative ventures adopted to provide affordable housing are mostly emergent and still do not cover the majority of residents. This is due to the high demand vis-à-vis the limited financial resources available.

Inclusive financing is therefore critical for the provision of affordable housing. This may entail the provision of credit facilities to low- and middle-income persons who would not have qualified for these under market rates charged by commercial banks (Lam & Feather, 2016, pp. 189–190, 198). Among the poor, incremental sources of funding are normally used to provide affordable housing. These may come from individual or community-based financing. Individual funding may come from personal resources, families or friends. Community-based financing emanate from both formal and informal sources. The formal mechanisms include credit unions and cooperatives. The informal approaches include rotating savings and credit associations, savings associations, as well as accumulating savings and credit associations (Smets, 2019, p. 586).

The production systems associated with housing provision are presently characterised by ad-hoc initiatives premised on a reactive planning approach. These endeavours emanate from the glaring reality of the housing backlog which begs for urgent responses to alleviate the plight of low-income residents. However, they tend to be carried out at a small scale and in incremental formats. Another reality of existing approaches is that they are primarily spearheaded by the central government and local authorities such as municipalities. The implication is that other actors whose resources and initiative can address issues are not provided equal opportunities.

To address the challenges associated with existing approaches to housing production, alternative approaches such as private non-commercial, cooperative and public sector housing production systems can be envisaged. Private non-commercial approaches entail interventions by individual and non-profit organisations with the aim of providing affordable housing. Moreover, cooperative systems involve collaborative or communal approaches to the production of housing units. Also, public sector housing production systems entail strategic interventions spearheaded by public actors such as central and local governments to provide social housing which addresses the needs of low-income residents. These interventions should be carried out in conjunction with other like-minded actors such as non-profit or cooperative organisations to maximise their impacts.

Empirical evidence indicates that low-income residents have the capacity to engage in the co-production of services. This however requires effective governance support structures and community visioning (Adewunmi et al., 2023, p. 1). Collaborative housing thus refers to a situation where housing is jointly produced by the end users in conjunction with the already established traditional producers. The development changes the dynamics associated with housing supply where users who were hitherto passive beneficiaries now play a more active role in ensuring the availability of this basic need. Moreover, the success of this endeavour depends on the roles of primary and secondary stakeholders, as well as actors in the wider environment (Czischke, 2018, p. 77).

Generally, the housing market in informal settlements is associated with low security of tenure due to unfavourable urban land use regulations. Although property owners are victims of this insecurity, renters appear to be the most vulnerable. These people are often subject to exploitation and abuse by their landlords leading to high rental charges and risk of eviction. The situation underscores the need for the provision of more affordable units in the formal housing market (Gunter & Massey, 2017, pp. 25 and 32). Moreover, due to the rapid growth of the informal rental sector, public policy reforms are required to proscribe unacceptable standards in terms of quality, rental charges and treatment of tenants by landlords (Scheba & Turok, 2020, pp. 109–112, 129–130).

In their present form, informal settlements are associated with emergent and unfavourable land use regulations. These may be partly explained by the fact that in Namibia, land governance generally is still defined by power vestiges emanating from the colonial and apartheid eras (Yankson, 2021, p. 75). These are underpinned by a modernist worldview which prioritises orderly physical development to the virtual exclusion of other considerations. Simultaneously however, the layout of these settlements epitomises a reactive approach to planning. This is due to their proliferation with little or no controls exerted by municipal authorities. To address these challenges, it is important for municipalities to start zoning for informal settlements. This entails acknowledging their existence and incorporating them into the existing zoning ordinance with special considerations made for their unique attributes. It is also important to underscore a more proactive approach to planning which anticipates the emergence of these settlements in advance so they can be planned for beforehand. This implies land allocation and servicing prior to the emergence of informal settlements.

Notwithstanding the potential benefits of rural–urban migration, the reality however does not always match the expectations of these migrants resulting in the emergence of informal settlements especially at the urban periphery. Thus, while rural–urban migration was meant to be a coping mechanism against rural squalor, it may end up creating a new form of urban deprivation (Niva et al., 2019, pp. 4–6). Similarly, rural–urban migration may perpetuate existing levels of urban inequality. This occurs due to differences in migrant agency and translocal networks available to them. The ramification is that while some migrants may experience belonging, others may become victims of social exclusion in their new places of abode (Brøgger, 2019, p. 97). Generally, the arrival of migrants on a large scale also puts pressure on the demand for urban housing, resulting in the need for urgent measures to deal with the situation (Montagna & Grazioli, 2019, pp. 577–579).

Currently, rural–urban migration may be regarded as an unplanned form of migration into informal settlements. To begin with, many of these migrants are compelled to move due to the need to improve their quality of life. Coupled with this is the unpreparedness of municipalities to attend to the needs of these migrants once they arrive or settle in their new places of residence. Going forward, it is therefore important for a new planning approach which anticipates arrivals of these migrants and makes provision for them. Part of the approach involves an enhanced system of data collection on migration trends to ensure better planning. Moreover, the migration dynamic must be linked with both current and future housing needs of residents to ensure better planning.

With respect to institutional and policy frameworks, informal settlements are presently associated with piecemeal interventions by municipalities and central government. This is evident in terms of their short-term nature and limited effectiveness in addressing the root causes of urban informality. Moreover, private interests are evident in home ownership which takes advantage of the high demand for housing to make profits. An alternative urban future for informal settlements must therefore emphasise long-term municipal and central government interventions. Also, the promotion of public–private partnerships would serve to leverage the synergies of both public and private actors towards simultaneously improving the quality of housing and delivering value for money by private investors.

Often, the policy choices adopted to deal with informal settlements are conceptualised in terms of entrepreneurial approach which seeks to regain land values for public and private actors. It is however important to note that nuances of these policy options may be shaped by the nature of intergovernmental relations in terms of decentralised or centralised decision-making, local development financing, politics and civil society capacity (Ren, 2018, pp. 79–88). Moreover, while much of contemporary urban policy making in terms of housing provision appears to be largely understood within the ambits of modernism and neoliberalism, state-led interventions nonetheless remain important in certain countries. These government policies on housing seek to improve supply and quality while simultaneously attaining political objectives such as allocating specific housing units to targeted socio-economic groups (Croese & Pitcher, 2019, pp. 401–404, 414–415).

Namibia's housing policy of 2009 underscores principles such as public–private partnerships, innovation, economic empowerment, sustainability, quality, affordability and a pro-poor orientation. Specific approaches and targets include: enhancing infrastructure provision to facilitate the land delivery process; promoting the use of alternative building materials; increasing the stock of housing to a maximum of 2,200 units per year to overcome the supply shortfall; enhancing the access to credit for the purposes of acquiring housing; attracting at least N$1.5 billion worth of resources in private funding for housing and infrastructure provision over five years; formalising 75% of informal settlements by the year 2030 (Republic of Namibia: Ministry of Regional and Local Government, Housing and Rural Development, 2009, pp. 16–20). Similarly, Namibia's housing policy of 2023 underscores the role of partnership building and a community-based approach to addressing the housing backlog. Thus, it is based on a participatory approach to informal settlement upgrading. Moreover, it emphasises the role of public–private partnerships towards the delivery of affordable housing based on a system of credit which can be accessed by low-income persons (Republic of Namibia: Ministry of Urban and Rural Development, 2023, pp. 11–12).

3.9 Conclusions

This chapter has applied the concept of contrapuntal urbanism to the analysis of urban housing. Contrapuntal urbanism is a largely emergent school of thought which examines the combination of factors which work together to configure certain outcomes in the urban landscape. By utilising this concept, the chapter has helped to shed light on how factors such as financing, cooperative production systems, land use regulations and rural–urban migration define urban housing. Moreover, it embodies how the interlinkages of top-down planning and bottom-up dynamics define access to and use of housing units.

The discourse on housing itself has been variously broached in existing literature. However, there appears to be little or no research on this within the lens of contrapuntal urbanism. By adopting this framework, the chapter has unpacked the nuances of relational geography in configuring this endeavour. It demonstrates the socio-spatiality of housing as an aggregation of fragmented and yet converging factors.

Similarly, the discourse enriches conceptual relationships between contrapuntal urbanism and postcolonialism by applying this in the Namibian context. While previous analyses of the dialectical interlinkages between the two terms have explored other parts of the Global South, the Namibian context appears to have been largely ignored. By undertaking this analysis, the chapter has brought alternative context-specific insights to bear on the existing literature.

References

Adewunmi, Y. A., Chigbu, U. E., Kahireke, U., Simbanegavi, P., Mwando, S., Issa, A. A., & Hayford, S. (2023). A multi-faceted approach to improving public services in low-income housing in Windhoek, Namibia. *Sustainability, 15*(6), 1–25.

Average, C. (2019). Low income housing problems and low-income housing solutions: Opportunities and challenges in Bulawayo. *Journal of Housing and the Built Environment, 34*, 927–938.

Brøgger, D. (2019). Urban diaspora space: Rural–urban migration and the production of unequal urban spaces. *Geoforum, 102*, 97–105.

Brown-Luthango, M., Reyes, E., & Gubevu, M. (2017). Informal settlement upgrading and safety: Experiences from Cape Town, South Africa. *Journal of Housing and the Built Environment, 32*, 471–493.

Cai, H., Wang, Z., & Zhang, Q. (2017). To build above the limit? Implementation of land use regulations in urban China. *Journal of Urban Economics, 98*, 223–233.

Centre for Affordable Housing Finance in Africa. (2024). *Housing finance in Namibia*. https://housingfinanceafrica.org/countries/namibia/

Cerutti, E., Dagher, J., & Dell'Ariccia, G. (2017). Housing finance and real-estate booms: A cross-country perspective. *Journal of Housing Economics, 38*, 1–13.

Chen, J., Yang, Z., & Wang, Y. P. (2014). The new Chinese model of public housing: A step forward or backward? *Housing Studies, 29*(4), 534–550.

Chiodelli, F., & Mazzolini, A. (2019). Inverse planning in the cracks of formal land use regulation: The bottom-up regularisation of informal settlements in Maputo, Mozambique. *Planning Theory and Practice, 20*(2), 165–181.

Chipungu, L., Kamuzhanje, J., Makonese, E. D., & Magidimisha-Chipungu, H. H. (2022). Leapfrog developments—Gaps, challenges and opportunities in urban development: Cases of Harare and Durban. *Journal of Urban Systems and Innovations for Resilience in Zimbabwe-JUSIRZ, 4*(1), 45–65.

Chipungu, L., & Magidimisha, H. H. (2021). *Housing in the aftermath of the fast track land reform programme in Zimbabwe*. Routledge.

Chirisa, I., & Matamanda, A. R. (2016). Addressing urban poverty in Africa in the post-2015 period: Perspectives for adequate and sustainable housing. *Journal of Settlements and Spatial Planning, 7*(1), 79–87.

Chiwara, P., & Lombard, A. (2020). Multidimensional poverty in informal settlements in Windhoek, Namibia, and Orange Farm, South Africa. *Social Development Issues, 42*(1).

Croese, S., & Pitcher, M. A. (2019). Ordering power? The politics of state-led housing delivery under authoritarianism—The case of Luanda, Angola. *Urban Studies, 56*(2), 401–418.

Czischke, D. (2018). Collaborative housing and housing providers: Towards an analytical framework of multi-stakeholder collaboration in housing co-production. *International Journal of Housing Policy, 18*(1), 55–81.

Fields, D., & Uffer, S. (2016). The financialisation of rental housing: A comparative analysis of New York City and Berlin. *Urban Studies, 53*(7), 1486–1502.

Goytia, C., Heikkila, E. J., & Pasquini, R. A. (2023). Do land use regulations help give rise to informal settlements? Evidence from Buenos Aires. *Land Use Policy, 125*(106484), 1–15.

Grubbauer, M. (2019). Postcolonial urbanism across disciplinary boundaries: Modes of (dis)engagement between urban theory and professional practice. *The Journal of Architecture, 24*(4), 469–486.

Gunter, A., & Massey, R. (2017). Renting Shacks: Tenancy in the informal housing sector of the Gauteng Province, South Africa. *Bulletin of Geography. Socio-Economic Series, 37*, 25–34.

Habitat International Coalition. (2024). *Saamstan housing cooperative: A case study in Windhoek, Namibia.* https://www.hic-net.org/saamstaan-housing-cooperative-a-case-study-in-windhoek-namibia/

Han, W., Zhang, X., & Zheng, X. (2020). Land use regulation and urban land value: Evidence from China. *Land Use Policy, 92*(104432).

Hanratty, M. (2017). Do local economic conditions affect homelessness? Impact of area housing market factors, unemployment, and poverty on community homeless rates. *Housing Policy Debate, 27*(4), 640–655.

Karuaihe, S. T., & Wandschneider, P. R. (2018). Limited access to services for the urban poor in Windhoek, Namibia. *Development Southern Africa, 35*(4), 466–479.

Kisiała, W., & Rącka, I. (2021). Spatial and statistical analysis of urban poverty for sustainable city development. *Sustainability, 13*(858), 1–18.

Kok, N., Monkkonen, P., & Quigley, J. M. (2014). Land use regulations and the value of land and housing: An intra-metropolitan analysis. *Journal of Urban Economics, 81*, 136–148.

Lam, A., & Feather, C. (2016). Promoting access to affordable housing finance: Morocco's fogarim guarantee fund and US housing finance. *Cityscape, 18*(2), 189–200.

Ma, X., Heerink, N., van Ierland, E., & Shi, X. (2014). Land tenure insecurity and rural-urban migration in rural China. *Papers in Regional Science*, 1–24.

Magidimisha, H. H., & Chipungu, L. (2011). Unconventional housing provision: Reflections on health aspects—A case study of Zimbabwe. *Journal of Housing and the Built Environment, 26*, 469–485.

Monkkonen, P. (2013). Urban land-use regulations and housing markets in developing countries: Evidence from Indonesia on the importance of enforcement. *Land Use Policy, 34*, 255–264.

Montagna, N., & Grazioli, M. (2019). Urban commons and freedom of movement: The housing struggles of recently arrived migrants in Rome. *Citizenship Studies, 23*(6), 577–592.

Mponzi, W. P., Msaky, D. S., Binyaruka, P., & Kaindoa, E. W. (2023). Exploring the potential of village community banking as a community-based financing system for house improvements and malaria vector control in rural Tanzania. *PLOS Global Public Health, 3*(11), 1–20.

Myers, G. (2014). From expected to unexpected comparisons: Changing the flows of ideas about cities in a postcolonial urban world. *Singapore Journal of Tropical Geography, 35*, 104–118.

Nakale, A. (2018). *40% of Namibians live in shacks.* New Era. https://neweralive.na/posts/40-of-namibians-live-in-shacks

Namibia Statistics Agency. (2015). *Namibia 2011 census: Migration report.*

Namibia Statistics Agency. (2019). *Sustainable development goals baseline report: Namibia 2019.* https://nsa.nsa.org.na/new_publications/sdg-baseline-report-2019/

Ndjavera, M. (2022). *World Bank calls on NHE to restructure.* New Era Publishing Corporation. https://neweralive.na/posts/world-bank-calls-on-nhe-to-restructure

Niva, V., Taka, M., & Varis, O. (2019). Rural-urban migration and the growth of informal settlements: A socio-ecological system conceptualization with insights through a "water lens." *Sustainability, 11*(3487), 1–16.
O'Callaghan, C. (2012). Contrapuntal urbanisms: Towards a postcolonial relational geography. *Environment and Planning A, 44*, 1930–1950.
Patel, K. (2016). Sowing the seeds of conflict? Low income housing delivery, community participation and inclusive citizenship in South Africa. *Urban Studies, 53*(13), 2738–2757.
Pendleton, W., Crush, J., & Nickanor, N. (2014). Migrant Windhoek: Rural–urban migration and food security in Namibia. *Urban Forum, 25*, 191–205.
Rahman, M. M., Hassan, M. S., Bahauddin, K. M., Ratul, A. K., & Bhuiyan, M. A. H. (2017). Exploring the impact of rural–urban migration on urban land use and land cover: A case of Dhaka city, Bangladesh. *Migration and Development*, 1–18.
Rasmeni, M. (2022). *City of Windhoek continues to bridge housing backlog: 100 affordable houses handed over to informal settlement dwellers.* Namibia Economist. https://economist.com.na/70927/general-news/city-of-windhoek-continues-to-bridge-housing-gap-100-affordable-houses-handed-over-to-informal-settlement-dwellers/
Ren, H., Guo, W., Zhang, Z., Kisovi, L. M., & Das, P. (2020). Population density and spatial patterns of informal settlements in Nairobi, Kenya. *Sustainability, 12*(7717).
Ren, X. (2018). Governing the informal: Housing policies over informal settlements in China, India, and Brazil. *Housing Policy Debate, 28*(1), 79–93.
Republic of Namibia. (2002). Communal land reform act 5 of 2002. *Annotated statutes.*
Republic of Namibia. (2018). Urban and regional planning act 5 of 2018. *Annotated statutes.*
Republic of Namibia: Ministry of Land Reform. (2016). *Guide to Namibia's flexible land tenure act, 2012 (Act no. 4 of 2012).* Directorate of Land Reform and Resettlement.
Republic of Namibia: Ministry of Lands and Resettlement. (2015a, March). *Baseline report (Vol. 1) for the Zambezi integrated regional land use plan.*
Republic of Namibia: Ministry of Lands and Resettlement. (2015b, March). *Integrated regional land use plan for the Kavango East Region: Baseline report* (Vol. 1).
Republic of Namibia: Ministry of Lands, Resettlement and Rehabilitation. (1998, April). *National land policy.*
Republic of Namibia: Ministry of Regional and Local Government, Housing and Rural Development. (2009). *Namibia national housing policy.*
Republic of Namibia: Ministry of Urban and Rural Development. (2023). National housing policy (second revision). Final draft submission to cabinet: 27 Sept 2023.
Republic of Namibia: Ministry of Urban and Rural Development. (n.d.). *National housing programme: Build together—Housing the poor of the nation.*
Robinson, J. (2011). Cities in a world of cities: The comparative gesture. *International Journal of Urban and Regional Research, 35*, 1–23.
Robinson, J. (2014). New geographies of theorizing the urban: Putting comparison to work for global urban studies. In *The Routledge handbook on cities of the Global South* (pp. 57–70). Routledge.
Roy, A. (2005). Urban informality: Toward an epistemology of planning. *Journal of the American Planning Association, 71*, 147–158.
Said, E. (1994). *Culture and imperialism.* Vintage Books.
Scheba, A., & Turok, I. (2020). Informal rental housing in the South: Dynamic but neglected. *Environment and Urbanization, 32*(1), 109–132.
Selod, H., & Shilpi, F. (2021). Rural-urban migration in developing countries: Lessons from the literature. *Regional Science and Urban Economics, 91*(103713).
Shack Dwellers Namibia. (2024). *Welcome to Shack Dwellers Federation of Namibia (SDFN) and Namibia Housing Action Group (NHAG): Improving the lives of poor people in Namibia.* https://www.shackdwellersnamibia.com/

Smets, P. (2019). Indian community-based housing finance systems: Potentials and pitfalls for urban development and housing improvement. *International Journal of Urban Sciences, 23*(4), 586–600.

Soyinka, O., & Siu, K. W. M. (2018). Urban informality, housing insecurity, and social exclusion; concept and case study assessment for sustainable urban development. *City, Culture and Society, 15*, 23–36.

Storper, M., & Scott, A. J. (2016). Current debates in urban theory: A critical assessment. *Urban Studies, 53*(6), 1114–1136.

Sullivan, E., & Olmedo, C. (2015). Informality on the urban periphery: Housing conditions and self-help strategies in Texas informal subdivisions. *Urban Studies, 52*(6), 1037–1053.

Sweeney-Bindels, E. (2011/10). *Housing policy and delivery in Namibia*. Institute for Public Policy Research.

Weber, B. (2017). Addressing informal settlement growth in Namibia. *Namibian Journal of Environment, 1*(Section B), 16–26.

Weber, B., & Mendelsohn, J. (2017). Informal settlements in Namibia: Their nature and growth. In *Exploring ways to make Namibian urban development more socially just and inclusive*. Development Workshop Namibia.

Wetzstein, S. (2017). The global urban housing affordability crisis. *Urban Studies, 54*(14), 3159–3177.

Wijburg, G., & Aalbers, M. B. (2017). The alternative financialization of the German housing market. *Housing Studies, 32*(7), 968–989.

World Bank. (2017). *Gini index: World Bank estimate*. https://data.worldbank.org/indicator/SI.POV.GINI

Wyly, E. (2023). Response to commentaries: Contrapuntal gentrifications. *Dialogues in Urban Research, 1*(1), 108–115.

Xie, Y., & Jiang, Q. (2016). Land arrangements for rural–urban migrant workers in China: Findings from Jiangsu Province. *Land Use Policy, 50*, 262–267.

Yankson, E., & Allotey, A. A. (2024). Rural-urban migration and the right to the city: Urban social movements in the informal settlements of Namibia and Ghana. In A. Domaradzka & R. Zajonc (Eds.), *Handbook on urban social movements* (pp. 148–166). Edward Elgar Publishing.

Yankson, E. (2021). Land governance as a restitutive mechanism for asserting ownership and tenure rights in postcolonial contexts: Insights from Namibia and Ghana. In H. Chitonge & R. Harvey (Eds.), *Land tenure challenges in Africa: Confronting the land governance deficit* (pp. 75–98). Springer.

Eric Yankson is an Associate Professor of Spatial Planning at the Department of Architecture, Planning and Construction of the Namibia University of Science and Technology (NUST). He is the current Ph.D. Programme Coordinator for Spatial Science and the immediate past Associate Dean for Research and Innovation of the Faculty of Engineering and the Built Environment of NUST. Prof. Yankson is also an Advanced Researcher affiliated with the Global Centre of Spatial Methods for Urban Sustainability at the Technical University of Berlin. He serves on the editorial boards of two international journals: Discover Sustainability and the International Journal of E-Planning Research. He holds a Ph.D. in Urban and Public Affairs from the University of Louisville.

Chapter 4
Land Assembly for Social Housing Developments and the Advancement of Spatial Transformation: Case of the Aloe Ridge Social Housing Project in Msunduzi Municipality

Fulufhelo Nyadzani and Lovemore Chipungu

Abstract Recognising that South Africa is one of the most unequal societies in the world, the government has a pivotal role in equalising the imbalance and unjust socio-economic conditions exposed to the marginalised and disadvantaged communities. The establishment of the social housing programme promised a change in the state of housing provision and an outlook towards spatial transformation for urban centres still staggered by the legacy of apartheid spatial planning and land dispossessions. The spatial transformation envisioned was to be anchored by restructuring zones entrusted with the determination of suitable areas for social housing projects. The study analysed factors impeding the ability of social housing programme in fulfilling a core government mandate of reversing the spatial imbalances in urban areas. The research utilised a mega social housing project 'Aloe Ridge Social Housing' located in Pietermaritzburg to assess the impact of the social housing in transforming urban space. The study draws form the theoretical perspective of right to the city, locational and smart growth theories as foundation for the provision adequate housing, locational choice and inclusive developments. The research literature explored the global comprehension and application of various social housing programmes and the commitment towards inclusive urban centres. A South African perspective on social housing is advanced while articulating the historical implication of colonial and apartheid rule towards housing provision and land availability for low-income households. The study research methodology employed a mixed-method approach towards data collection process, selected for its ability to account for demographics data analysis and the need to comprehend personal experiences of beneficiaries.

F. Nyadzani (✉)
Pelham, Pietermaritzburg, South Africa
e-mail: 221117143@stu.ukzn.ac.za

L. Chipungu
School of the Built Environment and Development Studies, Howard College, University of KwaZulu-Natal, Glenwood, Durban, Republic of South Africa
e-mail: chipungu@ukzn.ac.za

L. Chipungu et al. (eds.), *Contemporary Models for Pro-Poor Housing in Urban Sub-Saharan Africa*, https://doi.org/10.1007/978-3-032-05489-0_4

The research utilised purposive sampling method in locating the study area and acquiring data from the social housing beneficiaries and key informants. The data collected revealed that the social housing programme has performed a significant task in securing affordable housing to low-income households, however its impact towards spatial restructuring in urban centres has been minimal. The research outcome uncovered a policy gap in the link between social housing projects land assembly process and municipal development policies, impeding the charge towards spatial inclusion. The study advances that the social housing restructuring zone guidelines must be reconfigured to ensure compulsory alignment municipal spatial planning plans. It further proposes an increase in the development of inclusionary housing to counter the impact of limited suitable land and costly land assembly process while also sharing the social responsibility with the private sector.

Keywords Land assembly · Inclusive development · Spatial restructuring · Social housing · Spatial planning

4.1 Introduction

The 2030 Agenda on Sustainable Development echoes the importance of urban centres' performance in accommodating more than half of the world's population, and further notes that the figure will increase by two-thirds by the year 2050 (United Nations Development Programme [UNDP], 2015). The Integrated Urban Development Framework (IUDF) (RSA. Department of Cooperative Governance and Traditional Affairs [CoGTA], 2016) states that with such levels of urbanisation, urban centres are incubators of poverty, unemployment, homelessness, and infrastructure, and livelihood inequalities. The IUDF (RSA. CoGTA, 2016) estimates that over 60% of the South African population already resides in urban areas and suffers from the same urbanisation forces experienced across the globe. Continuous influxes to urban areas and prevalent urban poverty have increased the need for housing provision with an aim of addressing the adequate housing agenda and mitigating the negative socio-economic conditions linked to urbanisation. Turok (2015) advances that a critical discourse is vital to unravel the nature of spatial planning and public housing provision in South Africa.

South African Cities Network [SACN] (2004) notes that besides dealing with the impact of rapid urbanisation, South African cities have to contend with the remnants of Apartheid, delayed urbanisation and political transformation. The dual forces of urbanisation and the Apartheid legacy have contributed immensely to the implementation of housing policy and the attempts to bridge the socio-economic divide in existence within South African cities. McKenna (2019) posits that the scale of the housing demand due to the historical supply inadequacy and urban migration has placed immense pressure on land demand and the supply of basic service infrastructure. Aigbavboa (2016) argues that the pre-1994 Apartheid state worked to entrench its land tenure system and endeavoured to limit the availability of land for non-white citizens. Spatial inequality ramifications enforced by a draconian and oppressive legislative framework left unimaginable scars on the South African urban form.

Tomlinson (2002) concurs with the assertion that the current South African government inherited cities that perpetuated and mirrored the Apartheid era's inequalities. The National Development Plan (RSA, 2011) surmises that South African cities are highly fragmented, and that limited progress has been achieved in reversal of the inherent unsustainable growth patterns. The support for urban restructuring in South Africa carries a deeper sense of meaning due to the plight created by the Apartheid government's segregation mandate. The contestation of the urban form has immense sway on public housing provision and the efforts made to fulfil the mandate bestowed on the government by Section 26 of the South African Constitution to provide adequate access to housing to all citizens (RSA, 1996). According to Lee and Singh (2022), housing as a tool for social alleviation is complex and its development without consideration has economic, social, cultural, physical, environmental, and financial ramifications.

This chapter revisits the housing question through the window of social housing. It argues that if the provision of if social housing (and housing in general) is to be achieved, the issue of land must be tackled first. This is simply because land is the key input in the housing development process. Unfortunately, its limited supply has contributed to some of the challenges associated with the development of housing. Hence the idea of using restructuring zones for social housing production has emerged as an innovative approach to unlock strategic land that is in close proximity to opportunities. Housing provision and spatial restructuring are intertwined and as such, an integrated approach is necessary to address the continued lack of affordable quality housing and spatial equality. The government's position on redressing the inherited inefficient and unsustainable urban spatial form is executed through multiple restructuring tools, and among these is the Social Housing Programme. The Social Housing Programme introduced by the Breaking New Grounds Policy (RSA. The Social Housing Programme [DHS], 2004) is viewed as a housing delivery mechanism capable of addressing the provision of low-cost rental stock, performing a catalytic role in stabilising areas experiencing urban decay, and prompting urban regeneration and urban restructuring through economic, social, and spatial integration.

4.2 The Nature of Social Housing Provision

Pohl and Badenhorst (2001) define the primary purpose of the Social Housing Programme as the provision of essential housing services by non-profit social housing institutions at cost to people with limited financial means through the provision of rental stock and rent-to-buy stock. Kholodilin et al. (2022) note that social housing is distinct from rental housing and home ownership. Lundgren (2019) expands on the meaning of social housing by outlining five criteria associated with social housing systems, namely: the nature of the beneficiaries targeted, the nature of tenure, the main social housing provider, and financial assistance mechanisms. Demonstrating how a welfare state influences housing, Kholodilin et al. (2022) identify the determinants of social housing provision. The GDP dictates the scale of the social housing provision; social housing policies should be pro poor; it can be utilised as a complementary tool to rent regulation; and governments with stronger redistributive mandates can implement social housing on a larger scale.

Turok and Visagie (2021) identified three critical ways in which social housing promotes integration; through locating housing for the poor close to job opportunities and public facilities, and through the creation of a supportive living environment. The urban restructuring policy mandate for social housing is unique across all housing provision mechanisms established through the National Housing Code (RSA. DHS, 2009), but this programme is restricted to urban areas to facilitate urban settlement restructuring. To achieve the alluded social housing objectives, suitable urban land is a prerequisite. Manomano et al. (2016) add that the location of housing projects on such suitable land is critical for the overall social housing objective if it is to benefit the beneficiaries' by fostering their socio-economic empowerment.

Scheba and Turok (2021) note that social housing can be a potent tool with which to bridge the divide in cities through provision of quality rental accommodation for low- and middle-income households; yet there is also limited evidence of its impact on the surrounding neighbourhoods, including the attraction of complementary public and private investment on areas of development. The urban restructuring mandate endowed to the Social Housing Programme has been demoted to a secondary objective/priority behind the provision of affordable rental units. The increasing housing backlog has heightened the need to fast track affordable housing provision, and the spatial restructuring objective of social housing has been deemed secondary and now continues to receive less focus. Gerber (2018) states that land availability and administration continue to be a thorn in the side of facilitating the delivery of sustainable human settlements which include social housing development. The availability and provision of suitable land for housing is critical in the housing implementation value chain. The implementation of housing provision requires readily available land to address the continuous rising national housing backlog and the dependency of housing provision on land availability presents a conundrum for all stakeholders involved in the process. In responding to the land challenge, the role of land provision to achieve urban restructuring is becoming increasingly significant. The challenge in providing suitable land limits the potential of social housing to support urban restructuring.

Social housing intervention can manifest in three main forms: regulation; linkage to public policies; and subsidies and direct provision of social housing through public bodies or publicly owned companies. Zadeh et al. (2021) recognise the diversity of social housing across countries in terms of ownership models, the institutions managing construction, rental prices, the funding or subsidy stream, the type of tenure, and the objective of the social housing provision. Morris (2017) reports that globally, social housing has been funded and managed by local, state, or national governments, but it is increasingly becoming managed and owned by non-profit housing providers. The social housing is centred on cooperation between public and non-government sectors.

Fluchtmann et al. (2020) deduce that there are two social housing systems, classified as universal or targeted systems or models. The universal model caters for the broader population while the targeted model focuses on specific segments of the population, primarily low-income and vulnerable populations. Scanlon et al. (2015) observe that the application of social housing programmes has left little distinction between the universal and targeted systems. There has been a transition to targeted models for those social housing systems that were initiated as universal systems, reducing the distinction between the two predominant models. In dissecting the anatomy of social housing, Onatu (2018) cites three types of social housing institutions: Tier 1 umbrella institutions that primarily provide property management services, for a fee, to other smaller institutions. Their mission tends to be the servicing of collective tenant organisations. Tier 2 social housing institutions that develop, own, and manage their own rental stock without any intermediary associations. They deal directly with their tenants. Tier 3 social housing institutions that have been established by tenant groups as a mechanism for acquiring and managing their own housing.

4.3 Land Availability for Social Housing Projects

The first essential condition for a vibrant and well-functioning housing sector is the availability of residential land, in ample supply and at affordable prices (Marutlulle, 2021). Land is a vehicle for inclusive and adequate housing; however, the ability of the state to secure suitable land for housing has been a hindrance to the inclusive drive. Gerber (2018) asserts that land continues to be one of the major problems hindering the delivery of sustainable human settlements in our country. Slow and complex land identification, allocation, and development processes have resulted in insufficient land for housing development purposes (The Fuller Centre for Housing, 2014). Vital and valuable, land holds numerous values for competing stakeholders. The legacy which the country was built on renders ordinary methods ineffective and results in minimal impact on the low-income communities with regard to improving their livelihoods.

The discourses between free market and state interventionists are long and complex (Napier, 2009). Dowall (1993) observes that there are three accepted justifications for government intervention in land markets: elimination of market imperfections and failures to increase operating efficiencies; removal of externalities so that the social costs of land market outcomes correspond more closely to private costs; and to redistribute society's scarce resources so that disadvantaged groups can share in society's outputs. Napier (2009) recognises South Africa's history of exclusion of ownership of prime land and the dispossession of land and acknowledges the impact that this has had on the land market and the justification for the state to intervene. Brown-Luthango (2006) cautions that it cannot be left to the market to address the current development and spatial challenges, as the current operation of the market reproduces and reinforces marginality, exclusion, and poverty. Duncan (2008) posits that rampant land speculation, failure of the central government to regulate the land markets and failure of the municipal governments to provide workable urban development strategies exacerbate the challenges faced when assembling land for housing. The operation of the urban land market is such that it works progressively to displace the poor to increasingly marginal parts of the city over time (Huchzermeyer et al., 2019).

According to Lawson and Ruonavaara (2020), the ever-increasing obligation by the government to assemble land to implement social housing in urban areas continues to be a hindering factor in urban centres' housing provision. Myeni and Okem (2020) identify that one of the impediments restricting access to suitable land is the absence of a legislative framework that incentivise private owners to sell urban land and allow access, since the Housing Development Agency (HDA) cannot acquire many hectares of well-located urban land. The social housing sector is reliant on provinces and municipalities availing such strategic land and buildings through their developmental partnerships with the social housing institutes. The SACN (2016) conducted case studies on the municipal public land assembly process and concluded that municipal SDFs place limited emphasis on the integration of provincial and national spatial investment logic. The lack of coordination in land assembly between government entities further limits the effectiveness of social housing implementation. According to McKenna (2019), the problem with strategic land within restructuring zones is that most properties available for transformational development are often small portions in key nodes or large portions on the periphery of the cities.

Molefe and Nkhahle (2019) advance that when state-owned entities (SOEs) dispose of land, municipalities should have the right of first refusal to acquire the land. The *High-Level Panel Report on the Assessment of Key Legislation and the Acceleration of Fundamental Change* (National Assembly of South Africa, 2017) acknowledges the limitations of the *Government Immovable Asset Management Act No. 17 of 2007* in governing the land owned by state-owned enterprises (SOEs) since much of the well-situated and vacant urban land is owned by SOEs rather than by government departments. The panel further recommends the utilisation of expropriation where land is being held for speculative purposes. Lawson and Ruonavaara (2020) outline the mechanisms capable of supporting social housing development, including land banking or leasing, land re-adjustment, land value recapture, and various regulatory and planning tools. McKenna (2019) further infers that there is an inclination to focus on larger land portions that can deliver large scale projects located on the periphery of urban centres and to avoid the complexities associated with smaller portions of land.

4.4 Advancement of Social Housing with Smart Growth Theory

Shrivastava and Sharma (2011) categorise smart growth as concerned with managing growth in city centres to decrease the urban sprawl and creating compact, transit-oriented, walkable, bicycle-friendly land uses, including neighbourhood schools, complete streets, and mixed-use developments with various ranges of housing choices. Chen et al. (2017), and Litman (2022) allude that smart growth is underpinned by accessibility and mobility. It offers a higher degree of accessibility to the city in order to reach desired goods, services, and activities by reducing the distance between common activities such as home, work, schools, and services. Backhouse (2015) notes that utilisation of the smart growth theory leads to smart cities, and it employs human intelligence to address the challenges that emanate from cities to achieve better living conditions.

Observing smart cities, Backhouse (2015) articulates three criteria that distinguish the smart growth theory's approaches and functionalities from those of other planning theories: infrastructure-based services which employ information and communication technologies to improve governance; business-led urban development that focuses on providing infrastructure to attract specific businesses and industries; and social inclusion, learning and development orientated towards meeting community needs. Smart Growth Theory's applicability to housing developments pertains to the need for the convenient location of services in relation to the housing developments. In order for a housing development to serve its beneficiaries efficiently, supplementary services and amenities must be provided with some degree of accessibility. This will improve the livelihoods of those in the community while reducing the transport and environment related costs. Social housing can be utilised as a tool to achieve the principles of smart growth, and the social housing policy (RSA, 2003) advocates for social housing to be utilised to promote urban regeneration and spatial restructuring. Those elements are being promoted through infill development and densification and by promoting social housing developments along development corridors in order to capitalise on the concept of the agglomeration of the economy.

The South African Smart Cities Framework (RSA. CoGTA, 2016) describes certain conditions critical to the implementation of smart cities principles. The preconditions for becoming smarter require comprehension of the state's current ability and furthermore the existence of enablers to implement smart city technologies. Those enablers include institutional and organisational arrangements, existing digital infrastructure, the capacity of the government officials and community involvement to execute smart growth principles.

4.5 Spatial Inclusion of Urban Centres and Land Availability

Spatial dimension is a cornerstone of social inclusion, and it deserves more attention than it usually receives (Serageldin, 2016). Focusing the lens on spatial exclusion, the World Bank (2015) notes its close association with a myriad of intra-urban inequalities as it deals with accessibility to land for housing and infrastructure for social and basic services. Furthermore, land availability dictates the effects of spatial exclusion in the form of high prices and rapid cost appreciations which outperform income increases worldwide. Land effects have cascading implications on local governments' ability to address most pressing social needs, and this then erodes the inclusionary impact of land. The constraints in the supply of suitable land are also associated with non-progressive development regulations, lengthy administrative processes for land development and land speculation practices. In identifying potential catalysts for inclusion, the World Bank (2015) acknowledges that land is a fundamental building block that enables the urban poor to access other services and to build assets. The locational aspect of housing cannot be separated from the provision of infrastructure for social services.

When addressing the root cause of spatial exclusion, there are potential entry points for investments aimed at facilitating inclusion. These include affordable land and housing, access to basic services and infrastructure, the inclusion of marginalised groups, progressive economic development strategies, investment in data management, proactive investment in urban planning, and capacity building. Inclusion can manifest in a variety of traits that affect the social, spatial, and environmental dimensions. The understanding of the interdimensional influence of these attributes on the overall functioning of urban centres must not be viewed through a restrictive lens when focusing on the spatial dimension's contribution to inclusive cities. Rather, a holistic approach must be utilised to formulate effective and site-specific mechanisms to bring about inclusion. The World Bank (2015) states that as an institution, its financial support must be multidimensional to tackle the challenges of exclusion. The World Bank (2015) has recognised the importance of linking inclusion policies and strategies to financing tools such as result-based financing, which is aimed reaching out to the marginalised populations and eradicating their marginalisation. The influence that spatial configuration has on urban centres ability to support the inclusiveness of human settlements is becoming evident. Acioly (2018) notes that strong legal and urban planning provisions that acknowledge the right to adequate housing (as outlined by the New Urban Agenda) are critical for the cultivation of spatial inclusion and access to affordable and inclusionary housing. Such policy advancements have the potential to mitigate poor communities being driven to peripheral locations that are commonly underserviced to land that is not suitable for human habitation. According to the World Bank (2015) the application of inclusion strategies when formulating infrastructure plans and service provision has the potential to minimise spatial segregation and improve the quality of life of all urban residents.

4.6 Historical Account of South African Housing and Spatial Planning

Strauss (2019) avows that spatial segregation denotes a significant dimension of the historical development of the South African urban settlement patterns which are deeply rooted in colonial oppression. Ile and Makiva (2020) assert that the Apartheid system's exclusion of the black African population from public housing has contributed to the inequality in the country, and the repercussions of this persist today. The *High-Level Panel Report on the Assessment of Key Legislation and the Acceleration of Fundamental Change* (National Assembly of South Africa, 2017) notes that South Africa's statute framework is a narration of the country's history of conquest, domination and racial segregation. The panel concludes that the imposition of the former draconian statutes manifested in the form of a spatial outcome littered by segregation and deep inequities across space. Myeni and Okem (2020) claim that the aggressive approach taken to enforce the segregationist and suppressive laws by the former Apartheid government paved for the way for the racially and spatially segregated human settlement patterns still evident. The legal apparatus enabled the harsh treatment of the majority of the population by the government, and this manifested in a myriad of avenues in a number of sectors including housing, the economy, spatial planning, freedom of movement, and basic services provision.

The aggressive implementation of the various segregationist laws by the Apartheid regime led by the National Party facilitated the emergence of human settlement patterns characterised by different race groups in different locations (Myeni & Okem, 2020). Marutlulle (2021) states that the Apartheid state employed housing as a political tool to further their segregationist agenda. The Apartheid government resorted to preventing African people from accessing housing in towns and cities and simultaneously ensured tight control over increases in housing stock for Africans residing in cities (Thomas, 2010). Walker (2017) further substantiates that the Apartheid government's utilisation of repressive policies of urban influx control, population relocation, and the tribalisation of local administration in the reserves-maintained control of the urban spatial form. The Apartheid spatial planning reinforced the segregated nature of African townships with limited housing opportunities. Confined to mostly unsuitable land, the settlements were afforded limited public housing and were heavily policed in terms of the Native Urban Areas Act. The persistent need for control over the black population in urban centres led to the establishment of the Apartheid City Model that segregated the population along racial lines. The urban centres became synonymous with the physical barriers implemented to prevent cohabitation by different races. The Apartheid City Model further restricted access to socio-economic opportunities including affordable and adequate housing for the black majority who were located predominately on the periphery of the urban areas.

4.6.1 *Human Settlements Challenges in Post-1994 South Africa*

Urban development and the current human settlements policy continue to identify the persistent trio of problems namely, poverty, inequality and unemployment that arose as a result of the Apartheid geography (RSA. NDP, 2011). According to Turok (2011), South African cities have two distinctive identities, and this requires strategic resource allocation and direction on where and how to encourage density and the instruments required to facilitate such densification. Mtantato (2011), and Turok (2011) observe that South African population densities are still relatively low when they are compared to other developing countries, and this continuously limits the ability of the low-income population to improve their livelihood strategies.

Turok (2018) associates the struggle in addressing the spatial imbalances with the government's indecisiveness in its approach to spatial divides. The Centre for Development and Enterprise (CDE) (2017) reveals that the issues of space, place and territory have not received the attention they deserve since 1994. Minimal consideration has been given to internal population shifts, despite their significance in income mobility. Turok (2016) argues that when developing human settlements, minimal consideration has been given to the efficiency of the households (travel costs and distances to work); the wider economy's efficiency (transport costs and congestion from urban sprawl); and the fiscal implications for the urban municipalities. The Financial and Fiscal Commission (FFC) (2011) estimates that the resultant low-density urban sprawl costs the country's economy about R6.4 billion annually.

Dugard et al. (2016) concede that despite the Constitutional Court's reference to the importance of location in housing cases such as the Blue Moonlight case brought before the court in 2011, and the acknowledgement by housing policies of the significance of housing locations, inadequate attention to locational issues regarding public housing still exists. According to Turok (2018), a perceivable threat exists within South Africa's policy and regulatory framework as it continues to locate poor people on cheap and inaccessible land as opposed to places with stronger economies. The Sustainable Development Solutions Network (2014) notes that the right to adequate housing remains unrealised for a distressing number of urban dwellers, especially for the poor, the vulnerable, and special needs groups (migrants, persons with disabilities, older persons, women, youths, and other marginalised groups).

In realising the importance of spatial configuration for communities' livelihoods, Turok (2018) observed that service delivery is costlier in dispersed locations, and the spatial outcome reproduced in South African cities promoted the distortions of the Apartheid city. McKenna (2019) attributed the discrepancy in spatial restructuring efforts to contestation over priorities between scale versus strategic spatial interventions; construction led delivery versus facilitating redevelopment and densification; green versus brown developments; grant funding systems that are geared to support infrastructure development as opposed to mixed financing based on an emerging property market; and lastly, the availability of land and opportunities to deliver housing within the existing urban centres.

McKenna (2019) observes that weak strategic planning results in strategic plans either ignoring the basic economic trends or showing a limited understanding of the importance of key industries which significantly influence the effective functioning of human settlements. In examining the importance of spatial economic analysis, Du Plessis (2014) contends that the interplay and associated trade-offs between the need for economic growth and competitiveness and for socio-economic redress are highly contested within spatial planning and as such, urban spatial plans must provide a comprehensive analysis of the urban economic space and the market and investment trends.

4.6.2 Provision of Infrastructure and Service Access in South Africa

Van Heerden et al. (2022) offer a critical distinction between having access and the level of access in matters of service delivery. The former measures the ability of households to reach a point of interest, while the latter examines the choices available. This distinction is vital in matters of policy formulation and subsequent implementation, as an accurate depiction of the status quo creates the foundation to provide lasting solutions. In sprawling urban landscapes, private transport becomes the dominant form of transportation, further excluding the lower class which is dependent on public transport to access opportunities. The National Treasury Cities Support Programme (RSA. NTCSP, 2018) observes that though public transport is still used extensively in South Africa, it was designed using a monocentric model based on a ‘hub and spoke’ approach. With the evolving urban form and a new polycentric model, the current network is no longer efficient enough to service the working class and transport them to multiple economic nodes within a geographical area. The SACN (2016) acknowledges that due to the spatial composition of South African cities, investment in urban public transport networks and systems is vital to mitigate the spatial marginalisation of the urban poor residing on the periphery of large cities.

Klein et al. (2012) postulate that the urban spatial form is highly influenced by the provision of infrastructure and this further impacts the sustainability, efficiency and inclusiveness of cities and local areas. While analysing the influence of infrastructure in reshaping cities, Klein et al. (2012) further observe that infrastructure departments tend to ignore spatial plans, due to the state’s failure to provide sufficient infrastructure and the private sector subsequently taking charge of infrastructure provision. The National Treasury Cities Support Programme (RSA. NTCSP, 2018) discovered that the pressure on municipalities to expand network infrastructure to cater for urbanisation was massive, especially the provision of services to the indigent and lower-income households for free or at a reduced charge. Intensified by stagnant economic growth, the provision of the mandated basic services has become underfunded or unfunded.

4.6.3 The Drive for Land Reform in Urban Areas

Politically, 'land reform' implies that the state ought to be actively addressing land inequalities (Mkhize, 2015). South African land reform consists of three main pillars, namely restitution of land rights, redistribution of land rights, and improving the security of a wide range of tenure forms (van der Walt & Pienaar, 2012). Gibson (2009) defines land reform as a system comprised of interventions meant to transform the patterns of land ownership, utilisation, and tenure systems to empower those impacted by land dispossession after the enactment of the *Natives Land Act, No. 27 of 1913*. According to Mkhize (2015), land restitution and land redistribution in urban areas take two primary forms: government-subsidised housing (in lieu of redistribution) or financial compensation (in lieu of restitution of land). Mkhize (2015) recognises though that the application of urban land reform is missing the profound impact of the socio-economic status of the urban residents.

The SACN (2015) uncovered that the urban land debate is one of the most neglected but influential issues in post-Apartheid South Africa. Despite the existence of different views in the urban land-reform dialogue, there is a unified argument for an expedited urban land-reform programme across these positions (Huchzermeyer et al., 2019). Huchzermeyer et al. (2019) further state that the urban property market excludes the poor and the urban land-reform programmes need to establish how best to mitigate this imbalance.

Hendricks et al. (2013) note that property privilege still excludes the black majority, and as such land remains a political challenge. Orthopher (2016) observes that urban land reform is hampered by wealth inequality, as it is estimated that at least half of South Africa's wealth is owned by one percent of South Africans. In addition, another challenge for urban land reform is that wealth is partly vested in urban real estate, which ties the fate of urban land directly to the functioning of an important part of the economy (Huchzermeyer et al., 2019). Moyo (2004) describes the conflict that exists within the urban land debate due to converging and diverging interests as constituted by the competing demands for the control of urban land. It is further argued that competing demands on contested ownership arise from actors including the state, at the central and local urban municipality level, and traditional authorities and leaders associated with urban areas.

According to Huchzermeyer et al. (2019), society and business have a social obligation to contribute towards the process of urban spatial transformation, but developers and financial institutions have resisted government's attempts to enforce inclusionary housing requirements. Huchzermeyer et al. (2019) outline a possible approach to address urban land reform by noting that the programmes must focus on more than just land redistribution; emphasis must be placed on land along transit corridors, around development nodes, and in areas with a strong infrastructural base. Moreover, the urban land-reform programme can be complemented with the implementation of rates surcharges on strategically located vacant land and selective expropriation of well-located properties.

4.6.4 South African Housing Landscape and the Gap Market

The period of transition to democracy in South Africa saw drastic changes to the housing legislative and policy framework. Gardner (2017) states that the housing approach in South Africa is considered to be a government aided, private sector-driven approach with a heavy emphasis on low-cost housing for low-income households. The RSA NTCSP (2018) surmises that the South African public housing programme is dominated by large, subsidised housing projects typically located in undeveloped greenfield areas at the city's edge, and that it lacks incentives and subsidies to redevelop neighbourhoods within the city. The RSA NTCSP (2018) further acknowledges that the programme has fallen short as a vehicle for urban spatial and economic integration. The Centre for Development and Enterprise CDE (2018) queried the large disparity in terms of the demand and supply of affordable housing and concluded that the quantity-driven, cost-minimising approach of the government has been the main culprit behind the inefficiencies in drastically altering the dynamics of housing in South Africa. The Centre for Development and Enterprise CDE (2018) determined that the dispersal of the poor and unskilled in isolated and socially homogenous poor neighbourhoods has further amplified the impact of segregation and limited access to agglomeration positives for the economy.

The policy gap tied to income baskets discovered during the implementation of the housing tools enacted to combat the housing backlog became known as the gap market. In the course of housing provision, it became apparent that the government had perpetuated housing exclusion to a significant portion of the population as a result of the policies that it had adopted. The Fuller Centre for Housing (2014) defines the gap market as comprising of the grouping of households earning between R3,500 and R15,000, with earnings considered too high to qualify for low-cost subsidised housing and too low to qualify for mortgage bonds. When reviewing the RDP (1994), the White Paper on Housing (RSA. Department of Housing, 1994 and the National Housing Code of 2000), it was discovered that the programmes were not tailored to respond and assist the gap market. The private housing market does not match the demand for unsubsidised and partially subsidised affordable housing for households earning between R3,500 and R15,000 per month either (Gardner, 2018). The realisation of the unserved segment of the population lead to the reinforcement of the social housing programme.

4.6.5 Social Housing Provision Approach in South Africa

Cloete et al. (2009) trace the formulation of the South African Social Housing Programme to influences by Western European countries and the international housing policies, especially the Netherland's social housing policy and its funding model. The Netherland's social housing model was based on the premise of inclusivity of various income groups and the housing policy was considerate of various tenure options. Cloete et al. (2009) further acknowledge that social housing garnered limited consideration in the founding years of the housing policy in the early 1990s. Social housing found minimal representation in the White Paper on Housing in 1994 (RSA. Department of Housing, 1994), which mainly provided guidance on low-cost housing provision. Equipped with a better comprehension of the impact of diversifying the housing market, the government introduced a social housing subsidy with the National Housing Code of 2009, which followed on from the introduction of the expanded guidelines of the Social Housing Programme under the Breaking New Ground Policy in 2004 (RSA, DHS, 2004).

The Social Housing Policy for South Africa (RSA, 2003) defines South African social housing as a rental or co-operative housing option for low-income persons at a level of scale and built form which requires institutionalised management, and which is provided by accredited social housing institutions or in accredited social housing projects in designated restructuring zones. The Social Housing Regulatory Authority (SHRA) (2020) elaborates that South African social housing projects include high-rise, medium-rise (walk-up) and low-rise (single-storey) housing located on contiguous sites or scattered across urban areas with density ranges from 60 to 200 units per hectare. Scheba and Turok (2021) observe that the Social Housing Programme's focus was originally not directed towards the extreme poor but rather towards the working-class households by relocating them to better locations. Scheba and Turok (2021) note that when evaluating the progress of social housing, the Department of Human Settlements recognised the lack of policy guidance for the programme as projects were initiated with inadequate skills levels and uncertainty in financing.

Onatu (2018) notes that in South Africa, the funds to implement the Social Housing Programme are availed through the national government's budget apportion to the Department of Human Settlements on an annual basis, and such allocation to individual provinces is regulated by the *Division of Revenue Act, No. 1 of 2018* and the budget allocations for social housing projects have been decreasing yearly. The SHRA (2020) notes that on average, the unit costs of social housing have increased annually by 11%, and the escalation costs are also closely associated with the 6.4% annual inflation rate. Scheba and Turok (2021) have discovered that the current social housing financing mechanism hampers the ability of social housing institutions to assemble strategic land due to the assumption of the land price equating to R30,000 per unit, while the realistic price for well-located land is significantly higher.

The Social Housing Regulatory Authority and Department of Human Settlements (2019) outlined a set of guidelines intended to provide clear direction for the implementation process for social housing. Core to these guidelines is the need to improve beneficiaries' access to socio-economic amenities. Social housing projects must be located in close proximity (less than 500 m) to established public transport routes and stops and be within a 15-min walk (±800 m radius) from other amenities. The housing projects must have access to at least three of the following amenities: public transport, schools, established economic centres, retail and convenience stores, health care facilities, leisure and recreational facilities, and an open space. For new developments, the projects must demonstrate alignment to an adopted urban design framework and demonstrate that essential social amenities related to education, health, and security have been considered and included as key amenities in the planning stage or that the capacities that already exist will suffice.

4.7 Research Methodology

The research employed a mixed-method approach to the data collection process. The mixed-method approach was selected based on its ability to account for quantitative data related to demographics and the income analysis of the social housing beneficiaries. The data required to comprehend the personal experiences of the beneficiaries was acquired through qualitative data collection. Creswell (2009) asserted that the different data collection methods all have inherent flaws, so the utilisation of a mixed-methods approach is effective in neutralising the perceived biases of the other methods as it seeks convergence across the quantitative and qualitative methods.

The research was case study driven and focused on the Aloe Ridge social housing project to determine the forces acting against the success of the social housing project. The research dived into the experiences of the social housing beneficiaries, the municipal role in the development of the project, and the surrounding communities' experiences of interacting with the project. The case study approach enabled the researcher to analyse the specific deliverables of the social housing project without much generalisation. The data gathered employing these primary sources provided a clear indication of the project status quo, an understanding of how the theoretical framework can be efficiently implemented and the possible mechanisms and approach that are compatible with the South African social housing environment.

Purposive sampling was utilised during data collection. According to Creswell (2021), purposive sampling involves identifying and selecting individuals or groups who have intimate knowledge about the research study through their personal experiences. Utilising purposive sampling, the researcher chose the case study of the Aloe Ridge social housing project. It is the largest social housing project within the Msunduzi Local Municipality, and it was completed in 2017. At the time of its completion, it was the largest social housing project in the country. The project's characteristics provided a credible platform to test the hypothesis postulated by the research, as it was a recently implemented project and provided a relevant baseline against which to measure the current social housing policy framework and guidelines.

The sample size of 70 participants was chosen and they were selected from among the beneficiaries of the Aloe Ridge social housing project based on a variety of attributes. Firstly, the case study project consisted of 952 social housing units and to achieve viable representation, the sample size had to equate to 7% of the beneficiaries residing in the housing units. This sample of housing beneficiaries was supplemented key informants' interviews with municipal officials tasked with planning and housing developments.

4.8 Location of the Study Area—Aloe Ridge

The project is located approximately 8 km from the Pietermaritzburg CBD and 7 km from Scottsville shopping centre. Aloe Ridge Social Housing is located within Msunduzi Municipality and managed by Capital City Housing. The Aloe Ridge social housing project was initiated by the Municipal Human Settlement unit in a quest to supplement the existing social housing rental stock consisting of the Acacia Park and Signal Hill social housing project. The main objective of the project was the continuation of affordable housing provision by the municipality to curb the utilisation of informal backroom rentals which further promotes formalisation. The location identified as suitable was a vacant piece of land approximately 14 ha owned by the municipality. After the identification of the site, the municipality a undertook the land preparation process which comprised of the rezoning of the property to General Residential zone as to accommodate a high-density residential development. Project was design to accommodate 952 units (KZN Human Settlement, 2020), with each unit approximately 45 m^2 in size with 2 bedroom and open plan living space. It was designed to accommodate 4,000 people, and the cost was R353 million. Financed through the Social Housing Programme, the project was funded by the Department of Human Settlements, Social Housing Regulatory Authority, National Home Financing Corporation and Capital City Housing. Political interference has been one of the major problems after the completion of the project. In 2018, the project was thrown into chaos with the liberation struggle veterans being accused on hijacking 300 units.

4.8.1 Data Presentation

4.8.1.1 Msunduzi's Urban Challenges in Driving Inclusiveness

The municipality is a focal point for the district's commercial and social enterprises, and as such it has a higher service demand compared to other adjacent nodes. The municipality has recognised the rapid population growth, increased housing demand, and lack of land available to accommodate the increasing population and housing and the urgent need to develop the appropriate tools to manage this urban growth. The Msunduzi Urban Networks Strategy (Msunduzi Local Municipality, 2022a) and the Municipal SDF (Msunduzi Local Municipality, 2022b) unveiled the struggles of the municipality, and Pietermaritzburg in particular, to eradicate the spatial configuration of the Apartheid system. With black African townships predominantly located on the city's margins and the lack of quality socio-economic amenities in these townships, the effects of suppression still influence the livelihoods of these communities. The SACN (2016) defined the municipality's challenges to spatial restructuring as topography, an Apartheid urban structure, and the municipality's fear of disturbing its tax base by bringing development closer to the centre and upsetting the upper/middle class residents. Additionally, policy interference during the implementation of the SDF and minimal collaboration in the alignment of programmes and projects has limited the pursuit of spatial transformation.

4.8.1.2 Msunduzi Municipality Spatial and Housing Programme

The SDF provides a housing toolkit for the various housing typologies supported within the various zones. It further acknowledges the priority housing development area identified by the HDA in recognition of the high-pressure areas requiring immediate housing provision attention. However, the restructuring zone approved by the Department of Human Settlements does not find recognition throughout the SDF's analysis of the housing need and proposed strategies. The Msunduzi housing policy acknowledges that spatial restructuring strategies have been part of the discourse throughout the municipality's policies, indicating the municipality's commitment towards breaking down spatial barriers. However, instead of a compact urban form, the municipality has witnessed an increase in the size of the peripheral areas in the Greater Edendale and Northern Areas. The housing policy thus recognises the dire need to unlock strategically located land for high-density developments and housing opportunities for the poor.

The Msunduzi housing policy outlines the following performance indicators for social housing: the municipality has to start three social housing projects per annum and one thousand residential units have to be delivered per annum. Neither of these targets have been reached since the adoption of the policy. It is however noted that the municipality is committed to the development of social housing within the restructuring zones, through a phased approached. The municipality will prioritise the availability of state or municipal owned land and bulk infrastructure and limit any community pushback to the development of social housing.

A disjuncture is perceptible in the location of the social housing restructuring zones in relation to the spatial plan's objectives. The Pietermaritzburg CBD and Raisethorpe CBD are categorised as major nodes within the municipality, while the Edendale Corridor is in a state of transition. The remaining areas are located far away from the socio-economic activities in the municipality and are furthermore outside the areas earmarked for densification or classified as nodes and corridors for development, bringing into question the methodology utilised to identify the areas for spatial restructuring and development.

4.8.1.3 Aloe Ridge Social Housing Beneficiaries' Access to Services and Economic Opportunities

The social housing norms and standards (Social Housing Regulatory Authority & Department of Human Settlements, 2019) dictate that social housing projects must be located within 500 m of public transport routes and stops, and within a 15-min walk (±800 m radius) of other amenities. The project must have access to a minimum of three of these amenities: schools (preschools, and public junior and senior schools);

established economic activities and employers such as office parks, industrial areas, retail and convenience stores; pay points for municipal accounts; health care facilities; leisure and recreational facilities; and open spaces/parks. The aforementioned facilities must also be within 15 min of travel time when utilising affordable public transport (buses, taxis, and trains). The social units' beneficiaries require the provision of essential services, and these may be existing or new facilities. The planning and development of the project did not invest in supplementary social facilities and instead relied on existing ones. The Msunduzi Local Municipality guidelines for urban development sourced from the Urban Network Strategy (2022a) note that developments within priority nodes and corridors must promote mixed-use activities within an 800 m radius, and that translates to a 15-min walking distance. This developmental approach is supposed to be anchored by an accessible, efficient, and affordable public transport system.

The nature of the existing municipal public transport system is that it is subjected to traffic congestion during peak hours and the estimated travel time increases. Moreover, pick-ups and drop-offs, diversions, and utilisation of secondary routes to maximise coverage and profits increase the average time travelled. The public transport mode accessible in the Aloe Ridge social housing project is taxis; the area is not covered by the public bus network and there are no commuter rail services within the municipality. The taxi fare from the Pietermaritzburg CBD to Aloe Ridge is R15 per trip, and the Pietermaritzburg CBD is utilised as the hub due to its dominant economic output as it houses the sector departments and municipal offices. A beneficiary of this social housing project spends an average of R600 on transport fares to commute to town monthly. If a second taxi is required to access a secondary node, then the cost is higher. The cost is also higher if more than one household member is travelling on a daily basis. The trips undertaken by the respondents indicated that the main destinations were the Pietermaritzburg CBD, Edendale, and Scottsville respectively. Considered as the larger nodes within the municipality, they represent access to both social facilities and economic opportunities for the municipal residents. The respondents indicated that the major reason for travelling to those particular destinations was access to employment.

The existence of tuckshops within close proximity to the social housing project can be construed as adequate access to commercial facilities. Observations by the researcher indicate that the tuckshops/corner shops only supply perishable goods and the variety of products on offer is limited. The businesses are small scale, and mostly operated in conjunction with a dwelling as a mechanism to supplement the household's income. The limited products on offer compel beneficiaries to travel to commercial nodes in search of variety, quality, and affordable products.

4.8.2 Discussion

4.8.2.1 Inclusive Urban Development

The recognition of the responsibility that the local authority has to promote and facilitate an inclusive society is monumental in the pursuit of equal opportunities for every citizen. That responsibility was acknowledged by the municipal officials interviewed; however, the challenge has always been the implementation of an integrated and coordinated approach and investment in resources. Furthermore, the lack of an overall municipal approach to achieve inclusion has left each municipal section to focus on individual priorities that are, at times, contradictory. The current municipal policy framework does provide a vision of what inclusive growth might resemble in the future, but like all under-resourced municipalities, there is inefficient commitment to the implementation plan and monitoring of the projects implemented against the strategic objectives is non-existent.

4.8.2.2 Travel Destination and Distance Travelled

The extra distances that the social housing beneficiaries have to travel to access economic nodes symbolises a disjuncture between the social housing regulations and policy implementation realities because they are from the low-income group and cannot afford this. This policy gap indicates minimal adherence to inclusive urban principles which advance the need to locate the low-income population in suitable environments that reinforce and promote their social mobility. The limited consideration of the spatial proximity of the social housing project to the major economic nodes has a potential impact on the beneficiaries' access to amenities. The social housing location also provides minimal contribution to the efforts by spatial transformation to address the Apartheid spatial policies' injustices. The lack of observation of the social housing guidelines pertaining to the spatial restructuring mandate of the Social Housing Programme thus leads to compromising of the standards enshrined in the social housing guidelines.

4.8.2.3 Transportation Choice for the Aloe Ridge Social Housing Beneficiaries

The vehicle ownership data from the beneficiaries revealed an income band of R3,500 to R15,000 and most of the individuals in this band cannot afford to own private vehicles. Data indicates that 76% of the beneficiaries do not own private vehicles and 85% of them depend on public transport to commute to work or to social activities. The data indicated that the availability of reliable and affordable public transport is vital to the livelihoods of the beneficiaries. Furthermore, proximity to socio-economic activities and opportunities is paramount to minimise the costs associated with using public transport.

The smart growth theory notes that a vibrant and inclusive urban centre must be underpinned by an efficient public transport system. The high degree of public transport utilisation by the beneficiaries seems to support the application of the smart growth theory; however, the limited choice in the variety of public transportation and the cost associated with the distances travelled places a heavy burden on the social housing beneficiaries' livelihoods. The efficient operation of the social housing project will require supplementation of the existing public transportation (taxis) with other forms to ensure flexibility in the transport choices and affordability for the users.

4.8.2.4 Municipal Role in Social Housing and Spatial Planning

The municipality, together with the social housing institutions, bears the responsibility of identifying potential social housing sites that are fully compliant with the municipal regulations and social housing guidelines. The municipality fulfils this role by searching proactively for suitable land upon which to accommodate social housing opportunities. The municipal spatial development plans do outline the priority corridors earmarked for densification and intensification. Theses corridors are located along the major transport routes that connect the different nodes around Pietermaritzburg city. Although the Aloe Ridge social housing site is located along one of the gateways to the city, its an area that lacks pull factors to facilitate the agglomeration of activities required for full spatial integration.

The realisation of the spatial improvement of the urban system was never the main objective, hence the utilisation of this large site on the outskirts of the urban centre. The integration of the existing public transport system to support the development was not considered, nor were any resources directed towards improving the transportation system available to the beneficiaries. Further to that, limited attention was directed towards the provision of supplementary public/social amenities such as low-order commercial activities, preschools, primary and tertiary education facilities, and municipal services.

4.8.2.5 Linkage Between Restructuring Zone and Municipal Policies

The local authority noted that a link exists between the housing sector plans and the spatial plans. The Housing Sector Plan was adopted by the municipal council in 2020, and the Spatial Development Framework was approved in 2020. The SDF provides a breakdown of the various housing typology demands of the current population and the projected growth over the next five years. However, the SDF provides no reference to the existence of the approved restructuring zone and the importance it has in facilitating urban integration through housing and infrastructure projects. Furthermore, the location of the social housing project do not align with the densification guidelines of the SDF.

The disjuncture identified between these strategic municipal policies has hindered the municipality's drive towards an inclusive city with the continued utilisation of peripheral land for social housing projects, due the land being affordable or already owned by the municipality. Such an approach addresses the short-term goal of affordable housing but compromises the long-term social mobility goal of the Social Housing Programme to ensure that the urban centre's spatial form supports the livelihoods of its citizens. The existing strategic policies thus do not fully reinforce the principles of the New Urban Agenda and the drive towards compact development.

There seems to be confusion between the Town Planning unit and the Human Settlements unit on who is responsible for overseeing and monitoring the implementation of the restructuring zone's objectives. The existence of minimal coordination between these two critical units in matters of housing and planning has diminished the role and recognition of the restructuring in all municipal policy documents. Urban governance matters have been at the forefront of the inclusive cities approach and have additionally been linked to the Right to the City. Clearly defined roles and responsibilities of the local governance in matters of spatial transformation and housing provision are therefore critical. The inefficiencies in the administration of the adopted municipal restructuring zone have hampered the support required to transform the urban spatial form successfully. This has led to the continued placement of the marginalised in locations that do not offer them socio-economic emancipation.

4.8.2.6 Land Assembly Process for Social Housing by the Municipality

The Msunduzi Local Municipality officials noted that the land assembly process for the housing project was executed by the Human Settlements unit, and the Town Planning unit only advised on the potential suitability of the portions of land that were identified. It was noted that land portions owned by the municipality were the first priority because they minimised the land acquisition costs and limited the administration that would be required when dealing with privately owned properties. When looking for sizeable tracts of municipal owned land, it is evident that the large portions are mostly located on the outskirts of the urban centre. The policy for the disposal of municipal land has also complicated the land assembly process as land capable of accommodating social housing developments has been allowed to be sold or leased without consulting with the Human Settlements unit. The municipality has always been reluctant to acquire privately owned land due to the limited financial resources at its disposal, but the municipal officials noted that land owned by the government sectors and public enterprises can supplement the demand for housing development. Unfortunately, the administrative process required for the deals to secure this land makes it difficult to secure the funding required.

The location theory alludes to market forces playing a role in the availability and accessibility of prime property locations. The concept is clearly visible when dealing with land assembly for public housing, with the government only has limited financial resources for land acquisition. The result is housing projects that do not provide any assistance towards spatial transformation and partly fortify the old Apartheid planning doctrine. The municipal approach towards land assembly for social housing also circumvents the social housing regulations which emphasise that socio-economic and spatial integration must be a priority of the programme. The limited success in the assembly of land for social housing thus far highlights the need for a change in the approach to smart growth intiatives in terms of its advocacy for compact development and the application of technological analyses for the selection of potential sites to weigh up their spatial strengths and possible contribution towards the beneficiaries.

4.9 Recommendations

The study assessed the progression of the social housing policy's implementation and concentrated on the matters concerning land assembly and spatial restructuring. The analysis conducted uncovered a propensity to relax the social housing guidelines when dealing with mega projects, and which hampers the efforts to achieve true spatial transformation. The gaps identified within the existing social housing policy framework and the implications of these gaps have necessitated the need to articulate possible mechanisms, projects and approaches that can be implemented to improve the way in which the social housing programme is currently being implemented.

The municipality is facing a shortage of suitable land that is readily available (owned by it) and that can be utilised for the development of social housing projects. The municipality can thus institute an inclusionary housing policy to negate the need for a public land supply. This inclusionary housing policy can mandate private developers to provide a certain percentage of their housing developments for social housing for the low-income group The policy implementation process must be tailored to suit smaller municipality and secondary town conditions where smaller scale developments might be prevalent in comparison to the size of the developments in the larger municipalities. The policy must also seek to place the low-income population along strategic corridors or within vibrant nodes of development where access to resources can be maximised.

The urban policies pertaining to land administration, spatial planning and housing provision must be reviewed and aligned, as this is vital to combat the exclusionary forces within urban centres. These urban policies, once properly aligned and linked by a shared vision for the city, will form the foundation to combat the urban housing inefficiencies. Social housing provision is more than just providing rental units to tenants. It requires coordinated efforts to actualise the supplementary services, opportunities, and infrastructure necessary to make the projects sustainable. An integrated implementation framework must be designed that makes all service departments responsible for the end product. The requirements and responsibilities must be detailed methodically to ensure communication between and accountability by all service departments. This initiative will reduce the occurrence of unfunded mandates/projects being imposed on the municipality due to the sector departments' lack of commitment to the projects' outcomes, as has been observed with the Aloe Ridge social housing project.

There are multiple indications of gaps within the existing social housing regulations and within the norms and standards on spatial restructuring. These have resulted in social housing projects, including the Aloe Ridge project, having minimal impact on the realignment of the structure of the surrounding urban settlements. A policy review is warranted to supplement the existing guidelines, and the reviewed policy must define spatial restructuring more clearly and provide the measurable indicators necessary to assess whether or not a project has fostered a spatially efficient and inclusive urban centre. A review is required of the existing restructuring zones that have been identified. The locations of these restructuring zones must be reviewed for better alignment with the spatial development framework and the urban development framework. The current articulation of several restructuring zones situated in non-priority residential neighbourhoods and nodes will have limited capacity as developments in these areas stretches the resources available that could otherwise be directed towards prioritised areas identified by the recently reviewed spatial plans.

The McKinsey Global Institute (2014) cautions that it is extremely difficult to overcome the problems caused by the development of housing in the wrong location. A land assembly framework is required to enhance the identification, records, management, and release of the land available for social housing. The municipal land register must be updated as this is vital in this endeavour. The updated register must be shared with the municipal housing and planning units to ensure long-term planning coordination and comprehension of the land resources available. A taxation policy should be introduced to deter land speculation and the existence of dilapidated buildings. The tax policy can be used to stimulate development on suitable vacant land. It can further include property tax exemptions for new housing developments, and penalties in the form of idle-land taxes for property owners who choose not to sell their land. The sector departments and state-owned entities that are reluctant to cooperate with the municipalities by sharing information on land at their disposal, reduces the ability of the municipalities to source suitable and affordable land for housing. Therefore there is a need to adjust the inter-governmental guidelines on land disposal and make the municipalities the first priority when the government departments and state-owned entities are encouraged to dispose these properties.

The current social housing subsidy does not cater for the acquisition of privately owned land, thus limiting local governments' efforts to locate social housing in the best possible localities. The shift to peripheral locations was made to counter the escalating costs for housing development within the urban centres. The introduction of a subsidy mechanism to assist municipalities to acquire prime land to facilitate spatial restructuring will be beneficial as it will respond to some of the shortcomings of the Social Housing Programme. The powers and resources conferred to the Housing Development Agency to assist municipalities with land acquisition matters do not seem to be closing the gap between the demand for housing and the supply of suitable land. This subsidy mechanism must be linked to the updated restructuring zone guidelines to offer synergy and alignment during implementation. The approach will negate the misappropriation of the funds for projects that do not support spatial restructuring efforts. The long-term effect will be the reduction of social housing projects similar to Aloe Ridge being situated on the outskirts of urban centres in an effort to minimise the land acquisition costs.

4.10 Conclusion

Rydin (2013) observes that the desire to confront the urban order status quo and introduce a more inclusive approach does not cease. It does, however, fluctuate over time as it is influenced by the realisation of injustice; the belief that it can be addressed; the willingness to do what it takes to right the injustice; and the technical and human ability to achieve it comes together. The South African government has acknowledged the need to play an active role in the provision of housing to counter the lingering effects of Apartheid's segregation policies. While the polices adopted for housing provision and the resources committed symbolise the government's willingness and desire to change the status quo, it is evident that effective policy implementation continues to lag behind, prolonging the suffering of the low-income communities residing in inadequate housing and unsustainable urban settlements.

The results indicate that the attempt to provide mass numbers of rental units has been accompanied by a systemic shift to mega projects in major cities, aimed at expediting the rate of rental units' provision to the masses. This has resulted in the short-term success of the social housing programme and a political win for the state. However, when such projects are located on the periphery of city centres, it removes the benefits of a convenient location for the beneficiaries and thus limits the advantages that social housing is supposed to impart to communities. While the sacrifices made to locate and provide housing units on the outskirts of city centres has allowed local governments to increase the number of housing units provided, it has minimised the effects of spatial restructuring and diminished the impact of the social housing programme in reversing the spatial legacy of the Apartheid state. Fortunately, there is always space for improvement and policy implementation is a continuous cycle that requires inwards and outwards reflection to keep up with the ongoing socio-economic challenges encountered by communities.

References

Acioly, C. (2018). Affordable housing is key for the realization of an inclusive and human rights-based city.

Aigbavboa, C. (2016). A dialogue on the genesis of housing problems in South Africa. In *41st IAHS World Congress Sustainability and Innovation for the Future*, 13–16th September 2016, Albufeira, Algarve, Portugal.

Backhouse, J. (2015). Smart city agendas of African cities. University of the Witwatersrand.

Brown-Luthango, M. (2006). Capturing unearned value/leakages to assist markets to work for the poor. Urban LandMark.

Centre for Development and Enterprise (CDE). (2017). *The growth agenda priorities for mass employment and inclusion.* ISBN: 978-1-920653-21-7.

Centre for Development and Enterprise (CDE). (2018). Building better cities: new approach to housing and urban development. The Growth Agenda. https://www.cde.org.za/

Chen, G., Jue, J., Xu, L., & Wang, N. (2017). *Comprehensive modelling and planning of urban smart.* Atlantis Press. 978-94-6252-355-5.

Cloete, J., Venter, A., & Marais, L. (2009). *'Breaking New Ground', social housing and mineworker housing: The missing link.* University of the Free State.

Creswell, J. W. (2009). *Research design: Qualitative, quantitative, and mixed methods approach* (3rd ed.). SAGE Publications.

Creswell. (2021). *A concise introduction to mixed methods research.* SAGE Publications.

Dowall, D. E. (1993). *The role and function of urban land markets in market economies.* [Workshop on privatization of land in Ukraine]. Ukraine Ministry of Construction and Architecture, State Committee on Land Resources, and United States Agency for International Development. http://www.urbanlandmark.org.za/downloads/Dowall_Role_and_Function.pdf

Du Plessis, D. (2014). A critical reflection on urban spatial planning practices and outcomes in post-Apartheid South Africa. *Urban Forum, 25*, 69–88. https://doi.org/10.1007/s12132-013-9201-5

Dugard, J., Clark, M., Tissington, K., & Wilson, S. (2016). The right to housing in South Africa. Foundation for Human Rights.

Duncan, J. (2008). Causes of inadequate housing in Latin America and the Caribbean. Habitat for Humanity. https://www.habitat.org/lc/lac/pdf/causes_of_inadequate_housing_in_lac.pdf

Financial and Fiscal Commission (FFC). (2011). Submission for the 2012/13 Division of Revenue Technical Report. ISBN: 978-0-621-40083-0

Fluchtmann, J., Plouin, M., & Adema, W. (2020). *Social housing: A key part of past and future housing policy.* OECD.

Fuller Centre for Housing. Western Cape, South Africa. (2014). *Housing delivery in South Africa.* Draft report: Housing delivery 2014. https://fullercenter.org/wp-content/uploads/sites/default/files/Housing%20delivery%20-%20South%20Africa.pdf

Gardner, D. (2017). *South African urbanisation review. Analysis of the Human Settlement Programme and subsidy instruments.* World Bank.

Gardner, D. (2018). *Social housing structure in South Africa (inception of social housing in SA).*

Gerber, J. (2018, May 10). *R10bn budget slash for human settlements department* (p. 53). City Press.

Gibson, L. (2009). *Overcoming historical injustices: Land reconciliation in South Africa.* Cambridge University Press.

Hendricks, F., Ntsebeza, L., & Helliker, K. (2013). Land questions in South Africa. In F. Hendricks, L. Ntsebeza, & K. Helliker (Eds.), *The promise of land: Undoing a century of dispossession in South Africa.* Jacana.

Huchzermeyer, M., Harrison, P., Charlton, S., Klug, N., Rubin, M., & Todes, A. (2019). Urban land reform in South Africa: Pointers for urban policy and planning. *Town and Regional Planning, 750*, 91–103. https://doi.org/10.18820/2415-0495/trp75i1.10

Ile, I., & Makiva, M. (2020). Evaluating social housing intervention strategies in a South African municipality. *African Journal of Public Affairs, 6*(4), 2013.

Kholodilin, K., Kohl, S., & Müller, F. (2022). The rise and fall of social housing? Housing decommodification in long-run perspective. *Journal of Social Policy.* https://doi.org/10.1017/S0047279422000770

Klein, G., Klug, N., & Todes, A. (2012). Spatial planning, infrastructure and implementation: Implications for planning school curricula. *Town and Regional Planning, 600*, 19–30. https://journals.ufs.ac.za/index.php/trp/article/view/494

SA. KwaZulu-Natal Department of Human Settlements. (2020). Roll out plan for social housing (2018–2023) presentation. KwaZulu-Natal Department of Human Settlements. https://www.kzndhs.gov.za/Uploads/Social_HPL.pdf

Lawson, J., & Ruonavaara, H. (2020). *Land policy for affordable and inclusive housing: An international review*. RMIT University.
Lee, H. S., & Singh, A. (2022, September). *Adequate and affordable housing: Enhancing ADB's support to developing member countries*. [ADB sustainable development working paper series]. Asian Development Bank. https://doi.org/10.22617/WPS220353-2
Litman, T. (2022). *Evaluating criticism of smart growth*. Victoria Transport Policy Institute.
Lundgren, L. (2019). *A repeatable multi-criteria decision model for social housing asset intervention decisions*. akademin för teknik och miljö. https://www.diva-portal.org/smash/get/diva2:1322531/fulltext01.pdf
Manomano, T., Tanga, P., & Tanyi, P. (2016). Housing problems and housing programmes in South Africa: A literature review. *Journal of Sociology and Social Anthropology, 7*(2), 111–117.
Marutlulle, N. K. (2021). A critical analysis of housing inadequacy in South Africa and its ramifications. *Africa's Public Service Delivery and Performance Review, 9*(1), a372. https://doi.org/10.4102/apsdpr.v9i1.372
McKenna, L. (2019). *Integration and spatial transformation of South African cities*. Centre for Affordable Housing Finance Africa (CAHF), National Treasury.
McKinsey Global Institute. (2014). Tackling the world's affordable housing challenge. McKinsey Global Institute. www.mckinsey.com/mgi
Mkhize, M. (2015). *The politics of urban land and ownership: Locating spatial transformation in the urban land question*. [The urban land paper series volume 1]. South African Cities Network.
Molefe, N., & Nkhahle, S. (2019). Municipal urban land release and acquisition—SALGA's proposals to facilitate spatial transformation and inclusive economic growth through efficient urban land governance and management approaches. *Town and Regional Planning, 75*, 3–5.
Morris, A. (2017). *Social housing*. Sydney University of Technology. https://doi.org/10.1007/978-3-319-69892-2_756-1
Moyo, S. (2004). *African land questions, the state and agrarian transition: Contradictions of neoliberal land reforms*. African Institute for Agrarian Studies.
Msunduzi Local Municipality. (2022a). *Urban network strategy*. http://www.msunduzi.gov.za/site/search/downloadencode/nNyeaqSubNeqrYyx/index.pdf
Msunduzi Local Municipality. (2022b). *Spatial development framework*. http://www.msunduzi.gov.za/site/search/downloadencode
Msunduzi Local Municipality Housing Sector Plan. http://www.msunduzi.gov.za/site/search/downloadencode
Mtantato, S. (2011). *Impacts of current land-use patterns on public transport and human settlements*. In Submission for the Division of Revenue 2012/2013—Technical report. FFC.
Myeni, S., & Okem, A. (2020). *The political economy of government subsidised housing in South Africa*. Routledge. ISBN: 978-1-138-36491-2 (hbk).
Napier, M. (2009). *Making urban land markets work better in South African cities and towns: Arguing the basis for access by the poor*. Urban Land Markets Programme (Urban LandMark).
National Assembly of South Africa. (2017). *High level panel report on the assessment of key legislation and the acceleration of fundamental change*. National Assembly of South Africa. https://www.parliament.gov.za/storage/app/media/Pages/2017/october/High_Level_Panel/HLP_Report/HLP_report.pdf
Onatu, G. (2018).*The challenge facing social housing institutions in South Africa: A case study of Johannesburg*. University of Johannesburg, Johannesburg, South Africa.
Orthopher, A. (2016). *Wealth inequality in South Africa: Evidence from survey and tax data*. [REDI3X3 Working Paper series]. University of Cape Town.
Pohl, S., & Badenhorst, M. S. (2001). *Social housing: What lessons can be learnt from the international experience?* University of Pretoria.
RSA. (1994). *The reconstruction and development programme fund Act No. 7 of 1994*. Government Printers.
RSA. (1996). *Constitution of the Republic of South Africa, act no. 108 of 1996*. Government Printer.

RSA. (2011). *National development plan 2030*. https://www.gov.za/issues/national-development-plan-2030
RSA. Department of Cooperative Governance and Traditional Affairs [CoGTA]. (2016). *Integrated Urban Development Framework (IUDF)*.
RSA. Department of Human Settlements. (2000). *National housing code, 2000*. https://www.dhs.gov.za/content/national-housing-code-2000
RSA. Department of Human Settlements. (2004). *Breaking new ground. A comprehensive plan for the development of sustainable human settlements*.
RSA. (2003). Social Housing Policy for South Africa. Towards an enabling environment for social housing development. https://www.gov.za/sites/default/files/gcis_document/201409/social-housing-policy.pdf
RSA. Department of Human Settlements. (2009). *National housing code, 2009*. https://www.dhs.gov.za/content/national-housing-code-2009
RSA. National Treasury Cities Support Programme (NTCSP). (2018). Managing urbanisation to achieve inclusive growth. National Treasury. https://csp.treasury.gov.za/
Rydin, Y. (2013). *The purpose of planning: Creating sustainable towns and cities*. The Policy Press.
Scanlon, K., Fernández Arrigoitia, M., & Whitehead, C. M. E. (2015). Social housing in Europe. *European Policy Analysis, 17*, 1–12. http://eprints.lse.ac.uk/62938/
Scheba, A., & Turok, I. (2021). The role of institutions in social housing provision: Salutary lessons from the South. *Housing Studies, 38*(6), 1132–1153. https://doi.org/10.1080/02673037.2021.1935765
Serageldin, M. (2016). Inclusive cities and access to land, housing, and services in developing countries. [Urban Development Series Knowledge Papers]. World Bank Group. https://documents1.worldbank.org/curated/en/702761467995074482/pdf/103994-WP-Inclusive-cities-22-WEBpdf-PUBLIC.pdf
Shrivastava, R., & Sharma, A. (2011). Smart growth: A modern urban principle. *Architecture Research, 1*(1), 8–11. https://doi.org/10.5923/j.arch.20110101.02
South African Cities Network (SACN). (2004). *State of the cities report 2004*. First edition. ISBN: 620-31150-9.
South African Cities Network SACN. (2015). The urban land paper series volume 1. https://www.sacities.net/wpcontent/uploads/2020/03/Urban-Land-Series-Vol-1-Released-2015-3.pdf
South African Cities Network (SACN). (2016). *State of South African cities report 2016*. SACN.
Social Housing Regulatory Authority and Department of Human Settlements. (2019). Norms and standards for social housing. Social Housing Regulatory Authority. https://www.kzndhs.gov.za/Uploads/Norms.pdf
Social Housing Regulatory Authority (SHRA). (2020). *Annual report—2020/2021*. https://shra.org.za/download-category/annual-reports/
Strauss, M. (2019). *A historical exposition of spatial injustice and segregated urban settlement in South Africa*. Fundamina.
Sustainable Development Solutions Network. (2014). Indicators for sustainable development goals. United Nations. https://www.unsdsn.org/wp-content/uploads/2015/02/140522-SDSN-Indicator-Report-MAY-FINA
Thomas, C. (2010). Why the homeless rebel: Housing struggles in post-Apartheid South Africa. *African Historical Review, 42*(2). https://doi.org/10.1080/17532523.2010.517402
Tomlinson, R. (2002). Botshabelo: Coping with the consequences of urban apartheid. *International Journal of Urban and Regional Research, 21*(4), 691–705. https://doi.org/10.1111/1468-2427.00109
Turok, I. (2011). Deconstructing density: Strategic dilemmas confronting the post-apartheid city. *Cities, 28*, 470–477. https://doi.org/10.1016/j.cities.2010.10.003
Turok, I. (2015). South Africa's new urban agenda: Transformation or compensation? *Local Economy*, 1–19. sagepub.co.uk. https://doi.org/10.1177/0269094215614259
Turok, I. (2016). *Are South African cities becoming ungovernable again?*
Turok, I. (2018). *Worlds apart: Spatial inequalities in South Africa*.

Turok, I., & Visagie, J. (2021). Social inequality and spatial segregation in Cape Town. In *Urban socio-economic segregation and income inequality* (pp. 71–90). https://doi.org/10.1007/978-3-030-64569-4_4

United Nations Development Programme [UNDP]. (2015). *The new urban agenda*. United Nations.

Van der Walt, A. J., & Pienaar, G. J. (2012). *Introduction to the law of property*. Juta & Co.

Van Heerden, Q., Karsten, C., Holloway, J., Petzer, E., Burger, P., & Mans, G. (2022). *Accessibility, affordability, and equity in long-term spatial planning: Perspectives from a developing country*. Inclusive Smart Settlements & Regions, Council for Scientific and Industrial Research.

Walker, C. (2017). *The land question in South Africa: 1913 and beyond*. https://doi.org/10.1093/acrefore/9780190277734.013.79

World Bank. (2015). *Exclusion, in urban areas, is multidimensional and is manifested along a continuum*. World Bank.

Zadeh, V., Moulaert, F., & Cameron, S. (2021). Housing systems in the Global South: The relevance of the 'social housing' approach in meeting housing needs. *PlaNext—Next Generation Planning, 11*. https://doi.org/10.24306/plnxt/74

Fulufhelo Nyadzani holds a master's degree in Housing as well as a bachelor's degree (Honours) in Urban and Regional Planning. Ph.D. candidate in Urban and Regional Planning, focusing on urban developmental economics. He is an active member of the South African Council of Planners, registered as a Professional Town Planner. He is currently a Chief Town Planner within the local government sphere, specialising in strategic spatial planning. He has extensive experience in government policy formulation and implementation, project management in the development of spatial strategies and economic initiatives.

Lovemore Chipungu holds a Ph.D. in Town and Regional Planning, a Master's Degree in Rural and Urban Planning as well as a Bachelor of Science Degree (Honours) in Rural and Urban Planning. He is an active member of the Zimbabwe Institute of Regional and Urban Planners as well as a Corporate Member of the South African Planning Institute. He is currently an Associate Professor in the School of the Built Environment and Development Studies, at the University of KwaZulu-Natal. Professor Chipungu is both a practitioner and academic. In addition, he is an editorial board member of various journals and the Managing Editor of the Journal of Inclusive Cities and the Built Environment. He has published widely in peer-reviewed journals and books; and his areas of interest are housing policy, land-use planning and urban design. He is currently a co-principal investigator on a project on inclusive cities in Southern Africa.

Chapter 5
From Passive Victims of Elite Housing Policies to Formal Decision-Making Spaces for Impactful Delivery: Lessons from Efforts of Urban Social Movements in Zimbabwe

Kudzai Chatiza

Abstract The degree to which the urban poor's struggles have yielded policy and practical changes in Zimbabwe's housing sector is little understood let alone acknowledged adequately. Messaging about their struggles has overwhelmingly emphasised how they have experienced repeated demolitions and evictions. This focus has overshadowed their efforts towards assisting the Zimbabwean state to tame its penchant to steer elite ordering of urban spaces. Through dialoguing with and exposing state officials to alternative approaches inside and outside Zimbabwe, urban social movements have aided considerable urban transformation through mutual state–society capacitation. These successes have had limited reach resulting in relevant lessons not being available to comparable African urban jurisdictions. This paper draws on external evaluations of the work of Dialogue on Shelter Trust, Zimbabwe and the Centre for Community Initiatives, Tanzania. Both non-governmental organisations technically support movements of the urban poor in the two countries. The movements are affiliated to Slum Dwellers International through which they access good practices, solidarity for their efforts, and get empowered to act locally. The paper stitches the impacts of this category housing social movements within the context of Zimbabwe's recent experiences and snippets from Tanzania. It also uses experiences of housing cooperatives that, even with political manipulation, stretching and abuse of the core model, made an impact on housing delivery and broader urban policy. Further, insights from a three-year Inclusive Urban Infrastructure research project (A project funded by the UK Research and Innovations' (UKRI), Global Challenges Research Fund under the title 'Towards Trajectories of Inclusion: Making infrastructure work for the most marginalised' (Grant reference number ES/T008067/1) coordinated by the University of Sussex's School of Global Studies.) in 6 Zimbabwean settlements are drawn on to show efforts of the urban poor on which states can build while simultaneously value-adding to better guide expanded self-provisioning.

K. Chatiza (✉)
Waterfalls, Harare, Zimbabwe
e-mail: kudzai@mweb.co.zw

L. Chipungu et al. (eds.), *Contemporary Models for Pro-Poor Housing in Urban Sub-Saharan Africa*, https://doi.org/10.1007/978-3-032-05489-0_5

It concludes that housing social movements have made important advances towards realisation of the right to adequate housing from which other poor urban Africans and those governing them can learn for sustainably and inclusively scaling low-income housing.

Keywords Urban poor · Housing standards · Social mobilisation · Urban reforms · Right to adequate housing · State-society mutual capacitation

5.1 Introduction

The mystery or out-of-reach status of urban housing in Zimbabwe and other comparable African jurisdictions has waned over the years. Africa's urban poor have become increasingly aware of how housing is generally done, the specific parts they can play, the value chain points they can participate in, and approaches to utilise towards practically and creatively stacking a claim in urban spaces. The demystification of urban housing has been a product of rising spatial planning and general literacy among the urban poor. Additionally, an accumulation of experiences of interacting with public and private sector authorities involved in housing has led to considerable familiarity with the plural approaches to and actors in housing development and management. In the process the urban poor have been able to learn from both their successful and unsuccessful acting on aspects of urban housing. It is also critical that most African countries have adopted the right to adequate housing in their constitutions alongside applying enabling approaches. Some of this institutional versatility as well as considerable inclusiveness has been due to demonstrated public sector housing delivery inadequacies both qualitatively and quantitatively. Zimbabwe's Section 28 of the constitution provides as follows:

> The state and all institutions and agencies of government…must take reasonable legislative and other measures, within the limits of resources available them, to enable every person to have access to adequate shelter (Government of Zimbabwe, 2013a, 2013b: 23).

The constitution further provides in s74 that:

> No person may be evicted from their home, or have their home demolished, without an order of the court made after considering all relevant circumstances (ibid.: 37).

When read together with s13, which requires the state to take measures to promote private initiative and self-reliance alongside the emphasis on involving, and empowering Zimbabwean citizens through balanced urban and rural development, the role of citizens in delivering housing and implementing other aspects of development becomes evident. That the alliance and housing cooperatives whose experience this paper draws on predate Zimbabwe's 2013 constitution shows that the struggle to take direct part in inclusive urban housing started much earlier (Chitekwe-Biti, 2009; Chitekwe-Biti & Mitlin, 2001). For instance, the first two cooperatives were set up in 1985 while the alliance started in 1998. Cooperative numbers rose to 135 by 1999,

3500 by 2015 (Muchadenyika, 2020) and were more than 6000 by 2018 (Chatiza & Dube, 2018). Membership of the Federation side of the alliance rose to 55,000 families by 2008. These two categories of housing social movements got involved in urban housing, health, and other economic activities around Zimbabwe's informal and slum settlements. They also sought land on which they established new settlements. The dominant land use was housing with some exceptions being education and community centres, commercial activity, and building material production (Chatiza, 2016a; Chatiza & Dube, 2018).

Over the same period associations of residents increased in numbers and urban governance influence especially after the year 2000 (Musekiwa & Chatiza, 2015). Social movements broadly and housing social actors particularly opened significant urban social development and political spaces. Muchadenyika (2020) observes critical urban transformation elements of access to services, policies related to inclusive cities, spatial reconfiguration, and shifts in state–society relations of power. Contestations over and co-production of these elements altered urban physical form, land uses, socio-economic and political life as well as the approaches framing both urban form and life.

5.2 Zimbabwe: Governance and Development Context and Status

Zimbabwe has a three-tier government system of (i) national, (ii) provincial and (iii) local authority levels as provided for in the 2013 constitution. Its executive is supported by a two-chamber legislative arm,[1] a judiciary and independent constitutional commissions.[2] Zimbabwe's 15.2 million people (ZIMSTAT, 2023) live in 10 administrative provinces (8 rural and 2 metropolitan) with 38.6% of them living in urban areas (ibid.). With a land area of 390,757 km^2, Zimbabwe has a population density of 39 persons per km^2 (ZIMSTAT, 2022). Harare Province's density of 2,784 people per km^2 is the highest followed by Bulawayo at 1,340. In terms of age structure there are only two provinces (Manicaland and Masvingo) where the 0–17 age group (children) is higher than the 18 years and older category (ibid.). Though the population is generally youthful the population pyramid's 'child bulge' is clearer in rural than urban areas. On race and language, Zimbabweans are 99.6% African (ibid.) and speak 16 official languages, including English with Shona dominant dialect at 80% and Ndebele at 14%.

The constitution provides for devolution of governmental powers, responsibilities, and resources to provincial and metropolitan councils (tier two) and local authorities (tier three). Provincial and metropolitan institutions oversee economic development,

[1] 350-member strong Upper House with 80 Senators and 270 Lower House with 270 Members, https://www.parlzim.gov.zw/about-parliament/how-parliament-is-structured.

[2] These are the Human Rights, Gender, Electoral, Peace and Reconciliation, Media, and Anti-Corruption.

coordination and implementation of government programs, environmental protection, and promotion of tourism. Their work is mainly executed by deconcentrated national government institutions (field offices of national Government Ministries, Departments and Agencies, MDAs). The Office of the President and Cabinet (OPC) politically headed by a Minister for Provincial Affairs and Devolution with support a Permanent Secretary who heads a technical team oversees the work of the provincial and metropolitan MDAs. Their operations are funded through the national budget, allocations to individual MDAs and intergovernmental transfers provided in s301 of the constitution.

Local authorities (numbering 92 in 2023, 32 urban and 60 rural) fund their budgets through a suite of taxes, service charges and rates. They also receive intergovernmental transfers. Further, some MDAs deliver certain services in local government areas adding to funding sources for local spending. Local authority functions include water supply, sanitation, health (primary health care), housing, transport, education, environmental protection, and public safety.[3] Zimbabwe's urban local authorities have a longstanding tradition and capacity to deliver a wider range of services than rural local authorities. Local authority budgets have underperformed in the last 20 years due to protracted macroeconomic instability and governance regression. The ruling party's regaining of urban control through recentralised governance was mainly through the Ministry responsible for local government and involved taking powers and sources of revenue away from councils (Chatiza, 2010; McGregor, 2013). The serious political tensions associated with the fight to control urban local authorities (McGregor & Chatiza, 2019, 2020, 2023) between the political opposition voted for by urban citizens and the ruling party. For instance, between 2002 to 2008 the City of Harare was run by an appointed Commission (Kamete, 2009a, 2013). Just before the 2013 elections national government wrote off Harare's USD400 million debt that it was owed by ratepayers incapacitating the in-coming Council (Chatiza et al., 2013; City of Harare, 2018).

Both factors of macroeconomic instability and contested governance undermined local governance and service delivery affecting operation and maintenance of existing infrastructure let alone expansion to unserved or poorly served urban and peri-urban areas. In some of these settlements, officials of both national and local authorities were unable to effectively interact with residents (Chatiza & Gotora, 2021). Some of the fastest urban growth has occurred in these areas developed on land irregularly converted from commercial and communal farmland. The settlements' unstable governance systems have imperilled infrastructure planning, financing, emplacement, and management. Several audits confirmed land related corruption with some political actors identified as involved deeply (Chatiza & Gotora, 2021; City of Harare, 2010; Government of Zimbabwe, 2013a, 2013b, 2015, 2019, 2021; Muchadenyika, 2015). The associated corrupt and violent resource extraction (Mutongwizo, 2014) created serious spatial governance dilemmas (Chatiza, 2016a) related to what

[3] This covers firefighting, local policing, street lighting, maintenance of public open spaces, and parks.

Marongwe et al. (2011) observed to be negatively affecting Zimbabwe's urban land markets.

Despite the structural constraints Zimbabwe has seen rapid yet jobs-shy sprawling and fragmented urban growth in the face of climate change and financialisation of the housing sector. The latter pushed a previously robust national housing system to its limits creating a housing problem (UN, 2021). Zimbabwe is estimated to be short of 2 million housing units (CAHF, 2023a, 2023b). Marongwe et al. (2011) identified 1995 as a turning point in that formal supply of serviced residential land started to fall short of demand. This is also the year that Chatiza et al. (2013) suggest was when local authority performance started to decidedly slip and never recovered. Drivers of housing shortage include high cost of serviced land, building materials, construction, inadequate state support for low-cost housing programs, construction and lending inflexibility, and hyperinflation in an unstable economy with inadequate foreign currency (CAHF, 2023a, 2023b). Some 22% of urban residents live in informal settlements (CAHF, 2023a, 2023b). However, given that state quantification of informality remains a contested issue it is possible that the proportion of urban residents living in informal and slum areas is higher.

With reduced state capacity to deliver urban services (Chirisa et al., 2015; Musekiwa & Chatiza, 2015) residents began to lead responses to housing shortage (Chirisa et al., 2014). Slums and informal settlements began to increase (DOS, 2014). Slow state responses in terms of regulations for housing development saw slum and informal settlement clearance as a dominant response to *make-do* housing including backyard shacks and unapproved outbuildings (see Kamete, 2007, 2009a, 2009b; Marongwe et al., 2011; Muchadenyika, 2020; Mwonzora, 2022; Patel & Adams, 1981; UN, 2005). Operation Murambatsvina in 2005 was the most significant[4] in terms of numbers of those affected, the scale of demolitions and associated rights violations (Chatiza et al., 2021; Solidarity Peace Trust, 2005; UN, 2005). Despite slum clearance and post-2005 regulatory adjustments, land and housing informality, politicisation, and disregard of environmental law grew (Government of Zimbabwe, 2021).

Post-Murambatsvina delivery of land for housing deserves some strategic attention though. The responses by the state from criticisms of having negatively affected upwards of 700,000 residents (UN, 2005) saw a surge in state-led urban housing programs. First, was provision of state-built homes to displaced citizens and others needing housing. When financing and other technical aspects of the military-implemented program became overwhelming as hyperinflation spiralled the state adjusted the model to provision of unserviced land. A National Housing Convention in 2009 consolidated this land allocation policy position supported by guidance on incremental development previously adopted in 2005 (Government of Zimbabwe, 2005, 2009). Access to affordable urban state land at a large-scale unblocked housing delivery. Different urban housing movements and groupings that were allocated the

[4] Another massive urban displacement occurred on the eve of the 1991 Commonwealth Heads of Governments Meeting (CHOGM) held in Harare. Thousands were displaced with some being taken to a Holding Camp at Porta Farm towards Norton.

land did not have to immediately pay for it. State logic was that delayed payment for land would allow beneficiaries to build their homes and contribute towards the design and emplacement of infrastructure and services.

Even within the context of the politics of selectively using urban land audits to purge political opponents a 2019 national Commission of Inquiry into the sale of state land in and around urban areas since 2005 concluded that 170 farms had become available for urban development. Of these 91 had been formally acquired by the Ministry responsible for lands, handed over to the Ministry responsible for local government, planned and developed while 79 awaited handing over but had since been developed (Government of Zimbabwe, 2019). The estimated value of the state land was USD3 billion with 90% of the intrinsic land value not collected (ibid.). Those allocated land were given 5 years as a 'grace period' after which they were expected to make payments (Chatiza & Gotora, 2021). Unfortunately, most did not make the expected payments as established by the Commission. Interestingly, the Commission's estimate of the infrastructure costs on the acquired land was close to what the state should have collected in intrinsic land value. As it turns out, the lack of infrastructure that has made some of the settlements dysfunctional in the state's estimation (Government of Zimbabwe, 2021) arose from non-collection of the revenue which could have been applied for that purpose. This lack of services was also a product of misapplication of the resources mobilised by the various groupings that received state land.

Land that became available at scale for urban and peri-urban housing was accessed through invasion by home seekers, war veterans who later converted farming into housing land, formal allocations to cooperatives, trusts and land developers by the Ministry responsible for local government (Government of Zimbabwe, 2019). It also included land on which serving and aspiring Members of Parliament created new settlements as part of mobilising support creating considerable politicised control of land for housing (ibid.; Muchadenyika, 2015). Additional to the above factors (non-collection of intrinsic land value and misapplication of beneficiary contributions) was the politicised processes that made private sector participation difficult. On their part established private actors in housing generally stayed clear of urban land and housing in post-2005 settlements arguing that tenure arrangements, planning, and development procedures were too opaque to guarantee security of and return on their investments.

As noted, opposition-led urban local authorities were not involved in the land access and development processes. The ruling party considered that this would empower them. Consequently, land and housing delivery by urban local authorities dropped to a trickle in comparison to the model overseen by national government. For instance, on the edge of Masvingo City, 8165 residential stands were allocated to 15 developers on government acquired and planned land on Lot 1 and the Remainder of Victoria Ranch between 2007 and 2019 (GRZ, 2020). City records further show that a total of 12,873 residential stands were submitted to the city relating to peri-urban state land (Masvingo City Council, 2023). Masvingo City received administration fees for 8,328 of these stands (65%) suggesting 35% were yet to initiate processes towards formal land and housing tenure security. In comparison the city delivered

776 residential stands between 2011 and 2023 on land within its boundary, 6% of those on state land.

While the innovations, including the inevitably messy yet scaled and relatively inclusive urban land allocations, not enough consolidation of the lessons has been done. The balancing of state and societal competence in transformed urban land and housing delivery has not fully been engaged with. The question is whether this '*pay for urban state land later or not to pay at all*' can be protected from elite and politicised accumulation on one hand and how established public or private housing authorities envision financialisation of non-land assets and services in human settlements. Related, is what societal literacy and organisation is suited to mediating such an urban development model. It also requires exploring the kind of land and housing rights that can be defined and protected in such a framework. Essentially, the messiness had some positive outcomes. Zimbabweans accessed affordable land on which they built urban homes freeing themselves from some financially costlier lodgings.

5.3 Mutual State–Society Capacitation: A Conceptual Framework

Experiences of Zimbabwe's urban poor have involved seeking visibility for their ambitions and realising their right to work and live in urban areas. This has seen scaled efforts at avoiding being blocked and evicted from their houses and trading spaces. The ideas of the urban poor and the state have not always been in sync. Even in cases where alternatives were triggered in terms of adjusting building material and other housing standards momentum was not always sustained. However, as the housing social movement grew at a time of stressed state capacity and legitimacy some of their approaches gained traction allowing mutual society-state capacitation as depicted in the Fig. 5.1. In selected cases, this allowed deep relations and delivering practical housing solutions.

Outcomes of the intersection of state capacity and social mobilisation are yet to be fully explored in urban housing. Chitekwe-Biti (2009, 2014) and Muchadenyika (2020), among others make very competent attempts on which future efforts can build particularly with a focus on inclusive and sustainable urban governance. The opening of space from 2006 created some interconnected and interdependent systems

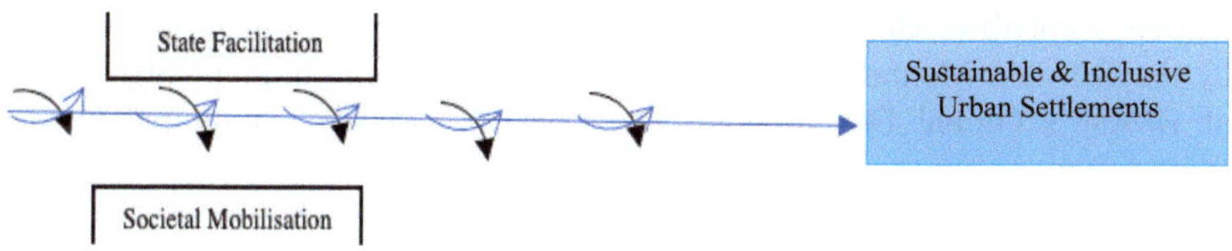

Fig. 5.1 Advantageous state–society intersectionality

to address previous sequences or experiences of disadvantaging and discriminating against the urban poor. Using their agency, residents of cities failing to deliver housing at a scale possible in peri-urban areas traded their support (faked and real) to the ruling party for access to land for housing. In a sense they individually and collectively mobilised their actual and prospective votes in exchange for secure access to a service otherwise unavailable from the opposition-led cities emaciated by Zimbabwe's ruling party (McGregor & Chatiza, 2020).

It is within this convenient enrolment into the ruling party's patronage system that land for housing was delivered at a scale that planners, steeped in authoritarian modernist planning ideals, couldn't cope with. This was at a time when cities couldn't sustain the myth that they lacked informal settlements. Further, a moratorium on large-scale demolitions had been replaced by a grudging shift towards in situ slum upgrading and regularisation. It was also at a time when standards had been lowered to help make housing affordable for the urban poor and 'parallel development'[5] was adopted allowing home seekers to inhabit stands before servicing (Government of Zimbabwe, 2005, 2009). We use evidence from the work of the alliances in Zimbabwe and Tanzania to pull out aspects of mutually advantageous state–society capacitation to illuminate policy and practice lessons with some potential for wider application in Southern Africa.

5.4 Mutual State–Society Capacitation in Housing Practice

The Harare Slum Upgrading Project (HSUP), 2010–2016

This section of the paper draws on a mixed method (qualitative and quantitative) evaluation of the Harare Slum Upgrading Project (HSUP) conducted in July 2016. Data collection involved 16 focus group discussion (FGD) sessions, 8 key informant interviews, a household survey that reached 100 respondents, 11 site/project visits, and visits to 3 other local authorities[6] impacted by HSUP. FGD sessions targeted different aspects of the project. These included beneficiaries of the Harare Slum Upgrading Finance Facility, HSUFF (4 sessions), groups outside Harare (4), 1 session each with HSUFF staff, Harare City officials, and a local and international group[7] at the Dzivarasekwa site, which sprawled from a state-established holding camp for victims of a 1991 mass eviction (Chatiza, 2016a, 2016b, 2016c). Because the site was outside the city boundary national government involvement was critical.

HSUP was a five-year City of Harare (CoH), Dialogue on Shelter Trust (DOS) and the Zimbabwe Homeless People's Federation (ZHPF) collaboration supported by the Bill and Melinda Gates Foundation (BMGF). Its implementation was part of BMGF's Global Development Program on Inclusive Municipal Governance

[5] This was a core of Circular No. 70 of 2004.

[6] The Cities of Bulawayo, Chinhoyi and Masvingo.

[7] The evaluation also involved representatives of SDI affiliates from Uganda and Zambia.

(GPIMG) launched in 2010. DOS and ZHPF are an alliance where the former offer technical support to the latter, which is a social movement affiliated to Slum Dwellers International (SDI). GPIMG targeted five African cities[8] to build city government and the urban poor partnerships to tackle challenges inclusively.

The Harare partnership entered an MOU on March 19th, 2010, following cycles of formal and informal, joint, and separate discussions both within and outside Zimbabwe. For instance, securing national government permission to work in Dzivarasekwa was negotiated with the Minister responsible for local government in Vancouver on the sidelines of the 2006 World Urban Forum. A second strategic meeting was held in Nanching, China, at the 2008 World Urban Forum between the alliance and Advocate Muchadeyi Masunda, who became Mayor between 2009 and 2013. Other meetings were held between him and the Foundation's representatives before a city-alliance meeting was convened in early 2009 in Harare. This last series of engagements signalled the beginning of negotiating the MOU finally signed in 2010. The cycles of engagement are consistent with the iterations involved in mobilising state facilitation (at local level in this case) through deployment of mobilised social groups (ZHPF) with local technical support (DOS) and underpinned by international financing (the Gates Foundation).

HSUP implementation was politically led by Harare's 2008–2013 and the 2013–2018 Councils guided by six MOU objectives which focused on:

- Conducting situational analyses of Harare settlements with inadequate shelter and services
- Selecting at least one settlement for implementing detailed needs assessment
- Creating an action plan to address priority needs in the selected settlement(s)
- Mobilising additional funding to further resource the action plan
- Documenting and disseminating lessons learnt from the experience
- Creating a work plan to replicate the process in other disadvantaged settlements in Harare

Over a 10-year period, with reduced intensity in the 2013–2018 period, Harare city pursued recovery of service delivery and restoration of the City's governance authority. HSUP implementation advanced inclusive housing by (i) defining and identifying slums in the city, (ii) engaging the urban poor in slum upgrading, (iii) creating joint platforms for innovative planning and implementation of integrated services in slum areas, and (iv) developing alternative financing models drawing on city and community resources. Project partners improved their relations and performance aided through exchanges within the global south. HSUP hosted and sent delegations from/to Namibia, Ethiopia, Botswana, Malawi, India, Zambia, Uganda, Kenya, Ghana, Tanzania, Swaziland, the Philippines, and South Africa. Officials and the urban poor were involved in the exchanges had opportunities to jointly learn and develop mutual respect.

USD3,879,358.03 of Gates Foundation funding enabled direct service delivery in Dzivarasekwa Extension to 501 families, enumeration of 63 slums and initiation of

[8] Harare, Monrovia, Luanda, Lilongwe, and Cairo.

city-led servicing of Council land in Mabvuku-Tafara to accommodate 121 families to be resettled from Gunhill Informal Settlement. Besides, the city was able to procure relevant equipment under the project. These included a van, an all-terrain utility truck, a 7-tonne truck and a tipper truck, Plan Plotter, AutoCAD 2012 software, copiers, laptop and desktop computers. The city was also transformed following slum area enumerations. It went beyond responding to health emergencies like cholera to actual service delivery and governance involvement in selected settlements.

HSUP also set up key institutions like the Harare Slum Upgrading Finance Facility (HSUFF) which delivered 228 loans worth USD374,595 to 959 urban poor people in 137 groups. Of these borrowers, 533 (55.6%) were women and 426 males. 577 (60.2%) borrowed for income generation purposes and the balance for housing. In terms of age distribution 877 (91.5%) were aged 21–50. The loan facility has initiated transformation of the lives of Harare's urban poor. Further, six Zimbabwean Councils and other non-Zimbabwean organisations learnt from and contributed to HSUP implementation.

Regarding the quality of services, 192 model houses were set up with resident participation, stormwater infrastructure reached 449 properties, water connections reached 285 houses with at least 50 connected to reticulated sewer and another 247 emplaced ecological sanitation facilities with Council approval. This alternative sanitation technology was also approved for intermediate use by the cities of Masvingo, Chinhoyi, and Bulawayo in their own peri-urban semi-formal settlements. Besides reticulated water supply, other HSUP-enabled service improvements were as shown in Table 5.1.

Harare City adopted a Slum Upgrading Strategy, initiated the setting up of a Slum Upgrading Unit, became clearer on its commitment to pro-poor housing including its financing, which saw Council resolving to authorise its housing department to approach the private sector for pro-poor housing finance and slum upgrading partnerships. City departments evolved and deployed some improved horizontal collaboration approaches. These were made possible through the HSUP Project Management Committee, created a platform for joint planning and implementation. Operationally, the city waived gravel, plan preparation and approval, individual water connection, and subdivision fees for HSUP recognising community input mobilised under the project for plan development and actual service emplacement through in-kind contributions.

Outside Harare, HSUP impacted other Councils through a consolidated definition of and acceptance of the existence of slums. The local authorities of Epworth, Bulawayo, Kariba, Masvingo, and Kadoma each identified sites where HSUP lessons were applied. The alliance helped mobilise additional resources and convinced each of these Councils to also make financial and technical contributions including regulatory changes to enable more inclusive delivery of relevant services.

Table 5.1 Pre- and Post-HSUP service delivery at Dzivarasekwa

Water and sanitation	Before HSUP	After HSUP
	60% borehole reliance, 30% unprotected wells, 52% using Ecosan toilets, 18% pit latrines and 30% unsafe sanitation	69% now use both reticulated and borehole water supply and all unprotected wells decommissioned, Ecosan (40%) and reticulated (60%) sanitation co-existing
Property road access	98% with informal and unimproved road access	86% with formal, improved % direct road access
Solid waste management	Partial compositing, illegal dumping and burning with minimal recycling (only 1% of sampled beneficiaries). No municipal collection	Municipal collection, community recycling initiated with burning and dumping dropping. City supplied litter waste bins for collection, sorting and separation. 30 women trained and spearheading the project
Energy sources	Firewood as only source for both cooking and lighting (64%) with no electricity	Firewood remains dominant (56%) for cooking with solar lighting as area is unserved with electricity (only 1% indicated use of electricity)
Standard of housing	64% of respondents rated pre-HSUP housing as poor, 15% good and excellent and 21% as fair. Also, no Resource Centre and ECD facilities	52% of the respondents rated housing as good and excellent, 12% as fair and 38% as poor. One Community—cum-ECD centre established with 10 computers & 1 printer and three sites identified for ECD by June 2016

Source Chatiza (2016b: 15)

5.5 Community-Led Water and Sanitation Delivery in Tanzania and Zimbabwe

The second case to further illuminate state–society capacitation in housing and related services is based on an evaluation of a Homeless International/Reall project (Chatiza, 2016c). Reall, a UK-based non-government organisation (NGO) was founded in 1989 focusing on addressing urban poverty in Africa and Asia through long-term partnerships with local organisations supporting poor communities' initiatives. The project evaluated was implemented in Tanzania and Zimbabwe with UK Global Poverty Action Fund (GPAF) funding. In country activities were conducted by the Centre for Community Initiatives (CCI) and Dialogue on Shelter (DOS), organisation with which Reall had relations since 2006 and 1999, respectively.

The project focused on water and sanitation targeting 30,680 slum dwellers in terms of water supply and 21,520 for sanitation. It aimed to increase the proportion of urban population with sustainable access to improved water supply and sanitation based on building community capacities, administering revolving loan funding, supporting direct community engagement in delivering services, and improving local authority-infrastructure provider collaborations.

Like DOS in Zimbabwe, CCI is a technical partner of the Tanzania Urban Poor Federation, an affiliate of Slum Dwellers International. In both Zimbabwe and Tanzania urban poor funds (Gungano and Jenga respectively) supported project activities through loans totalling 567,040.01 British Pounds, being 376,002.47 for Gungano and 191,037.54 for Jenga. GPAF capitalisation of the two funds allowed them to increase loan disbursement that tangibly impacted poor people's lives through service delivery, empowerment, and accountability within the context of conflict management, enhancing and housing tenure security and social justice from neighbourhood to city levels. Project implementation originally covered 10 local authorities[9] and 17 settlements over three years between October 2013 and September 2016. However, other local authorities and settlements were added during implementation including sites where the two local Reall partners had implemented GPAF-supported activities between 2010 and 2011. This brought the total to 21 local authorities.

The evaluation examined the impact of CCI and DOS activities. It involved reflecting on approaches used and drew lessons of good (bad) practices based on verified project records, assessments of the value for money of activities conducted, validating the result chains, and reviewing the overall effectiveness of the project strategy. Qualitative methodology was mainly adopted with secondary literature analysis, FGD sessions, key informant interviews, facility/service observations, and administration of a household questionnaire. The evaluation reach was 10 local authorities, 19 settlements, 28 key informants, 24 FGD sessions, 324 questionnaire respondents, and 18 facility visits.

The project reached 30,940 for water supply and 30,335 for sanitation. Results generated by the project related to service availability, council-community relations, urban poor movement visibility, social profiles of beneficiaries, and enhanced capacities of the technical partners (CCI and DOSS). Councils confirmed appreciating and benefiting from the approaches used, noted that some of the work done would have been impossible without mobilised societies, acknowledged resident behaviour change, and local authority level reforms regarding facility emplacement, facility operation and maintenance. Implementation of activities, developing infrastructure, and managing services provided scope for women's active and meaningful participation considering the centrality of water and sanitation services to women and girls. The meaningfulness of participation related to actual technical input and the decision-making power over service location, design, and management. Respectively, 70–80% of Federation members working with DOS and CCI are women. Consequently, most of the leaders of the urban housing social movements are females.

In Tanzania, loan application and administration involved government officials at local level while in Zimbabwe savings prior to borrowing by members of self-selected and trained groups improved loan performance. Further, local builder training, hiring them at discounted labour costs, negotiating for the waiver of certain fees by public institutions in Zimbabwe and allocation of free plots for construction of public toilets in Mwanza, Tanzania increased the volume of local resources availed for project implementation. In the Zimbabwe case a total of USD 2.2 million of Council and

[9] Dar-es-Salaam, Dodoma, Mwanza, and Morogoro in Tanzania, and Harare, Bulawayo, Makonde Rural District Council, Chinhoyi, and Hwange in Zimbabwe.

community contributions became available for project activities. In Bulawayo, Chinhoyi and Masvingo (Zimbabwe) water and sanitation access under the project significantly progressed issues of tenure security which would have been difficult without household/property level services.

Elderly beneficiaries who would have been unable to access services let alone approach the local authorities were among the most appreciative. WASH in schools (e.g. Keko Mwanga) improved menstrual hygiene management. In Mhangura, DOS brought together the local authority, the national water authority, ZINWA, and the community together despite years of community mistrust of the two public institutions over fruitless shunting of the water functions between the two. Recognising their importance to sustainable water, sanitation, and hygiene (WASH) services in Mhangura DOS facilitated ZINWA's involvement. Staff of the agency were already technically involved in redesigning aspects of the project with the authority providing some of the needed materials at the time of the evaluation.

5.6 Discussion

The political economy within which housing and associated services for the urban poor are delivered in most of Southern Africa is fragmented, characterised by mistrust, and often leaves the poor to their own devices. This disconnection reduces prospects for mutual reinforcement of state and societal capacities needed for inclusive and sustainable housing. For instance, local and national governments are often absent in slums and informal areas. In the case of Zimbabwe through the 2005 mass evictions that absence is at times due to denying the existence of such grossly underserved areas while at the same time itching to demolish them. At other times it is about lack of entry points and capacities especially where communities are organised enough to engage with the state and vice versa.

When left to their devices or left to self-provide the urban poor opt for technologies that they can afford, which invariably tend to be ill adapted to densely populated urban spaces posing health and other human security risks. In other instances, the urban poor fall prey to corrupt urban actors resulting in loss of resources and time. Where this happens, state actors do not always treat such losses as negatively affecting wider society and economy. Resolution of such matters does not always receive adequate state attention. The two case clusters used in this paper show how involvement of non-governmental organisations (DOS and CCI) to support the urban poor (Federations) to mobilise communities, identify technical solutions, and engage state actors. Essentially, the NGOs acted to facilitate state–society interaction. This approach allows connected performance in terms of technical solutions, resources or effort, and political trust. Where transformed into a pro-poor orientation and effectively deployed, the local and national state's aggregative and regulatory capacities inspire community confidence allowing additional contributions. At the same time critical urban reforms that matter in housing broadly and for the urban poor particularly were made leveraging service access and sustainable inclusion.

As service complexity increases in both scope and scale with financial and technical requirements beyond the means of mobilised poor communities, state action becomes more critical. This suggests a need for deploying state capacity in parts of the housing value chain that require strategic leverage for purposes of sustainable scaling given the huge service backlogs. The example of the Zimbabwe government's de facto provision of urban state land for free is critical to revisit here. It allowed scaled accessing of land by mostly the urban poor even where political actors facilitated access and abused beneficiary contributions. However, inadequate attention to or diminished state capacity to monitor or oversee parallel development, itself a complex model that could not be left to home seekers, created the current settlement dysfunctionality the Zimbabwean state is having to attend to. Yet dysfunctionality or incompleteness does not invalidate the land access and significant housing development. It only exposes the inadequate deployment of relevant state competences. Further, securing of land and housing tenure after the fact provides scope for financial models that can allow capturing of land value for service and infrastructure development. With the political heat subsiding and residents of some of these settlements motivating for services, additional capacity can be mobilised including from the private sector.

5.7 Conclusion

Mutual state–society capacitation for joint urban land and housing sector performance is critical. The positive outcomes it potentially generates are varied. They include mutual trust and respect, resource mobilisation and efficient use, appropriate urban reforms, informed public sector restructuring, and state–service provider partnerships. These results require active, committed, and competent facilitation over the medium to long-term. Additionally targeted financial, solidarity, and technical support to both sides (of social mobilisation and state facilitation) is also very critical. The two cases used in this paper benefited from such external support. However, they also show how the project opportunities were expanded on to activate local capacities in terms of actual financing including creating streams of stable sources of micro-loans availed to the urban poor complemented by Municipal innovations to ease and expand access. The Government of Zimbabwe's capacity to postpone payment for land is another key lesson to draw. It suggests possibilities of reducing the cost burden associated with purchasing land and then building any costs of land the state (local or national) needs to recover as part of post-construction levies and charges including property tax. In Zimbabwe's case, there has been a rush to sell land, both serviced and unserviced, by public institutions largely to finance operational costs rather than the services expected on the land in question. Once sold the thrust turned to acquiring more land without properly linking urban land resources to other functions critical to sustainable urban management. That trajectory fed a land corruption beast nurtured by political actors to shield it from legal and administrative sanctions. One net result has been serious regression of state facilitation capacity. Another is loss of trust in state land management by citizens and relevant private sector actors. More research work is needed on the issues of (i) balancing state and societal competences where relations of the two sides have been positively transformed, and (ii) articulating the place of urban land and how it can be accessed in ways that allow expanded, just, and sustainable urban housing development.

References

Centre for Affordable Housing Finance. (2023a). *Housing finance in Africa yearbook: 14th edition—2023*. https://housingfinanceafrica.org/resources/yearbook

Centre for Affordable Housing Finance. (2023b). *Dashboard: Housing and construction in Africa: 2010–2023*. https://housingfinanceafrica.org/

Chatiza, K. (2010). Can local government steer socio-economic transformation in Zimbabwe? Analysing historical trends and gazing into the future. In J. de Visser, N. Steytler, & N. Machinguta (Eds.), *Local government reform in Zimbabwe: A policy dialogue* (pp. 1–30). Community Law Centre, University of the Western Cape.

Chatiza, K. (2016a). *Analysing policy responses to the dilemmas of peri-urbanity and land management in Zimbabwe: A case study of Harare, in peri-urban developments and processes in Africa with special reference to Zimbabwe* (pp. 129–142). Springer.

Chatiza, K. (2016b). *End evaluation of the Harare Slum Upgrading Program (HSUP), September 2016*. Dialogue on Shelter, City of Harare and the Zimbabwe Homeless People's Federation.

Chatiza, K. (2016c). *End evaluation of the community-led water delivery for 30,680 slum dwellers and sanitation delivery for 21,529 slum dwellers in Tanzania and Zimbabwe (GPAF-IMP-099)*. Homeless International/Reall.

Chatiza, K., & Dube, T. (2018). *Zimbabwe national human settlements policy: Background and issues paper*. Ministry of Local Government, Public Works and National Housing.

Chatiza, K., & Gotora, P. (2021). The varying shades of settlement informality in Zimbabwe's urban areas and how this impacts public sector regularisation attitudes and responses. *Journal of Urban Systems and Innovations for Resilience in Zimbabwe-JUSIRZ, 3*(2), 38–67.

Chatiza, K., Makanza, V., Musekiwa, N., Paradza, G., Chakaipa, S., Mukoto, S., Kagoro, J., Ndlovu, K., & Mushamba, S. (2013). *Capacity building of local government for service delivery*. Government of Zimbabwe and UNDP.

Chatiza, K., Zivhave, M., Nyevera, T., Marewo, M., & Chatiza, T. M. (2021). *Demolitions and evictions in Harare, Zimbabwe: An analysis of drivers and some policy options, women and law in Southern Africa.*

Chirisa, I., Bandauko, E., & Mutsindikwa, N. (2015). Distributive politics at play in Harare, Zimbabwe: Case for housing cooperatives. *Bandung Journal of Global South, 2*, 15.

Chirisa, I., Gaza, M., & Bandauko, E. (2014). Housing cooperatives and the politics of local organization and representation in peri-urban Harare, Zimbabwe. *African Studies Quarterly, 1*, 15.

Chitekwe-Biti, B. (2009). Struggles for urban land by the Zimbabwe Homeless People's Federation. *Environment and Urbanization, 21*(2), 347–366.

Chitekwe-Biti, B., & Mitlin, D. (2001). The urban poor under threat and in struggle: Options for urban development in Zimbabwe 1995–2000. *Environment and Urbanization, 13*, 85–102.

City of Harare. (2010). *Special investigation committee's report on city of Harare's land sales, leases and exchanges from the period October 2004 to December 2009*. City of Harare.

City of Harare. (2018). *State of the city address by Mayor, Councillor Bernard Manyenyeni*, 11 April 2018.

Dialogue on Shelter. (2014). *Harare slum profiles report* (2nd ed.). Dialogue on Shelter and Zimbabwe Homeless People's Federation.

Government of Zimbabwe. (2005). *National housing: Circular number 70 of 2004-new housing standards*, 14 January. Government of Zimbabwe.
Government of Zimbabwe. (2009). *Second national housing convention report*, 26–30 October 2009, Elephant Hills Hotel, Victoria Falls.
Government of Zimbabwe. (2013a). *Constitution of Zimbabwe, Amendment No. 20.*
Government of Zimbabwe. (2013b). *Audit report on issues of land management and allocation: Chitungwiza Town and Seke District, November 2013.* Ministry responsible for Local Government, Public Works and National Housing.
Government of Zimbabwe. (2015). *Harare north housing cooperative compliance audit report, 16 July.* Ministry responsible for Local Government, Public Works and National Housing.
Government of Zimbabwe. (2019). *Summarising findings and recommendations of the commission of inquiry into the matter of the sale of state land in and around urban areas since 2005.* Government of Zimbabwe.
Government of Zimbabwe. (2021). *Update on identification and quantification of all irregular and dysfunctional settlements including flood-prone areas: Term of Reference No. 3*, 11th March 2021.
Kamete, A. (2007). Cold-hearted, negligent, and spineless? Planning, planners, and the (r)ejection of "filth" in urban Zimbabwe. *International Planning Studies, 12*(2), 153–171.
Kamete, A. (2009a). In the service of tyranny: Debating the role of planning in Zimbabwe's urban "clean-up" operation. *Urban Studies, 46*(4), 897–922.
Kamete, A. (2009b). For enhanced civic participation in local governance: Calling tyranny to account in Harare. *Environment and Urbanization, 21*(1), 59–67.
Marongwe, N., Mukoto, S., & Chatiza, K. (2011). *Scoping study: Governance of urban land markets in Zimbabwe.* Urban LandMark.
Masvingo City Council. (2023). *Data on housing delivery within the city and on its periphery, 2011 to 2023.*
McGregor, J. (2013). Surveillance and the city: Patronage, power-sharing, and the politics of urban control in Zimbabwe. *Journal of Southern African Studies, 39*(4), 783–780.
McGregor, J., & Chatiza, K. (2019). Frontiers of urban control. Lawlessness on the city edge and forms of clientelist statecraft in Zimbabwe. *Antipode, 51*(5), 1554–1580.
McGregor, J., & Chatiza, K. (2020). Partisan citizenship and its discontents: Precarious possession and political agency on Harare City's expanding margins. *Citizenship Studies, 24*(1), 17–39.
McGregor, J., & Chatiza, K. (2023). Geographies of urban dominance: The politics of Harare's periphery. In T. Goodfellow & D. Jackman (Eds.), *Controlling the capital: Political dominance in the urbanizing world* (pp. 172–202).
Muchadenyika, D. (2015). Land for housing: A political resource—Reflections from Zimbabwe's urban areas. *Journal of Southern African Studies, 41*(6), 1219–1238.
Muchadenyika, D. (2020). *Seeking urban transformation: Alternative urban futures in Zimbabwe.* Weaver Press.
Musekiwa, N., & Chatiza, K. (2015). Rise in resident associational life in response to service delivery decline in urban councils in Zimbabwe. *Commonwealth Journal of Local Governance, 16/17.*
Mutongwizo, T. (2014). Chipangano governance: Enablers and effects of violent extraction in Zimbabwe. *Africa Peace and Conflict Journal, 7*(1), 29–40.
Mwonzora, G. (2022). Demolitions of informal business structures in Harare, Zimbabwe under the guise of COVID-19 city clean up. *Journal of Asian and African Studies, 57*(7), 1446–1460.
Patel, D. H., & Adams, R. J. (1981). *Chriambahuyo: A case study in low income housing.* Mambo Press.
Solidarity Peace Trust. (2005). *Discarding the Filth/Operation Murambatsvina: Interim Report on the Zimbabwean Government's 'Urban Cleansing' and Forced Eviction Campaign*, May/June 2005.
UN. (2021). *Housing 2030.* United Nations Publications.

UN. (2005). *Report of the fact-finding mission to Zimbabwe to assess the scope and impact of Operation Murambatsvina by Special Envoy on Human Settlement Issues in Zimbabwe, Mrs. Anna Kajumulo Tibaijuka.*

ZIMSTAT. (2022). *Zimbabwe census preliminary findings report* (Vol. 1). Zimbabwe National Statistics Organization.

Kudzai Chatiza is a local governance, and spatial planning researcher, and student. He has over 30 years' experience in decentralization, development and strategic planning, citizen participation, housing policy and practice. He studied at the Universities of Zimbabwe, UZ (Bachelor, 1992 and Master, 1995 in Rural and Urban Planning) and Swansea (Ph.D. in Development Studies, 2008). He teaches and supervises research students at UZ and the Midlands State University. Dr. Chatiza also advises state and non-state development organizations on spatial planning, devolved governance, pro-poor land, agrarian, water and sanitation, urban infrastructure and housing governance in his current role as Director at Development Governance Institute www.degi.co.zw based in Harare. He has managed grant-funded research and published on these themes. Grant-funded work has been in partnership with the Universities of Sussex and Manchester (UK), Free State (South Africa) and UN agencies in Zimbabwe. Kudzai teaches on a temporary-full time basis at the University of Zimbabwe's Architecture and Real Estate Department, Faculty of the Built Environment. He also serves as a Deputy Editor of the Journal of Urban Systems and Innovations for Resilience in Zimbabwe-JUSIRZ, University of Zimbabwe.

Chapter 6
Best Practices for Designing and Building Alternative Models of Smart Affordable Housing for Low-Income Families

Lawrence Babatunde Ogunsanya

Abstract The affordable housing crisis is a pressing issue facing many communities around the world. The high cost of housing, particularly in urban areas, has left a significant portion of the population unable to afford suitable living arrangements. Traditional models of affordable housing for low-income families have faced numerous challenges, including high operating costs, inadequate infrastructure, stigmatization, and a lack of flexibility to meet the changing needs of residents. Smart affordable housing refers to residential developments that incorporate technology and design elements to reduce the cost of living for residents. It involves the provision of affordable housing which explores alternative models of smart technology with minimal gadgetry and integrated design elements to reduce the cost of living for residents. A few of these alternative models include sustainable buildings created to reduce their adverse effects on the environment and use resources efficiently by employing an integrated design process, climate adaptable buildings, the use of smart building materials and construction methods, smart waste management systems, and the use of smart microgrids: A small-scale energy system that is designed to operate independently of traditional power grids. In addition, the exploration of low-tech solutions that facilitates energy efficiency and comfort, without the need for expensive gadgets, such as: Insulation and weather-stripping and natural ventilation systems to help regulate indoor air quality. This chapter proposes to cover new and innovative technical advances in the building industry in production of affordable housing systems made up of case studies predominantly from Africa and various countries around the world. The intention is to contribute to the knowledge within the housing sector about how alternate models of smart technology can be implemented to improve the quality of life for low-income families, reduced housing costs and maintenance, enhanced security and privacy, and provide better housing flexibility and adaptability to changing needs of its users.

Keywords Smart technology · Affordable housing · Architecture

L. B. Ogunsanya (✉)
School of Engineering, Howard College, University of KwaZulu-Natal, Glenwood, Durban, Republic of South Africa
e-mail: ogunsanya@ukzn.ac.za

L. Chipungu et al. (eds.), *Contemporary Models for Pro-Poor Housing in Urban Sub-Saharan Africa*, https://doi.org/10.1007/978-3-032-05489-0_6

6.1 Introduction

The need for affordable housing continues to rise, necessitating novel approaches in meeting the demands of low-income households. Alternative kinds of smart affordable housing have evolved as a possible answer in recent years, merging technology, sustainability, community engagement, and developing housing polices.

Smart affordable housing offers an alternate solution to rising housing demand, particularly for low-income populations (Awad et al., 2017). It strives to create cost-effective, sustainable, and technologically advanced housing solutions by incorporating smart technology and innovative design ideas. Among these alternative innovative ideas are sustainable buildings that use an integrated design process to minimize their environmental impact and use resources efficiently; climate-adaptable buildings; the use of smart building materials and construction methods; smart waste management systems; and the use of smart micrograms, a small-scale energy system designed to operate independently of traditional power grids. Furthermore, the use of low-tech solutions that promotes energy efficiency and comfort without the use of costly gadgets. Smart affordable housing generally intends to meet the housing needs of low-income communities while promoting energy efficiency, community engagement, and long-term affordability by utilizing smart technology and adopting sustainable practices. This article investigates the best practices for planning and building such affordable housing models in order to provide low-income families with affordable and efficient living spaces. It identifies and discusses diverse models of providing and implementing affordable housing with smart and sustainable initiatives. South Africa has a critical need for affordable housing, particularly in impoverished communities. The current model utilized by the government is primarily based on a 'Reconstruction and Development Programme' house (RDPs). This programme involves the government providing beneficiaries with a fully built house for free. Unfortunately, this form of housing delivery has limitations in terms of design quality, sustainability, and community engagement (Nokulunga et al., 2018). Alternative models that combine smart technologies and community-oriented approaches offer viable solutions (Löw & Antalovsky, 2018). They improve the quality, sustainability, and inclusivity of low-income housing developments. By embracing innovative technologies, community engagement, and flexible policies, the government can create more equitable and resilient housing landscapes.

6.2 Theoretical Framework

The theoretical foundation of this article is based on the following theories and models: Technology Acceptance Model, Diffusion of Innovation Theory and Systems thinking. These theories and models provide a theoretical lens through which the subject matter of this article is conducted, examined, and interpreted. It establishes the theoretical foundation of how smart technology is infused and used in affordable housing, it also tries to identify research gaps and promotes the interpretation and the application of the theories in the context of housing. Finally, the framework assists in framing the phenomena of smart affordable housing within a broader academic or practical perspective, enhancing its contribution to knowledge on housing provision for low-income families.

6.2.1 Technology Acceptance Model

Davis' (1987) Technology Acceptance Model focuses on users' acceptance and adoption of technology. It implies that perceived usefulness and perceived simplicity of use are the major elements affecting an individual's intention to use technology (Davis, 1989). This theory explains how users perceive the benefits and usefulness of smart technology and how these views influence their propensity to accept and use it (Park et al., 2018; Shin et al., 2022). The Technology Acceptance Model (TAM) has evolved into one of the most useful instruments for analysing technological acceptability in a variety of industries (Nikou & Bouwman, 2014; Shin et al., 2022), including education (Huang et al., 2007), business and construction (Davies & Harty, 2013).

This model is beneficial in the empirical investigation of user perceptions of new technologies and for explaining the adoption process. Its theoretical origins were initiated in the psychological theory of reasoned action, developed by Fishbein (1967). TAM postulates that the actual use of technical systems is influenced by people and how useful the technology is for their tasks and how easy it is, to use the technology is (Nikou, 2019). According to TAM, the actual use of the technology is determined by Behavioural Intent to Use (BIU). In turn, behavioural intentions depend on the attitudes used by both parties of technology and Perceived Usefulness (PU), so Perceived Usefulness does not only influence behavioural intentions, but also directly affects attitudes towards it, by the use of technology. Additionally, Perceived Ease of Use (PEU) also has a direct impact on Perceived benefits and attitudes towards the use of technology (Davis, 1989). Several TAM studies have proven the importance of PU and PEU in predicting user intention and actual usage, either directly or indirectly (Chen & Barnes, 2007; Venkatesh & Davis, 2000). Therefore, using the constructs of TAM in the context of this article's subject matter; smart affordable housing's Perceived Usefulness (PU) will be an enhanced housing performance which is effective and productive. The Perceived Ease of Use (PEU) will be the ease of use, control and understanding of smart affordable housing and the Behavioural Intent to Use (BIU) will be the expectation, predictions, and strategies to employ the smart systems of affordable housing. In the framework of smart home technology, TAM has proven to be an effective model in explaining the adoption of the technology (Park et al., 2018).

6.2.2 *Diffusion of Innovation Theory*

Diffusion of Innovation Theory was established by Everett Rogers (1983) established, which investigates how new technologies spread and are embraced by individuals and civilizations. It classifies adopters into innovators, early adopters, early majority, late majority, and laggards. According to Yuen et al. (2020, p. 4) five factors can be used to measure an innovation's attributes or characteristics: Relative advantage, compatibility, complexity, experimentality, and observability. These factors are related to the innovation's acceptance and the decision-making process. This theory helps to explain the aspects that influence the acceptance and proliferation of smart technology and affordable housing, such as perceived benefits, compatibility with current systems, complexity, trialability, and observability.

Applying the identified attributes of the Diffusion of Innovation Theory into smart affordable housing can start from the various categories of adopters. Early adopters and innovators are often the first to accept new ideas or technologies (García-Avilés, 2020). In the context of affordable housing, these individuals or groups may include forward-thinking developers, social entrepreneurs, or government agencies that pioneer the implementation of affordable housing solutions. They frequently take risks, try out new techniques, and serve as role models for others. The degree to which an innovation matches established systems, values, and practices is referred to as compatibility. The difficulty of understanding and implementing an innovation is referred to as its complexity (Yuen et al., 2020). To ensure conformity with local rules, construction codes, and cultural norms, it is critical to assure compatibility with cheap housing technologies, including smart technology. Furthermore, reducing the complexity of smart technology through user-friendly interfaces, clear instructions, and training programmes might increase its adoption among affordable housing stakeholders. In the diffusion process, the theory emphasizes the importance of trialability and observability (Ali et al., 2019). Providing potential adopters with the opportunity to test and experience the benefits of smart affordable housing through pilot projects or demonstration sites can boost their desire to adopt these advances. Making the accomplishments and benefits of affordable housing programmes apparent and observable to others can also inspire and motivate additional adoption. Finally, the theory suggests that the perceived relative advantage of an innovation is a key determinant of its adoption (Yuen et al., 2020). The expected benefits of smart technology integration in affordable housing could include cost savings, energy efficiency, sustainability, increased quality of life, and community development potential. Demonstrating the advantages and positive outcomes of smart, affordable housing can encourage potential adopters to embrace these novel solutions.

6.2.3 Systems Thinking

Systems thinking is the ability to consider a system as a whole rather than just the pieces individually. Systems thinking entails employing a variety of methodologies to investigate various types of systems. Examples of systems thinking in nature include ecosystems, in which numerous elements such as air, water, movement, plants, and animals work together to exist or perish (Sanneh, 2018). The 'systems thinking approach' can be used to gain an understanding of both the qualitative and quantitative components of development trends for sustainability (Moldavska & Welo, 2015). Several scholars have defined systems thinking. The systems thinking paradigm views complex systems as interconnected and interdependent. It acknowledges the interdependence and consequences of ecological, social, and economic systems. To encourage long-term solutions, this theory emphasizes the importance of understanding feedback loops, dynamics, and unintended consequences of actions (Jackson, 2019).

Senge (1990) defined systems thinking as a 'framework for seeing interrelationships rather than things, for seeing patterns of change rather than static snapshots'. According to Davidz and Nightingale (2008), the concept of systems thinking is about interrelationships and interdependencies. They describe systems thinking as 'the analysis, synthesis, and understanding of technical, social, temporal, and multi-level interconnections, interactions, and interdependencies'. Systems thinking is also perceived as the ability to recognize, understand, and synthesize the interactions, and interdependencies in a set of components designed for a specific purpose (Moore et al., 2010). This involves the ability to notice patterns and repeats in interactions. Systems thinking is a means of seeing the connections, ties, or interconnections between different disciplines or sectors. Instead of viewing items in isolation, it allows the collective to be appreciated. Systems thinking focuses on how the entity being researched interacts with the other members of the system. It is about interdependence and interrelationships with linkages (Sanneh, 2018).

The application of systems thinking theory to smart affordable housing requires an understanding of the complex and interconnected structure of housing systems. Systems thinking promotes a holistic approach, considering the complete housing system rather than focusing simply on individual components. In the case of smart affordable housing, this involves the interactions and interdependencies of numerous factors such as technology, design, affordability, community engagement, and sustainability. Understanding these relationships aids in identifying possible leverage areas for increasing the overall effectiveness and impact of smart affordable housing efforts (Ammara et al., 2022; Driscoll et al., 2022). Systems thinking recognizes the involvement of several stakeholders in a system and their influence on each other. In the context of smart affordable housing, stakeholders include housing developers, legislators, residents, technology suppliers, financial institutions, and community organizations. Understanding the relationships and interests of the various stakeholders assist in identifying possible conflicts, aligning incentives, and fostering collaboration for the successful implementation of smart affordable housing programmes (Gosain et al., 2022; Lespagnard et al., 2023). Systems thinking understands that the system of smart housing displays emergent qualities that are not directly due to individual components of housing and technology but originate from their interactions. In the context of smart affordable housing, this involves understanding how the integration of smart technologies, sustainable design, and community engagement can create novel and beneficial qualities such as improved energy efficiency, improved quality of life, and higher social cohesion. Recognizing and harnessing these emerging qualities can lead to more effective and transformational smart affordable housing solutions (Driscoll et al., 2022; Nakamori, 2019).

6.3 Methodology

The research methodology of this study included a qualitative literature analysis and the incorporation of selected case studies. This strategy resulted in a holistic comprehension of the subject matter through the analysis of current literature and real-world examples. The procedure required rigorous analysis, and interpretation of existing literature on smart technology, smart houses, affordable housing, sustainability with construction materials and methods. The research aimed to provide significant insights and recommendations for the successful implementation of alternative smart affordable housing models by merging current information with real-life scenarios. The relevant information from the literature was extracted, including theories, concepts, frameworks, and empirical findings. The data was appraised based on relevance to the study's objectives, credibility of authors, publication dates and the impact of the findings. The gathered data was structured and classified into themes with supporting case studies. The case studies were chosen based on their relevance to the research topic and goals, diversity of their contexts, data availability, and the potential to generate adequate insights and knowledge about smart technology and affordable housing.

6.4 Literature Review

There is no definitive definition for 'smart affordable housing'; however, a number of scholars have described it in various ways, linked to smart technology and affordable or low-income housing. Smart affordable housing refers to residential developments that incorporate technology and design elements to reduce the cost of living for residents. It involves the provision of affordable housing which explores alternative models of smart technology with minimal gadgetry and integrated design elements to reduce the cost of living for residents (Ge et al., 2020; Kozubaev et al., 2019; Marikyan et al., 2019; Moghayedi et al., 2021; Saidu & Yeom, 2020; Simbanegavi & Ijasan, 2022; Wells & Vermeer, 2020). An affordable smart house is frequently constructed using environmentally friendly materials and building processes which may include features such as green roofs, solar panels, and water-saving fixtures to further decrease their environmental impact and lower living costs. Overall, a smart, affordable house, achieves a balance between modern elegance and practical cost (Simbanegavi & Ijasan, 2022; Wells & Vermeer, 2020). This type of housing model is an important component in the development framework of any smart city to provide housing access and efficiency to a majority of families in an urban area (Awad et al., 2017). For a housing scheme to achieve smartness and affordability, there is a need to understand the concepts of smart housing and affordable housing. Before discussing the concepts of smart affordable housing, it is important to briefly elucidate an overview and evolution of the housing policy frameworks in South Africa.

6.4.1 Overview of South Africa's Housing Policy

The post-apartheid era of South Africa made massive strides to rectify historical inequalities through various housing policies and programmes. The Reconstruction and Development Programme (RDP), through the 'Housing White Paper' initiated in the mid-1990s, was a key policy aimed at providing housing for disadvantaged households. Free-standing units were the products of housing delivery to the poor. The government distributes batches of capital subsidies per household to which, developers provide estates with tiny dwellings. These houses are allocated to qualifying beneficiary households for free (Huchzermeyer, 2014). However, this type of housing delivery faced several challenges, including slow implementation, inadequate infrastructure, and issues related to affordability and sustainability (Nokulunga et al., 2018). These estates lack creativity and end-user design, making the built environment unlivable. These estates are located far away from economic opportunities, community engagement, and participation are also minimal in the provision of the houses. The goal of the policy was mass production of housing, which led to unachievable targets, substandard building, and dissatisfied beneficiaries (Sethole & Sebola, 2020). The government subsequently introduced the Breaking New Ground (BNG) policy in 2004, emphasizing a more inclusive and sustainable approach to housing. BNG aimed to address the housing backlog, promote integrated urban development, and enhance community participation. It acknowledged the need for alternative housing models to better cater to the diverse needs of the population (Mkuzo et al., 2019). The products of the BNG policy include upgrading of informal settlements, self-help housing, social housing, new housing models, and subsidies for different household income levels within the disadvantaged communities. There was also the allocation of land close to the urban areas for affordable housing (Bredenoord et al., 2014).

Regardless of all the efforts and policies from the government, the housing challenges persisted, including a growing urban population, emerging informal settlements, corrupt state officials, and insufficient funding (Huchzermeyer, 2014). However, the government remained resolute and explored more innovative approaches to respond to the housing problem. Such approaches include; the changing role of the government from being a sole provider to a facilitator and partner in housing development initiatives. The adoption of inclusive housing, the concept for this type of housing is the mixing of individuals and households from diverse races and social economic statuses to facilitate densified sustainable developments (Karam et al., 2022; Onatu, 2012; Osman et al., 2011). Inclusive housing is an amalgamation of the three sub-markets (social rental housing, subsidized mortgage housing, and the market rate housing) with diverse unit typologies in one housing development, which translates into combining the poor and the rich in one housing scheme (Khadduri & Martin, 1997). This type of housing requires the government to have active partnerships with the private sector to promote affordability, poverty deconcentration, and racial integration. Another notable approach of responding to the housing challenge is the government's drive to implement affordable rented housing with smart technology. A smart house enables people to achieve numerous activities away from the physical places of employment or learning. This aspect reduces transportation costs and saves funds for low-income households (Chegut et al., 2016; Cilliers & Gaba, 2014). The implementation of these approaches with supportive policies, further facilitate the integration of housing solutions into the broader South African housing policy landscape.

6.4.2 Smart Housing

Smart housing refers to the integration of innovatory technology and systems within residential buildings to promote efficiency, comfort, and sustainability. It combines the use of smart devices, automation, and data-driven solutions to optimize energy use, improve security, and enable remote control and monitoring (Agee et al., 2021). The components of smart housing include energy management systems, automation and control systems, Internet of Things (IoT) connectivity, sensor technologies with security and safety systems (Gunawan et al., 2017). These components have been reported to assist in minimizing operational expenses of owning a house such as utility expenses and contribute to long-term affordability of providing affordable housing (King & Perry, 2017). A smart building is more than an automated entity which is controlled by its inhabitants but it is a building which assimilates from its inhabitants, conforms to their life patterns, and makes all decisions on its own (Batov, 2015). According to various existing literature, the essential features of a smart home or building include: Energy management and efficiency, automation, occupant comfort and security, multi-functionality, adaptability, and interactivity (Al Dakheel et al., 2020; Batov, 2015; Coskun et al., 2018; Gunawan et al., 2017; King & Perry, 2017; Le et al., 2012).

- Energy management and efficiency: Energy is prioritized in smart homes to reduce energy consumption and environmental effects. Numerous technologies are employed to monitor and control energy usage in real time, such as smart meters and sensors. Smart homes can optimize energy efficiency by adjusting the systems of home appliances based on occupancy, weather conditions, and energy demand.

- Automation: Smart homes use integrated building automation systems to control and manage numerous building functions and services such as lighting, plumbing, security, fire safety, access control, and ventilation equipment. Automation systems can be configured to respond to predetermined events or scenarios, thereby increasing operational efficiency and occupant comfort (Le et al., 2012; Lobaccaro et al., 2016).
- Occupant comfort and security: By utilizing sensor technologies and data analytics, smart homes give precedence to human comfort and well-being. Sensors monitor parameters such as temperature, humidity, air quality, lighting levels, and occupancy to produce the best indoor conditions possible. The information gathered can be evaluated to find patterns, reveal abnormalities, and make changes to improve occupant comfort, productivity and health. In addition, smart homes systems may include access control, video surveillance, intrusion detection, fire detection, and emergency response systems. To provide a secure and safe environment for occupants, who stay in the house (Gunawan et al., 2017).
- Multi-functionality: Multi-functionality in smart homes refer to the ability to provide numerous functions and services concurrently or interchangeably. This enables the homes to meet the varying demands of tenants, modify the internal space, optimize energy efficiency, and improve the overall building performance. The dynamic spaces created can support a myriad of activities and adapt to changing requirements by the user.
- Adaptability and interactivity: Adaptability of smart homes enables effective resource allocation, responsive environments, and smooth transitions between functions and modes of operation. Interactivity is the capability of the home to deliver a more engaging and user-centric environment, permitting occupants to have control, customization, and personalization of devices and systems, such as touchscreens, mobile apps, voice commands, and virtual reality (VR) technologies.

In the context of this study, smart housing can be useful and beneficial in sustainable and cost-effective ways. Smart housing encourages sustainable practices by utilizing energy-efficient technologies, incorporating renewable energy, and reducing resource use (Chel & Kaushik, 2018; Solanki & Nayyar, 2019). Incorporating smart technology with architecture creates a sustainable design approach that combines intelligent design, advanced technologies, and sustainable practices to create buildings that are energy-efficient, environmentally responsible, and responsive to the needs of occupants. This is achieved by optimizing resource usage, reduce environmental impact, enhance occupant well-being, and promote long-term sustainability (Hu & Hu, 2021). Another benefit of smart housing is its role in creating smart communities and neighbourhoods. This benefit is attained through as shared energy systems, smart mobility solutions, and community engagement platforms, highlighting their potential to enhance community resilience, social cohesion, and liveability. Smart housing effectively manages energy and resources used in the house via smart water and waste management systems. Smart water management systems provide water-efficient fixtures, leak detection systems, and automated irrigation systems help conserve water resources. Smart waste management systems promote recycling, waste segregation, and composting, reducing landfill waste and promoting circular economy practices (Hemidat et al., 2022). Other resource management benefits include the use of devices (i.e. smart meters) to monitor the use of energy, permitting homeowners to identify inefficiencies and make informed decisions to reduce consumption in their homes. This process is achieved through data-driven and decision-making capabilities of smart technology which leverages data analytics to provide insights for informed decision-making, leading to efficient predictive maintenance procedures and improved sustainability outcomes (Sarker et al., 2020). Smart housing also promotes the generation of renewable energy and lowering carbon footprints of buildings by subscribing to renewable energy sources like solar panels or wind turbines, generating clean energy, and reducing reliance on fossil fuels (Tan et al., 2021).

In the framework of affordability and cost savings, there a number of benefits facilitated by smart housing. Smart technology enables energy-efficient systems and appliances that can significantly reduce utility bills. Automated controls, smart thermostats, energy monitoring devices, and smart lighting systems help optimize energy usage and lower energy costs over time. By reducing energy consumption, residents can save money on their monthly utility bills, making housing more affordable. Another cost-saving benefit is the capability of smart technology to allow for remote management and control of various aspects of housing, such as lighting, heating, and security. This feature enables more efficient use of resources and reduces operational costs, hence making housing more affordable (Domb, 2019). Smart home security systems provide affordable housing with advanced security features such as video surveillance, motion sensors, and smart locks. These systems enhance the safety of residents and their belongings, potentially reducing the need for additional security measures and expenses (Prathibha et al., 2020). Access to high-speed internet connectivity is increasingly essential in today's digital world. Smart technology facilitates connectivity in affordable housing, enabling low-income families to access educational resources, job opportunities, telehealth services, and other online services that can improve their quality of life. Smart technology continually evolves, offering opportunities for innovation and adaptability in affordable housing design and management. As technology advances, new solutions emerge, providing cost-effective options for improving housing affordability for low-income families.

It is pertinent to acknowledge that the affordability impact of smart technology may vary depending on factors such as initial costs, access to technology, and ongoing support especially in low-income communities. These initial costs are usually mitigated by balancing initial investments with long-term cost savings and considering the specific needs of low-income households to make housing more affordable. Part of implementing these specific needs is the provision of smart technology devices that are affordable with low-tech automation systems which need minimal maintenance requirements. One of the most significant benefits of smart technology in low-income housing is its ability to improve energy efficiency. Smart thermostats, for example, can learn residents' temperature preferences and adjust accordingly, saving energy and money. Energy-efficient lighting, such as LED bulbs and smart lighting systems, can reduce energy consumption and extend the life of bulbs. Smart power strips can automatically turn off electricity to electronics when not in use, significantly minimizing energy waste. Another significant application of smart technology in low-income housing is to improve security. Smart locks, which can be operated remotely via smartphone apps, can increase access control while lowering the danger of break-ins. Video doorbells enable people to view and communicate with visitors without opening the door, adding an extra layer of protection. Security cameras can deter crime while also providing evidence in the event of a break-in. Smart technology can help improve connectivity in low-income dwellings, where internet access is often limited or costly. Wi-Fi routers can give wireless internet connectivity throughout the building, allowing residents to connect using their own devices.

The challenges and barriers associated with smart housing include: Cost, interoperability, data security, and user adoption (El-Azab, 2021; Maalsen, 2020). It is essential to acknowledge potential challenges and considerations related to smart technology and housing affordability:

- Initial Costs: The upfront costs associated with installing smart technology can be a barrier to affordability, especially for lower-income households or affordable housing projects. Efforts should be made to ensure that initial investments are balanced against long-term cost savings and potential benefits.
- Technological Accessibility: The usability and accessibility of smart technology can vary, and not all residents may have the necessary digital literacy or access to devices required for optimal utilization. Ensuring inclusivity and providing support for residents in adopting and using smart technology is crucial.
- Equity and Affordability Gap: There is a risk that the deployment of smart technology in housing may exacerbate the affordability gap if the benefits primarily accrue to higher-income households. Efforts should be made to ensure that smart technology is accessible and affordable for a wide range of residents, including those in low-income communities.
- Privacy and Data Security: Smart technology collects and processes data, raising concerns about privacy and data security. Measures must be in place to protect the personal information and ensure that data is handled responsibly and ethically

6.4.3 Affordable Housing

The goal of affordable housing is to provide low-income families with housing options that are both accessible and inexpensive. It focuses on lowering housing costs, preserving long-term affordability, and ensuring adequate living conditions for people (Keiser, 2019). The expression 'affordable house' signifies a myriad of things to a variety of people depending on their socio-economic status, their employment, their income, accommodation costs, family size, the size and location of their houses, access to necessary amenities, etc. (Doloi & Donovan, 2019). The general definition of an affordable house is by calculating the percentage of the household's monthly income spent on housing operational costs. This might be rent or mortgage payments, utility bills, and maintenance costs. The percentage should not be more than 30% of the household's monthly income (Luffman, 2006). This benchmark has been used for measuring housing affordability over the years; however, the definition is currently considered obsolete and irrelevant in today's housing landscape. The 30% income affordability criteria ignore regional variations in housing costs. Housing prices vary substantially among different regions, with certain locations facing extravagant housing costs in comparison with household earnings. Using a fixed percentage of 30%, fails to convey the genuine affordability issues encountered by households in high-priced areas (Aurand et al., 2021). In addition, housing costs have exceeded income growth in many regions, making the 30% criterion insufficient for capturing the real burden of housing prices. As property costs continue to rise, the 30% criterion becomes increasingly out of reach for many households, particularly low-income individuals and families. The 30% income affordability criterion excludes additional critical expenses such as utilities and ongoing maintenance, healthcare, transportation, education, and sustenance. It fails to consider the entire financial impact on households, resulting in a misleading representation of affordability. Given these issues, politicians, researchers, and housing advocates are increasingly realizing the limitations of the 30% income housing affordability criterion. Efforts are being made to produce more comprehensive and context-specific metrics of affordability that consider geographical variances, overall household spending, income disparities, and housing quality. A more inclusive approach entails identifying the percentage of households within each income group that are burdened by housing costs that surpass a given threshold. This aspect allows for the development of successful housing policies, initiatives, and interventions to meet the complex issues that individuals and families confront in today's housing context.

To achieve housing affordability, a combination of these techniques must be implemented, each adapted to the unique requirements and conditions of the community. One of the primary strategies is to expand the number of affordable housing units. This can be accomplished by building new affordable housing developments, enticing developers to incorporate affordable units into existing projects, and converting underutilized or vacant structures for affordable housing purposes. Collaborations between the public and private sectors can be helpful in improving housing affordability. Public–private partnerships (PPPs) can combine both sectors' resources, experience, and financial capacities to construct and operate affordable housing developments. These collaborations can lead to innovative finance approaches, faster construction procedures, and the use of underutilized public lands for affordable housing purposes. Another method of achieving housing affordability is the incorporation of sustainable design principles in affordable housing developments can help to minimize energy costs, lower maintenance costs, and increase overall affordability. Energy-efficient features like insulation, solar panels, and energy-saving appliances help residents save money on their electricity bills. Furthermore, creating affordable housing near public transportation hubs improves accessibility and lowers transportation expenses for tenants. Transit-oriented development promotes mixed-use development and provides chances for inexpensive housing in close proximity to employment, services, and amenities. Finally, government action is critical to improving housing affordability. This involves giving developers or people subsidies, grants, and tax breaks to support the creation and upkeep of affordable housing. Governments can also enact rent control measures and create housing trust funds to help with affordable housing projects (Baqutaya et al., 2016; Mueller & Tighe, 2022).

Housing affordability also has its fair share of challenges when it comes to implementation, ranging from high costs of construction, land and income disparities to regulatory constraints and funding limitations (Meen & Whitehead, 2020; Mueller & Tighe, 2022).

- High Construction and Land Costs: The cost of construction materials, labour, and land acquisition has been steadily rising, making it difficult to develop affordable housing. High construction and land costs can limit the feasibility of affordable housing projects and hinder their affordability.
- Insufficient Supply: There is often a shortage of affordable housing units relative to the demand, especially in densely populated areas. Limited supply results in higher competition and increased housing costs, making it challenging for low-income individuals and families to find affordable options.
- Income Disparities: Income disparities and stagnating wages can make it difficult for individuals and families to afford suitable housing. The gap between housing costs and income levels is widening, resulting in housing becoming increasingly unaffordable for a significant portion of the population.
- Gentrification and Displacement: Gentrification can lead to the displacement of low-income residents as neighbourhoods undergo revitalization and housing prices increase. This can disrupt communities and further exacerbate the affordable housing shortage, pushing vulnerable populations out of their homes.
- Regulatory and Zoning Constraints: Complex regulations, zoning restrictions, and lengthy approval processes can hinder the development of affordable housing projects. Restrictive land-use policies, density restrictions, and parking requirements can increase development costs and limit the potential for affordable housing options.
- Financing and Funding Limitations: Securing financing for affordable housing projects can be challenging. Traditional lending institutions may have stringent requirements, making it difficult for developers to obtain loans. Additionally, the availability of public funding and subsidies for affordable housing may be limited, further constraining the development of affordable housing units.
- Social Stigma and Discrimination: There can be social stigma and discrimination associated with affordable housing, perpetuating negative perceptions, and limiting the acceptance and integration of affordable housing developments within communities. Overcoming these biases is essential for creating inclusive and equitable housing opportunities.

- Infrastructure and Transportation Access: Affordable housing often faces challenges related to inadequate infrastructure and limited access to transportation, services, and amenities. Limited access to public transportation, schools, healthcare facilities, and employment opportunities can further increase the burden of housing costs for low-income households.

Addressing the difficulties of housing affordability necessitates a holistic approach that includes public–private sector partnerships, legislative reforms, community engagement, and novel financing options. It entails enacting supportive laws, streamlining regulations, increasing funding for affordable housing projects, and encouraging sustainable and inclusive development methods. Addressing these difficulties allows us to increase the supply of affordable housing, provide new alternative means of housing delivery, and ensure that individuals and communities have access to safe, decent, and inexpensive houses.

6.5 Alternative Models of Smart Affordable Housing

The motivation and need for best practices in designing and building alternative models of smart affordable housing for low-income families stem from the urgent need to address the housing affordability crisis, ensure access to safe and decent housing, promote financial stability, leverage technology for efficiency and cost savings, enhance sustainability, foster community development, and highlight innovation and replicability. Prioritizing these best practices enables the development of affordable housing solutions that improve the lives of low-income families while also contributing to more equitable and sustainable communities. The alternative models to be discussed further are climate smart housing, smart building materials, additive manufacturing (three-Dimensional printing), Internet of Things, smart mini-grids, and adaptable housing.

6.5.1 Climate Smart Housing

Climate-smart buildings and architecture have gained traction in an era of increased environmental concerns. As global society seeks new ways to address climate change, the construction and design industries play a critical role in lowering the environmental impact of buildings. Climate-smart buildings are built to use less energy, emit fewer greenhouse gases, and adapt to changing weather patterns. This comprehensive study examines the essential ideas, tactics, and technologies that characterize climate-smart buildings and architecture, emphasizing the relevance of sustainable practices in designing a greener future (Roy et al., 2021). Climate-smart buildings are designed to minimize their carbon footprint and optimize resource efficiency throughout their lifecycle. They are resilient structures that integrate sustainable design principles, utilize renewable energy sources, and prioritize occupant comfort and well-being (Hirsch, 2009). Examples of climate smart solutions are solar panels on the roof provide electricity to the building, a remote-controlled device which turns off ventilation and electricity in the building, and glass curtain walls in the balconies in each apartment with space to cultivate vegetables and herbs. Furthermore, the building features various communal rooms where residents can collaborate on shared initiatives (Wester, 2022). The primary aim of climate-smart buildings is to improve the climatic resilience and thermal comfort of affordable housing stock by using low-impact design, materials, and construction technologies. The core principles of climate-smart architecture include passive design tactics such as optimizing building placement and orientation to capture natural light and ventilation, thereby decreasing dependency on artificial lighting and HVAC systems. The use of high-performance insulating materials and designs to reduce heat transmission while increasing energy efficiency. Encourage natural ventilation, including cross-ventilation, moveable windows, and shading devices, in order to improve airflow and reduce the need for mechanical cooling. The intentional integration of daylighting in buildings by optimizing natural light penetration to reduce energy consumption for lighting reasons while producing a visually pleasing and productive indoor environment (Monis & Rastogi, 2022). Another important principle is the integration and optimization of renewable energy in building design. The type of design usually creates 'nearly-zero energy buildings' which are buildings that produce as much renewable energy as they consume over the course of the year. Smart technology provides effective energy management and optimization in net-zero buildings through the use of advanced automation, data analytics, and networking. Almost no buildings rely extensively on renewable energy sources to meet their energy requirements. Smart technology is important for the effective integration and optimization of renewable energy systems like solar panels and wind turbines. Smart systems that monitor renewable energy generation, energy storage, and building energy demand can dynamically control energy flows, maximize self-consumption of renewable energy, and assure optimal exploitation of clean energy resources (D'Agostino & Mazzarella, 2019).

There are a number of benefits to climate-smart buildings, firstly, the efficient use of energy by using advanced technologies and efficient systems to reduce energy consumption, resulting in lower operational costs and decreased reliance on fossil fuels. Secondly, by employing renewable energy sources and adopting eco-friendly materials and construction techniques, climate-smart buildings significantly reduce greenhouse gas emissions. Thirdly, these structures are designed to withstand the impacts of climate change, such as extreme weather events, rising temperatures, and rising sea-levels, ensuring long-term functionality and occupant safety. Finally, climate-smart buildings prioritize occupant health and comfort by improving the interior spaces of buildings and enhancing natural lighting, optimizing air quality, and promoting sustainable materials that minimize off-gassing (Wester, 2022).

Climate-smart buildings and architecture are essential components of a sustainable future. By incorporating passive design strategies, renewable energy integration, and innovative technologies, these buildings can significantly reduce energy consumption, lower emissions, and enhance occupant well-being. Integrating sustainable and energy-efficient features into affordable housing models benefits the environment and reduces long-term costs for residents. The incorporation of passive design strategies minimizes energy consumption with renewable energy and environmentally friendly building materials provide affordable and clean energy to residents. The collaboration among architects, engineers, policymakers, and the construction industry is crucial in fostering a transition to climate-smart practices globally. Embracing the principles and techniques outlined in this article will enable us to build a resilient and environmentally responsible built environment that mitigates climate change impacts and supports a sustainable future for generations to come. In Beira, Mozambique, a developing company named 'Casa Real' provides access to appropriate and sustainable housing for poor and middle-income families by delivering creative and high-quality affordable housing alternatives. The buildings are considered climate smart and resilient due to the concerted efforts made in the construction of the housing units to withstand Mozambique's adverse weather and climate.

6.5.2 Smart Building Materials

The most significant portion of the cost of a new house is more commonly the building materials and construction labour. Modern building materials in particular such as fired bricks, concrete, glass, and steel, are expensive to manufacture. They are usually manufactured at a central location far away from areas where affordable houses are situated, which usually results in substantial transport costs.

In the drive to alleviate the worldwide housing issue, the research and deployment of smart building materials have emerged as a game-changing answer. These novel materials not only provide cost-effective alternatives for building affordable housing, but they also have various other advantages, such as improved energy efficiency, increased durability, and a lower environmental impact to revolutionize affordable housing. As the demand for affordable housing continues to rise, the construction industry is exploring sustainable smart materials as a means to address this pressing issue. By combining sustainability and innovation, these materials offer a promising solution for creating cost-effective, environmentally friendly, and high-performance building materials for affordable housing.

Recycling building materials is an important method of sustainable construction. Materials such as reused wood, recycled steel, and recycled concrete lessen the demand for virgin resources while diverting waste from landfills. These materials can be obtained from dismantled buildings, salvaged from demolition sites, or purchased from recycling operations. Construction costs can be reduced while the environmental impact of resource extraction and waste generation is minimized by utilizing recycled materials in affordable housing developments. Cross-laminated timber (CLT) is an environmentally friendly and cost-effective alternative to standard building materials. CLT, which is made up of layers of wood panels glued together in a transverse pattern, has exceptional strength, stability, and fire resistance. Because of its prefabricated design, this material enables faster construction, lowering labour costs, and construction time. Furthermore, CLT structures have a lower carbon footprint than standard concrete and steel structures, making them a viable option for inexpensive housing developments. Insulated concrete forms (ICFs) have grown in popularity as an effective building alternative for low-income housing. These structures are composed of interlocking panels filled with concrete and made of materials such as expanded polystyrene (EPS) or extruded polystyrene (XPS). ICFs have outstanding insulation qualities, resulting in high thermal efficiency and low energy usage. They also have improved soundproofing qualities, increased longevity, and resistance to natural disasters such as hurricanes and earthquakes.

Traditional insulating materials, such as fiberglass and foam, are not only costly but also have negative environmental consequences. Smart building materials provide long-term solutions that are both affordable and environmentally beneficial. For example, cellulose insulation is created from recycled paper and has good thermal performance. Furthermore, natural fibre insulation materials such as hemp, wool, and cotton are renewable and provide good insulation while minimizing environmental impact. Smart glass, often known as switchable or electrochromic glass, is changing the way windows and facades work in buildings. This glass technology enables the control of transparency, colour, and heat transfer qualities. Smart glass can control the amount of sunlight that enters a building by varying its tint, decreasing the need for artificial lighting and saving energy. Smart glass also provides improved insulation, reducing heating, and cooling costs in affordable housing projects.

In Uganda, the technology of interlocking stabilized soil blocks (ISSB) is a notable climate smart building material being used for affordable housing. This material and construction technology have the advantages of low cost and low environmental impact while giving equivalent quality to traditional burned brick construction. The research and application of sustainable smart construction materials are critical to the creation of environmentally friendly, cost-effective, and high-performing affordable housing solutions. Policymakers, developers, and industry professionals must prioritize the use of these materials and processes to ensure that future housing initiatives match sustainability goals and satisfy the requirements of communities around the world.

6.5.3 Additive Manufacturing (Three-Dimensional Printing)

Additive manufacturing is one of the modern techniques that are emerging in the industry of building construction. It is characterized by being fast, low cost, and an environmentally friendly building process (Alkhalidi & Hatuqay, 2020). The development is the emergence of 3D printed affordable houses. By leveraging advanced technology and automation, 3D printing offers the potential to revolutionize the way we build homes, making them faster, cost-effective, and environmentally friendly. 3D printing, also known as additive manufacturing, involves creating three-dimensional objects by layering materials based on a digital model. In construction, 3D printing technologies utilize specially formulated concrete or other construction materials to build structures layer by layer, guided by computer algorithms and robotic arms. This automated process eliminates the need for traditional formwork and manual labour, resulting in faster construction times and reduced costs. 3D printing offers a green building technique. Compared to conventional building techniques, it eliminates a great amount of CO_2 emitting and energy-consuming processes. It is also flexible and opens the door for many architectural designs to be constructed at lower costs. The execution flexibility of 3D printing provides designers with the opportunity to create suitable elements for each project eliminating further treatments for energy reduction purposes. Supportive passive and renewable energy systems can be applied afterward to reach even lower, or zero-energy buildings. The use of 3D printing for construction project delivery is believed to be the way forward in the quest for sustainable construction projects—housing inclusive (Sakin & Kiroglu, 2017). With its project cost reduction benefits in terms of labour and materials (Lim et al., 2012; Mohd-Tobi et al., 2018), along with its ability to reduce the cost of transporting materials to site (Sakin & Kiroglu, 2017), economically sustainable housing projects that are affordable for low-income earners can be delivered. Furthermore, social sustainability can be achieved through 3D printing as the technology offers the creation of new jobs and the reduction of site accidents and fatalities (Fonseca, 2018; Pîrjan & Petroşanu, 2013). Aside from its economic and social attributes, the use of 3D printing promises the reduction in construction material wastage (Ellis, 2018) which over time has been a source of environmental concern for most construction experts (Ametepey & Aigbavboa, 2014).

Among the numerous recognized benefits of employing 3D printing for housing and aiding affordability is the potential to dramatically lower building costs by improving material consumption, minimizing waste, and reducing labour requirements. The shortened approach can help close the affordability gap in housing. Compared to traditional construction methods, 3D printing allows for faster construction by greatly expediting the total building process. Another advantage of 3D printing is its extraordinary design versatility, which enables the production of complex and personalized architectural shapes that are both aesthetically pleasing and structurally efficient. This flexibility opens up new avenues for inexpensive house design. 3D printing allows for the use of environmentally acceptable construction materials, lowers waste formation, and increases energy efficiency. Furthermore, 3D-printed constructions can include elements like insulation, energy-efficient windows, and smart technology to further enhance sustainability and affordability (Cascone et al., 2020; Rimmer, 2021). The Mvule Gardens development in Kenya, is an ideal example of how a housing developer is offering affordable, climate-resilient homes in Kenya to help bridge the country's infrastructural gap. The project is still under construction with a focus on healthy and comfortable living, addressing the local climate, and effective material use, made possible through structural design, 3D printing, and local material procurement.

To adopt 3D printing for sustainable low-income housing in developing countries, there is the need to promote the technology among relevant stakeholders and create some form of training for construction practitioners. This can be done through seminars, workshops, and conferences organized by both the vendors of these technologies and respective professional bodies responsible for building delivery in these countries. Developing long-lasting, high-performance 3D printing construction materials is an ongoing challenge. Researchers and manufacturers are constantly optimizing material compositions to ensure strength, longevity, and resilience to environmental conditions. There is the need for government support through legislation and the creation of policies to support and implement 3D printing as viable alternative to traditional construction methods. As technology advances, 3D-printed, inexpensive buildings have enormous promise for alleviating the worldwide housing crisis. Their potential influence extends beyond affordability to include disaster assistance, distant housing, and the creation of sustainable communities. Collaboration among architects, engineers, policymakers, and investors is critical for furthering the technology, driving innovation, and ensuring the widespread acceptance of 3D-printed affordable dwellings.

6.5.4 Internet of Things

The Internet of Things (IoT) has emerged as a significant force driving innovation in a variety of areas, including housing. IoT technology has enormous potential to make housing more inexpensive, efficient, and sustainable by linking everyday objects and gadgets to the internet and enabling them to communicate and exchange data (Greengard, 2021). One of the significant contributors to housing costs is energy consumption. IoT-enabled smart energy management systems provide real-time monitoring, analysis, and control of energy usage. Smart thermostats, energy-efficient appliances, and automated lighting systems optimize energy consumption, reduce wastage, and lower utility bills for residents. Additionally, IoT devices can integrate with renewable energy systems like solar panels, maximizing energy efficiency and further reducing dependence on the grid (Sepasgozar et al., 2020). IoTs are effective to use and implement quickly because they are mostly wireless and do not need extensive cable installations. Wireless technology reduces the need for expensive infrastructure installation and maintenance, such as cable laying or landline connections. This makes it more affordable for low-income families with limited financial resources. Wireless technology also enables quick access to the internet and communication services without the need for physical connections. It can be accessed via cell phones, tablets, or low-cost wireless devices, making it more accessible to low-income folks who may not have access to traditional wired services. This is advantageous for low-income housing or even informal structures which often lack permanent infrastructures. IoT technology enables efficient resource management in affordable housing. Smart water meters and leak detection systems help identify water wastage, enabling timely repairs and reducing water bills. Additionally, IoT-enabled waste management systems optimize garbage collection routes, reducing operational costs and enhancing environmental sustainability (Ismail, 2019).

There are a variety of affordable smart devices associated with IoTs which can be employed in low-income households to monitor and manage energy systems. These include thermostats, water heaters, lighting systems, security systems, and general electrical appliances integrated with the internet. IoT-enabled thermostats allow residents to control the temperature of their homes remotely through smartphone applications. These devices learn residents' behaviour patterns and adjust heating and cooling settings accordingly, maximizing energy usage and reducing utility bills. IoT-connected water management solutions, such as smart water meters or leak detection sensors, assist in measuring water consumption, identifying leaks, and offer real-time notifications to homeowners through smartphone applications. These technologies improve water management and prevent the waste of water. Water heaters enabled by IoT systems can be remotely regulated and scheduled to save energy. Residents can warm water before using it or change settings based on occupancy patterns, increasing energy efficiency and cost savings. IoT-connected lighting solutions allow residents to remotely manage and schedule lighting in their homes. They can use smartphone apps or voice commands to change the brightness, colour, and turn on/off lights, increasing energy efficiency and ease. Residents can benefit from economical security options such as smart door locks, video cameras, and motion sensors. These gadgets can be remotely monitored, send notifications to smartphones in the event of suspicious activity, and allow residents to permit entry to visitors even while they are not at home (Ghayvat et al., 2015). Finally, general smart electrical appliances which are IoT enabled such as refrigerators, televisions, washing machines, and geysers. Can be remotely controlled and monitored for efficient energy consumption and inform the users about maintenance and the replacement of consumables.

The Internet of Things (IoT) has the potential to change and enhance affordable housing by boosting energy efficiency, security, maintenance, resource management, and community engagement. By embracing IoT technology, housing providers can cut operational costs, improve the quality of life for residents, and develop sustainable and affordable communities. However, protecting data privacy, addressing security concerns, and incorporating IoT infrastructure into affordable housing projects require coordination among stakeholders, including housing developers, lawmakers, technology providers, and inhabitants, as stated earlier the process is easier and more accessible with use of wireless systems. Embracing the IoT revolution in affordable housing can lead to increased efficiency, decreased costs, and higher living standards for individuals and families in need of inexpensive housing options (Sepasgozar et al., 2020).

6.5.5 *Smart Mini/Micro-grids*

A smart mini-grid is a localized electrical distribution system that incorporates renewable energy sources, energy storage, and advanced monitoring and control technology. These grids are often smaller in scale than traditional power networks and are intended to service localized areas, such as low-income neighbourhoods, rural communities, and informal settlements. They deliver dependable, cheap, and sustainable energy solutions by utilizing renewable energy sources and optimizing energy generation, distribution, and consumption (Rajvanshi, 2022). Smart Grid is a stable, efficient, and reliable system (Tuballa & Abundo, 2016). Among the definitions, the European Technology Platform defines a Smart Grid as: 'An electricity network that can intelligently integrate the actions of all users connected to it—generators, consumers, and those that do both to efficiently deliver sustainable, economic, and secure electricity supplies' (Dileep, 2020).

Smart mini-grids can be sourced from a various procedure such as micro-hydroelectricity systems, solar, wind, and biomass. Micro-hydro is the power of running water as an energy source for a mini-grid. This system is usually on a favourable site and has the considerable advantage of producing electricity continuously all year round with diverse outputs. Hydro-electric systems generally do not require the need for backup devices such as batteries and diesel generators. However, they do require a continuous filtration and monitoring system for the effective flow of water, which is susceptible to seasonal debris and trash (Zebra et al., 2021). The abundant availability of solar energy makes it an attractive resource for a mini-grid; however, its seasonal and daily variations lead to the need for energy storage devices which are usually back up batteries. Another renewable, abundant energy resource is the wind. It can be accessed everywhere; however, the volatile relationship between wind speed and the energy that can be used for generating electricity is dependent on location. In addition, the seasonal movement of prevailing winds in an area makes this system require back up storage equipment. Biomass is an organic matter used as fuel to generate electricity. It can be processed in different forms; solid material, such as wood and crop residues, which can be subjected to direct combustion, or converted into a gas via a pyrolysis process. Wet material, such as food waste, animal dung and urine, which provides a good basis for anaerobic digestion to produce biomethane. Oils obtained from crops, such as rice bran and jatropha, by pressing. This is readily converted into an oil suitable for a diesel engine through a chemical process known as transesterification (Bhattacharyya & Palit, 2014).

There is a myriad of benefits from smart micro-grid systems, these include; smart mini-grids provide a stable and consistent electricity supply, replacing the erratic and often dangerous methods of energy generation used in low-income households. By reducing dependence on expensive fossil fuels and connecting to renewable energy sources like solar, wind, or biogas, these grids can offer affordable electricity tariffs, ensuring reliable access to power for households, businesses, and essential services. By leveraging renewable energy sources, smart mini-grids contribute to reducing greenhouse gas emissions and combatting climate change. Traditional methods of energy generation in informal and low-income settlements, such as burning fuels like kerosene or wood, release harmful pollutants and contribute to indoor air pollution. Smart mini-grids offer a cleaner and more sustainable alternative, reducing environmental health risks and fostering a greener and healthier environment. Access to electricity plays a vital role in economic development. Smart mini-grids enable the establishment and growth of small businesses within low-income settlements. Powered by reliable electricity, local entrepreneurs can engage in productive activities, such as running shops, operating communal charging stations for mobile phones, or operating small-scale manufacturing units. This economic empowerment helps create jobs, generate income, and uplift the overall living standards of the community. The cost of building a new optimized distribution infrastructure for the microgrid is also a small fraction of that of expanding the central grid to remote locations (Kuffar, 2020). Smart mini-grids also improve education and healthcare of communities by supply power to schools and local clinics.

The smart mini-grid implemented at the informal settlement of Qandu-Qandu in Cape town, South Africa focuses on disruptive innovation by installing solar towers that are linked to the shacks and managed by social companies. The four solar towers at Qandu-Qandu produce enough clean energy for night lighting and power up electrical appliances such as televisions for approximately 30 people (Bobbins et al., 2023). While smart mini-grids hold tremendous potential for empowering low-income settlements, several challenges must be addressed for its successful implementation: Firstly, establishing and maintaining smart mini-grids require significant investment. Public–private partnerships, innovative financing mechanisms, and grant funding can play a crucial role in making these projects economically viable, ensuring affordable tariffs for the residents. Secondly, the installation and management of smart mini-grids require technical expertise, especially in remote areas with limited infrastructure. Training and capacity-building programmes can empower local technicians to handle grid operations, maintenance, and troubleshooting, ensuring the long-term sustainability of the systems. Thirdly, the success of smart mini-grids heavily relies on community involvement and government supportive frameworks. Engaging the community in decision-making processes, awareness campaigns, and capacity-building initiatives is vital to ensure acceptance, ownership, and sustainable operation of the grids. Governments should provide incentives, streamline licensing processes, and facilitate partnerships to attract private sector participation and investment in these projects.

Smart mini-grids are a viable solution to the energy poverty that exists in informal settlements. These grids empower neglected communities by facilitating economic growth, boosting education and healthcare, and promoting environmental sustainability. Addressing the issues through community engagement, financial sustainability, technical knowledge, and supportive legislation will unlock the actual potential of smart mini-grids, providing better and more prosperous futures for individuals living in informal settlements (Bhattacharyya & Palit, 2014; Tlhatlha, 2020).

6.5.6 Adaptable Housing

Adaptable housing has emerged as a transformative solution to address the pressing need for affordable and sustainable housing. The design and construction of residential units that can be easily adjusted or expanded to fit changing occupant demands and preferences is referred to as adaptable housing. These homes provide flexibility by letting residents change the layout, size, and usefulness to meet their changing needs. Adaptable housing focuses on longevity, adaptation, and sustainability, allowing occupants to live comfortably throughout their shifting life stages (Gushen et al., 2021). Smart adaptable housing refers to residential buildings or homes that use sophisticated technologies and flexible design concepts to adapt and respond to tenants' changing demands over time. Intelligent technologies and features boost comfort, convenience, energy efficiency, and accessibility in these houses (Gushen et al., 2021; Kolarevic & Parlac, 2015).

According to existing literature, adaptable housing has a number of recognized diverse typologies, namely core and shell design, modular construction, convertible spaces, and adaptive reuse. In the core and shell design typology, there is core structure that supplies necessary utilities such as water and electrical systems, while the remainder of the area is flexible. Residents can adapt and partition the open area to suit their needs (Van Noorloos et al., 2020; Wainer et al., 2016). The modular construction typology involves the use of prefabricated modules that may be assembled to produce a range of layouts. These modules are readily added and withdrawn, allowing for growth or downsizing, as needed (Arvaniti, 2020; Magdziak, 2019). Convertible spaces include multi-functional spaces that can be easily altered to serve multiple needs. A living room, for example, can be changed into a bedroom or an office during the day. Incremental adaptable housing focuses on incremental development, allowing occupants to gradually increase the size of their homes as their demands and financial capacities rise. This typology enables the creation of fundamental structures that can be developed over time (Van Noorloos et al., 2020). The practice of changing one building typology into another from its initial intended use or function is known as adaptive reuse. Adaptive reuse is commonly utilized to modify old buildings to fulfil a new purpose. It is more environmentally friendly because the building is not demolished and old construction components are reused (Hebbert, 2021; Stone, 2019).

The influence and benefits of adaptable housing include a cost-effective solution for affordable housing. Residents can invest in a basic building and gradually extend and improve their homes as their financial situation allows by permitting incremental construction and customization (Femenias & Geromel, 2020). Adaptable housing makes the most use of available space, ensuring resource efficiency. Residents can optimize their living conditions based on their unique demands by modifying the layout and repurposing areas, avoiding the need for superfluous square footage. By adapting to shifting family dynamics, adaptable housing provides long-term affordability. The capacity to alter the space as families expand or decrease reduces the need for relocation and the accompanying costs (Manewa et al., 2016; yn der Voordt, 1990). Housing adaptation improves sustainability by lowering the environmental impact of demolition and new building. Adaptable housing promotes longevity while reducing waste and increasing resource efficiency (Estaji, 2017; Kinnane et al., 2016).

An example of prefabricated residential units found in Algeria displays the effective use of cutting-edge technology to increase the efficiency housing affordability and adaptability. Adaptable housing has enormous potential to alleviate Africa's affordable housing concerns. The housing sector may accommodate inhabitants' different requirements and preferences while maintaining long-term affordability and sustainability by embracing adaptable housing typologies and promoting flexibility. Adaptable housing, with additional research, funding, and policy backing, has the potential to be a catalyst for sustainable and affordable housing solutions across the continent, improving living conditions and boosting communities (French, 2016; Schmidt & Austin, 2016).

6.6 Conclusion

In conclusion, the alternative models of smart affordable housing present viable options to solve the pressing concerns of housing affordability, sustainability, and adaptation. These models cover several creative approaches, including climate-smart housing, smart building materials, additive manufacturing (three-dimensional printing), the Internet of Things (IoT), smart mini-grids, and adaptable housing. By merging these alternative models, it is possible to develop smart, affordable housing that is environmentally sustainable, technologically modern, and adaptive to changing conditions. These approaches have the ability to create inclusive and resilient communities, ensuring that safe and affordable housing is available to all. The successful implementation of alternative smart affordable housing models requires overcoming challenges such as initial investment costs, regulatory barriers, and the need for skilled labour. It is imperative for government, private sector, and communities to collaborate to create an enabling environment that supports the adoption of the models discussed in this chapter. Alternative models of smart affordable housing represent a promising avenue for addressing the complex challenges of housing in the twenty-first century. As the global population continues to urbanize, and the demand for housing rises, it is imperative to embrace innovative, sustainable, and community-centric solutions. The convergence of technology, sustainability, and community engagement not only redefines the concept of affordable housing but also lays the foundation for more inclusive, resilient, and environmentally conscious urban development. The journey towards widespread adoption of these models requires collective commitment, collaboration, and a shared vision of housing as a fundamental right and a catalyst for positive societal transformation.

References

Agee, P., Gao, X., Paige, F., McCoy, A., & Kleiner, B. (2021). A human-centred approach to smart housing. *Building Research & Information, 49*(1), 84–99.

Al Dakheel, J., Del Pero, C., Aste, N., & Leonforte, F. (2020). Smart buildings features and key performance indicators: A review. *Sustainable Cities and Society, 61*, Article 102328.

Ali, M., Raza, S. A., Puah, C. H., & Amin, H. (2019). Consumer acceptance toward takaful in Pakistan: An application of diffusion of innovation theory. *International Journal of Emerging Markets*.

Alkhalidi, A., & Hatuqay, D. (2020). Energy efficient 3D printed buildings: Material and techniques selection worldwide study. *Journal of Building Engineering, 30*, Article 101286.

Ametepey, S. O., & Aigbavboa, C. (2014). Practitioners' perspectives for the implementation of sustainable construction in Ghana. In *Proceedings of the DII-2014 conference on infrastructure investments in Africa* (pp. 114–124).

Ammara, U., Rasheed, K., Mansoor, A., Al-Fuqaha, A., & Qadir, J. (2022). Smart cities from the perspective of systems. *Systems, 10*(3), 77.

Arvaniti, E. (2020). *Ensuring adaptability in modular construction.* Aalborg University.

Aurand, A., Emmanuel, D., Threet, D., Rafi, I., & Yentel, D. (2021). *A shortage of affordable homes.* National Low Incoming Housing Coalition. Retrieved May 25, 2021.

Awad, J., Hyder, A., & Irfan, A. (2017). Development of smart cities from fiction to reality in member states of the Gulf cooperation council. In *Smart city networks: Through the internet of things* (pp. 43–63).

Baqutaya, S., Ariffin, A. S., & Raji, F. (2016). Affordable housing policy: Issues and challenges among middle-income groups. *International Journal of Social Science and Humanity, 6*(6), 433.

Batov, E. I. (2015). The distinctive features of "smart" buildings. *Procedia Engineering, 111*, 103–107.

Bhattacharyya, S. C., & Palit, D. (Eds.). (2014). *Mini-grids for rural electrification of developing countries: Analysis and case studies from South Asia* (pp. 55-e4sv). Springer International Publishing.

Bobbins, K., Caprotti, F., de Groot, J., Pailman, W., Moorlach, M., Schloemann, H., Densmore, A., Finlay, K., Fischat, E., Siwali, S., & Links, J. (2023). Smart and disruptive infrastructures: Re-building knowledge on the informal city. *Urban Studies*, 00420980231172582.

Bredenoord, J., Van Lindert, P., & Smets, P. (Eds.). (2014). *Affordable housing in the urban global south: Seeking sustainable solutions.* Routledge.

Cascone, P., Laddaga, M., & Forestiero, F. (2020). Digital construction: 3D printing for performative houses. In *Second RILEM International Conference on Concrete and Digital Fabrication: Digital Concrete 2020 2* (pp. 804–813). Springer International Publishing.

Chegut, A., Eichholtz, P., & Holtermans, R. (2016). Energy efficiency, and economic value in affordable housing. *Energy Policy, 97*, 39–49.

Chel, A., & Kaushik, G. (2018). Renewable energy technologies for sustainable development of energy efficient building. *Alexandria Engineering Journal, 57*(2), 655–669.

Chen, Y. H., & Barnes, S. (2007). Initial trust and online buyer behaviour. *Industrial Management & Data Systems, 107*(1), 21–36.

Cilliers, L., & Gaba, S. (2014). Improving the utilisation of smart technology in South African homes. *DDR, 2014*, 604.

Coskun, A., Kaner, G., & Bostan, İ. (2018). Is smart home a necessity or a fantasy for the mainstream user? A study on users' expectations of smart household appliances. *International Journal of Design, 12*(1), 7–20.

D'Agostino, D., & Mazzarella, L. (2019). What is a nearly zero energy building? Overview, implementation and comparison of definitions. *Journal of Building Engineering, 21*, 200–212.

Davidz, H. L., & Nightingale, D. J. (2008). Enabling systems thinking to accelerate the development of senior systems engineers. *Systems Engineering, 11*(1), 1–14.

Davies, R., & Harty, C. (2013). Measurement and exploration of individual beliefs about the consequences of building information modelling use. *Construction Management and Economics, 31*(11), 1110–1127.

Davis, F. D. (1987). *User acceptance of information systems: The technology acceptance model (TAM).*

Davis, F. D. (1989). Technology acceptance model: TAM. In *Al-Suqri, MN, Al-Aufi, AS: Information seeking behavior and technology adoption* (pp. 205–219).

Dileep, G. A. (2020). Survey on smart grid technologies and applications. *Renewable Energy, 146*, 2589–2625. Retrieved September 25, 2023, from https://doi.org/10.1016/J.RENENE.2019.08.092

Doloi, H., & Donovan, S. (2019). *Affordable housing for smart villages*. Routledge.

Domb, M. (2019). Smart home systems based on internet of things. In *Internet of Things (IoT) for automated and smart applications*. IntechOpen.

Driscoll, P. J., Parnell, G. S., & Henderson, D. L. (Eds.). (2022). *Decision making in systems engineering and management*. Wiley.

El-Azab, R. (2021). Smart homes: Potentials and challenges. *Clean Energy, 5*(2), 302–315.

Ellis, G. (2018). The truth about 3D printing in construction. Construction productivity blog. Available online: https://blog.plangrid.com/2018/11/the-truth-about-3d-printing-inconstruction/

Estaji, H. (2017). A review of flexibility and adaptability in housing design. *International Journal of Contemporary Architecture, 4*(2), 37–49.

Femenias, P., & Geromel, F. (2020). Adaptable housing? A quantitative study of contemporary apartment layouts that have been rearranged by end-users. *Journal of Housing and the Built Environment, 35*, 481–505.

Fishbein, M. E. (1967). *Readings in attitude theory and measurement*. Wiley.

Fonseca, L. M. (2018). Industry 4.0 and the digital society: Concepts, dimensions and envisioned benefit. In *Proceedings of the 12th international conference on business excellence* (Vol. 34, pp. 386–397).

French, H. (2016). *Adaptable housing: Accommodating change* (Doctoral dissertation, University of Brighton).

García-Avilés, J. A. (2020). Diffusion of innovation. In *The International Encyclopedia of Media Psychology* (pp. 1–8).

Ge, J., Zhao, Y., Luo, X., & Lin, M. (2020). Study on the suitability of green building technology for affordable housing: A case study on Zhejiang Province, China. *Journal of Cleaner Production, 275*, Article 122685.

Ghayvat, H., Liu, J., Babu, A., Alahi, E. E., Gui, X., & Mukhopadhyay, S. C. (2015). Internet of things for smart homes and buildings: Opportunities and challenges. *Journal of Telecommunications and the Digital Economy, 3*(4), 33–47.

Gosain, P., Zhang, L., & Ganapati, N. E. (2022). Understanding multisector stakeholder value systems on housing resilience in the City of Miami. *International Journal of Disaster Risk Reduction, 77*, Article 103061.

Greengard, S. (2021). *The internet of things*. MIT Press.

Gunawan, T. S., Yaldi, I. R. H., Kartiwi, M., Ismail, N., Za'bah, N. F., Mansor, H., & Nordin, A. N. (2017). Prototype design of smart home system using internet of things. *Indonesian Journal of Electrical Engineering and Computer Science, 7*(1), 107–115.

Gushen, M., Murphy, C., Valenta, L., Bertram, N., & Maxwell, D. (2021). *Adaptable housing for people with disability in Australia: A scoping study*. Australian Human Rights Commission.

Hebbert, A. L. (2021). *Renewed purpose: Exploring adaptive reuse for quality, affordable housing in Kansas City, Kansas*.

Hemidat, S., Achouri, O., El Fels, L., Elagroudy, S., Hafidi, M., Chaouki, B., Ahmed, M., Hodgkinson, I., & Guo, J. (2022). Solid waste management in the context of a circular economy in the MENA region. *Sustainability, 14*(1), 480.

Hirsch, T. (2009). *A case for climate neutrality: Case studies on moving towards a low carbon economy*. UNEP/Earthprint.

Hu, M., & Hu, M. (2021). Smart building and current technologies. In *Smart technologies and design for healthy built environments* (pp. 75–91).

Huang, J. H., Lin, Y. R., & Chuang, S. T. (2007). Elucidating user behaviour of mobile learning: A perspective of the extended technology acceptance model. *Electronic Library, 25*(5), 585–598.

Huchzermeyer, M. (2014). Changing housing policy in South Africa. In *Affordable housing in the urban Global South: Seeking sustainable solutions* (pp. 336–348).

Ismail, Y. (Ed.). (2019). *Internet of Things (IoT) for automated and smart applications*. BoD—Books on Demand.

Jackson, M. C. (2019). *Critical systems thinking and the management of complexity*. Wiley.

Karam, A., Viruly, F., Kariuki, C., & Akujuru, V. (Eds.). (2022). *Understanding African real estate markets*. Taylor & Francis.

Keiser, Z. (2019). *Revitalized public housing* (Doctoral dissertation, University of Cincinnati).

Khadduri, J., & Martin, M. (1997). Mixed-income housing in the HUD multifamily stock. *Cityscape: A Journal of Policy Development and Research, 3*(2).

King, J., & Perry, C. (2017). *Smart buildings: Using smart technology to save energy in existing buildings*. American Council for an Energy-Efficient Economy.

Kinnane, O., Grey, T., & Dyer, M. (2016, April). Adaptable housing design for climate change adaptation. In *Proceedings of the Institution of Civil Engineers-Engineering Sustainability* (Vol. 170, no. 5, pp. 249–267). Thomas Telford Ltd.

Kolarevic, B., & Parlac, V. (Eds.). (2015). *Building dynamics: Exploring architecture of change*. Routledge.

Kozubaev, S., Rochaix, F., DiSalvo, C., & Le Dantec, C. A. (2019, May). Spaces and traces: Implications of smart technology in public housing. In *Proceedings of the 2019 CHI Conference on Human Factors in Computing Systems* (pp. 1–13).

Kuffar, H. (2020, November). The profitability and affordability of smart renewable microgrid in Sub-Saharan Africa. In *2020 IEEE Electric Power and Energy Conference (EPEC)* (pp. 1–4). IEEE.

Le, Q., Nguyen, H. B., & Barnett, T. (2012). Smart homes for older people: Positive aging in a digital world. *Future Internet, 4*(2), 607–617.

Lespagnard, M., Galle, W., & De Temmerman, N. (2023). Visualizing equitable housing: A prototype for a framework. *Sustainability, 15*(5), 4110.

Lim, S., Buswell, R. A., Le, T. T., Austin, S. A., Gibb, A. G. F., & Thorpe, T. (2012). Developments in construction scale additive manufacturing processes. *Automation in Construction, 21*, 262–268.

Lobaccaro, G., Carlucci, S., & Löfström, E. (2016). A review of systems and technologies for smart homes and smart grids. *Energies, 9*(5), 348.

Löw, J., & Antalovsky, E. (2018). *Why Vienna gets high marks* (Vol. 2).

Luffman, J. (2006). *Measuring housing affordability* (pp. 75–001). Statistics Canada.

Maalsen, S. (2020). Revising the smart home as assemblage. *Housing Studies, 35*(9), 1534–1549.

Magdziak, M. (2019, February). Flexibility and adaptability of the living space to the changing needs of residents. In *IOP Conference Series: Materials Science and Engineering* (Vol. 471, no. 7, p. 072011). IOP Publishing.

Manewa, A., Siriwardena, M., Ross, A., & Madanayake, U. (2016). Adaptable buildings for sustainable built environment. *Built Environment Project and Asset Management, 6*(2).

Marikyan, D., Papagiannidis, S., & Alamanos, E. (2019). A systematic review of the smart home literature: A user perspective. *Technological Forecasting and Social Change, 138*, 139–154.

Meen, G., & Whitehead, C. (2020). *Understanding affordability: The economics of housing markets*. Policy Press.

Mkuzo, T. Z., Mayekiso, T., & Gwandure, C. (2019). An observational examination of houses built under the "Breaking New Ground" housing policy of South Africa. *Ergonomics SA: Journal of the Ergonomics Society of South Africa, 31*(1), 1–16.

Moghayedi, A., Awuzie, B., Omotayo, T., Le Jeune, K., Massyn, M., Ekpo, C. O., Braune, M., & Byron, P. (2021). A critical success factor framework for implementing sustainable innovative and affordable housing: A systematic review and bibliometric analysis. *Buildings, 11*(8), 317.

Mohd-Tobi, A. L., Omar, S. A., Yehia, Z., Al-Ojail, S., Hashimi, A., & Orhan, O. (2018). Cost viability of 3D printed house in the United Kingdom. *IOP Conference Series: Materials Science and Engineering, 319*(1), 1–7.

Moldavska, A., & Welo, T. (2015). Development of manufacturing sustainability assessment using systems thinking. *Sustainability, 8*(1), 5.

Monis, M., & Rastogi, A. (2022). A review of passive design strategies for improving building energy performance. *International Journal of Multidisciplinary Innovative Research, 2*(1), 43–50.

Moore, S. M., Dolansky, M. A., Singh, M., Palmieri, P., & Alemi, P. (2010). *The systems thinking scale* (Unpublished manuscript).

Mueller, E. J., & Tighe, J. R. (Eds.). (2022). *The affordable housing reader*. Taylor & Francis.

Nakamori, Y. (2019). *Knowledge construction methodology: Fusing systems thinking and knowledge management* (Vol. 20). Springer Nature.

Nikou, S. (2019). Factors driving the adoption of smart home technology: An empirical assessment. *Telematics and Informatics, 45*, Article 101283.

Nikou, S., & Bouwman, H. (2014). Ubiquitous use of mobile social network services. *Telematics and Informatics, 31*(3), 422–433.

Nokulunga, M., Didi, T., & Clinton, A. (2018, September). Challenges of Reconstruction and Development Program (RDP) houses in South Africa. In *Proceedings of the International Conference on Industrial Engineering and Operations Management* (pp. 1695–1702).

Onatu, G. O. (2012). Sustainable land use and development: Perspective on Cosmo City, Johannesburg, South Africa. *Sustainable Development and Environmental Protection, 2*(1), 67–77.

Osman, A., Sebake, N., & Davey, C. (2011). Focus on medium-density mixed housing: An important component in the transformation of South African housing environments. In *3rd CSIR Biennial Conference*, 31 August and 01 September 2010, CSIR International Convention Centre Pretoria (pp. 1–22). Retrieved September 20, 2023, from https://researchspace.csir.co.za/dspace/handle/10204/2516

Park, Y., Son, H., & Kim, C. (2018). Investigating the determinants of construction professionals' acceptance of web-based training: An extension of the technology acceptance model. *Automation in Construction, 22*, 377–386.

Pîrjan, A., & Petroşanu, D. M. (2013). The impact of 3D printing technology on the society and economy. *Journal of Information Systems and Operations Management, 7*(2), 163–173.

Prathibha, P., Kumar, A., Singh, A., Parul, K., & Jaiswal, S. (2020). Smart home security system using IoT. *International Journal of Research in Engineering, Science and Management, 3*(5), 752–755.

Rajvanshi, N. (2022). The smart city. In *The climate city* (pp. 320–334).

Rimmer, M. (2021). Automating fab cities: 3D printing and urban renewal. In *Automating cities: Design, construction, operation and future impact* (pp. 255–272).

Roy, J., Some, S., Das, N., & Pathak, M. (2021). Demand side climate change mitigation actions and SDGs: Literature review with systematic evidence search. *Environmental Research Letters, 16*(4), Article 043003.

Saidu, A. I., & Yeom, C. (2020). Success criteria evaluation for a sustainable and affordable housing model: A case for improving household welfare in Nigeria Cities. *Sustainability, 12*(2), 656.

Sakin, M., & Kiroglu, Y. C. (2017). 3D printing of buildings: Construction of the sustainable houses of the future by BIM. *Energy Procedia, 134*, 702–711.

Sanneh, E. S. (2018). *Systems thinking for sustainable development*. Springer International Publishing.

Sarker, M. N. I., Khatun, M. N., Alam, G. M., & Islam, M. S. (2020, September). Big data driven smart city: Way to smart city governance. In *2020 International Conference on Computing and Information Technology (ICCIT-1441)* (pp. 1–8). IEEE.

Schmidt, III, R., & Austin, S. (2016). *Adaptable architecture: Theory and practice*. Routledge.

Senge, P. (1990). *The fifth discipline, the art and practice of the learning organization*. Doubleday/Currency.

Sepasgozar, S., Karimi, R., Farahzadi, L., Moezzi, F., Shirowzhan, S., Ebrahimzadeh, M., Hui, F., & Aye, L. (2020). A systematic content review of artificial intelligence and the internet of things applications in smart home. *Applied Sciences, 10*(9), 3074.

Sethole, F. R., & Sebola, E. M. (2020). *Reconstruction and development programme: An Afrocentric analysis of the historical developments, challenges and remedies.*

Shin, J., Moon, S., Cho, B. H., Hwang, S., & Choi, B. (2022). Extended technology acceptance model to explain the mechanism of modular construction adoption. *Journal of Cleaner Production, 342*, Article 130963.

Simbanegavi, P., & Ijasan, K. (2022). Inclusive, affordable, and smart housing in Africa. In *Understanding African real estate markets* (pp. 122–134). Routledge.
Solanki, A., & Nayyar, A. (2019). Green internet of things (G-IoT): ICT technologies, principles, applications, projects, and challenges. In *Handbook of research on big data and the IoT* (pp. 379–405). IGI Global.
Stone, S. (2019). *UnDoing buildings: Adaptive reuse and cultural memory.* Routledge.
Tan, K. M., Babu, T. S., Ramachandaramurthy, V. K., Kasinathan, P., Solanki, S. G., & Raveendran, S. K. (2021). Empowering smart grid: A comprehensive review of energy storage technology and application with renewable energy integration. *Journal of Energy Storage, 39*, Article 102591.
Tlhatlha, P. V. (2020). *Renewable energy strategies for alleviating energy poverty in informal settlements: A case study of Diepsloot-South Africa.* University of Pretoria.
Tuballa, M. L., & Abundo, M. L. (2016). A review of the development of smart grid technologies. *Renewable and Sustainable Energy Reviews, 2016*(59), 710–725. https://doi.org/10.1016/J.RSER.2016.01.011
Van Noorloos, F., Cirolia, L. R., Friendly, A., Jukur, S., Schramm, S., Steel, G., & Valenzuela, L. (2020). Incremental housing as a node for intersecting flows of city-making: Rethinking the housing shortage in the global South. *Environment and Urbanization, 32*(1), 37–54.
Venkatesh, V., & Davis, F. D. (2000). A theoretical extension of the technology acceptance model: Four longitudinal field studies. *Management Science, 46*(2), 186–204.
Wainer, L. S., Ndengeyingoma, B., & Murray, S. (2016). *Incremental housing, and other design principles for low-cost housing*. Retrieved from Rwanda.
Wells, W., & Vermeer, K. (2020). *Blueprint for greening affordable housing*. Island Press.
Wester, M. (2022). My fault or default—Household behaviors living in a climate-smart building. *Buildings, 12*(2), 245.
yn der Voordt, T. J. (1990). *Building adaptable housing—From theory to practice.*
Yuen, K. F., Wong, Y. D., Ma, F., & Wang, X. (2020). The determinants of public acceptance of autonomous vehicles: An innovation diffusion perspective. *Journal of Cleaner Production, 270*, Article 121904.
Zebra, E. I. C., van der Windt, H. J., Nhumaio, G., & Faaij, A. P. (2021). A review of hybrid renewable energy systems in mini-grids for off-grid electrification in developing countries. *Renewable and Sustainable Energy Reviews, 144*, Article 111036.

Lawrence Babatunde Ogunsanya is a qualified professional architect with over twenty years of experience in the built environment industry. He holds two bachelor's degrees in Architecture, a master's degree in Housing, and a doctoral degree in Town, Regional, and Urban planning. With over twelve years of teaching experience, he has expertise in architectural design, housing, project management, and urban design. His research interests include architecture, sustainability, urban design, housing, and planning. He is currently a senior lecturer and the academic leader of the discipline of Architecture at the University of KwaZulu-Natal. His professional expertise is complemented by his active participation in various academic roles and activities in advancing the education and practice within the built environment.

Part II
Housing Finance

This part on housing finance will include various financing models that have been put in place in different countries to drive housing delivery for the poor. The typology of housing finance for the urban poor will depend on prevailing housing policies and housing production systems. Housing financing models can be both formal and informal such as: Private sources of housing finance; Community based sources of housing finance; Public sources of housing finance and international sources of housing sources. It is expected that these housing finance models will be discussed within prevailing contextual frameworks. Beyond this theoretical overview, this part delves into individual case studies and unpacks unique approaches used by households to finance housing production largely from their own coffers This mainly revolves around their meagre resources which they mobilise through remittances and savings of their hard-earned cash. In addition, this part interrogates challenges which contemporary governments south of the Sahara experience to finance adequate housing for the urban poor. Using empirical evidence provided through case studies from various countries, the authors illustrate other strategies that can be used to alleviate the financial burden since given that finance is a key resource which runs through the housing production process from inception to the final product.

Chapter 7
Uncovering the Inadequacies of South Africa's Housing Finance System: A Deep Dive into the Persistent Housing Challenges

Fortune Mangara and Hangwelani Hope Magidimisha-Chipungu

Abstract The housing finance system in South Africa has mainly been inadequate in addressing the country's housing challenges. A shortage of affordable housing, high levels of debt and financial stress among homeowners, and a lack of access to housing for many characterise the housing finance system. These challenges are a legacy of South Africa's historical inequalities, economic constraints, and a lack of political will to provide low-income housing, especially for the urban poor. Despite various government initiatives and policies like the Reconstruction and Development Programme (RDP), many South Africans struggle to access adequate housing. Against this background, the chapter argues that government initiatives and policies have not been sufficient to address the scale of the housing problem. Employing a desktop approach, this chapter documents the housing statistics in South Africa and how the government responded to housing delivery for the urban poor. To this end, the chapter analyses the policy decisions made by the government to curb the housing challenges. The study's findings show that the current housing finance system is inadequate with providing accessible, affordable housing for all South Africans. The government's housing subsidies and grants have often been insufficient or poorly targeted, leading to a lack of meaningful impact on the overall housing situation. Therefore, this chapter urges the South African government to improve the effectiveness, efficiency, and accessibility of the existing housing finance systems, create and devise innovative housing finance mechanisms and instruments, and promote equal and affordable access to housing finance for all people. This chapter suggests reforms to the housing finance system in South Africa and advocates for alternative financing options and approaches to prioritise the needs and rights of marginalised

F. Mangara (✉)
Unit for Environmental Sciences and Management, African Centre for Disaster Studies, North West University, Potchstroom, South Africa
e-mail: mangarafortune@gmail.com

H. H. Magidimisha-Chipungu
School of the Built Environment and Development Studies, Howard College, University of KwaZulu-Natal, Glenwood, Durban, South Africa

L. Chipungu et al. (eds.), *Contemporary Models for Pro-Poor Housing in Urban Sub-Saharan Africa*, https://doi.org/10.1007/978-3-032-05489-0_7

communities and address the housing crisis's underlying causes, including historical injustices and economic inequality.

Keywords Urbanisation · South Africa · Housing finance system · Housing challenges · Government initiatives

7.1 Introduction

Urbanisation is increasing rapidly, especially in African and Asian countries, with South Africa projected to have the highest rate of urbanisation at more than 70% by 2050 (UN, 2019). According to Kuddus et al. (2020), rapid urbanisation has resulted in increased poverty levels, unplanned urban settlements and slums, lacking basic infrastructure and services, and substandard housing. The UN-Habitat (2022) mentioned that housing is a fundamental human right and is essential to social and economic development. However, South Africa's housing finance system has been unable to effectively address the country's significant housing challenges (Marutlulle, 2021). This has led to a situation where millions of people cannot access affordable and decent housing, and there is a considerable shortage of housing units.

Several factors contribute to the inadequacy of the housing finance system in South Africa. One of the main factors is the lack of access to credit for many low-income households. Many people in South Africa do not have the necessary collateral or credit history to qualify for a mortgage, which makes it difficult for them to purchase a home. Additionally, high-interest rates and fees associated with home loans make it difficult for many people to afford the monthly payments.

Structural issues also contribute to the housing finance system's inability to address the housing challenges in South Africa effectively (Bah et al., 2018). For example, the government's previous policies of subsidising housing for low-income households have not effectively addressed the housing shortage. Additionally, the high cost of land and construction materials has made it difficult for developers to build affordable housing units. Given these challenges, it is clear that the current housing finance system in South Africa is not adequate to address the country's housing needs (Ganiyu et al., 2017). To address the housing crisis, it will be necessary to implement a more comprehensive and inclusive approach that considers the needs of low-income households and addresses the structural issues that have contributed to the housing shortage (Bah et al., 2018). This may involve a combination of measures such as improving access to credit, increasing funding for affordable housing development, and reforming policies that have contributed to the housing shortage. This chapter will unpack the challenges faced in the housing finance domain, considering the South African government policy interventions to address the country's housing challenges.

7.2 Challenges in the South African Housing Finance System

In recent years, the South African housing finance system has been the topic of extensive inquiry and controversy (Marutlulle, 2021). South Africa has a long history of housing problems, including inadequate housing, overcrowding, and informal settlements (Marutlulle, 2021). The South African government has attempted to solve this situation through a variety of policy actions, with varying degrees of success. The lack of access to affordable housing for low-income households is indeed a challenge for both the housing sector and the finance sector in South Africa. According to a South African Reserve Bank (SARB) survey, most South African households cannot afford to buy a home, with only 20% earning enough to qualify for a bond (SARB, 2022). This is due to high-interest rates, high transaction costs, and low wages.

A further issue is the lack of a unified and effective housing policy. The housing problem has persisted despite many governmental measures, such as the Rehabilitation and Development Programme (RDP) and the Breaking New Ground (BNG) policy (Trusler, 2009). According to research conducted by the Human Sciences Research Council (HSRC) (2018), these policies have had little success due to poor implementation, a lack of coordination among several government ministries, and a lack of money. A fundamental difficulty is the lack of a regulated and accessible housing finance system. Most home finance in South Africa comes from informal sources such as family and friends rather than formal institutions (UN-Habitat, 2008).

This research's theoretical foundation is based on institutional economics, which examines the role of institutions in shaping economic outcomes. According to institutional economists, the South African housing finance system is determined by the institutional environment, which includes government policies, rules, and cultural norms (Schirmer, 2017). According to Mbandlwa (2021), the institutional context has created impediments to access to cheap housing and formal finance, which has contributed to the continuance of South Africa's housing problem. Some other reasons have also contributed to the shortcomings of South Africa's housing finance system. One example is the absence of a well-functioning mortgage market. The mortgage market in South Africa is underdeveloped, with insufficient mortgage providers to match the demand for housing finance. This has resulted in high borrowing costs and limited access to credit for low-income households.

The absence of a secondary market for mortgages is another issue. A secondary market is a market where mortgages are bought and sold, hence increasing the availability of housing finance. Lenders are less likely to originate mortgages in the absence of a secondary market since they cannot sell them to other investors. This makes obtaining financing more challenging for households with low incomes. The question of land ownership is also a significant barrier in the context of South African home finance. According to a research conducted by the Development Bank of South Africa (DBSA) (2022), white South Africans continue to control the majority of land in the country, and there is a lack of available land for house development. This has resulted in a scarcity of land for housing development, driving up land costs and making it more difficult for low-income individuals to obtain affordable housing.

Other factors, in addition to those mentioned above, have contributed to the shortcomings of South Africa's housing finance system. One example is the absence of a well-functioning rental market (Scheba & Turok, 2020). Despite the great demand for rental property, South Africa's rental industry is underdeveloped and unregulated, resulting in high rents and poor living conditions (Gilbert et al., 2011). As a result, low-income households have found it difficult to find a reasonable rental property.

Lenders are less willing to give credit to persons with poor credit history without a comprehensive credit reporting system, making it more difficult for low-income households to receive home finance (Brevoort et al., 2016). The issue of informal settlements is also a serious challenge in the Framework of home finance in South Africa (Williams et al., 2019). According to a HSRC report, informal settlements are a major source of housing for low-income households but often lack access to essential utilities such as water and sanitation and are frequently threatened with eviction (SERI, 2018). This has made it difficult for low-income people to find affordable accommodation, contributing to the housing crisis's continuation.

Another major aspect is a lack of capacity and expertise in the housing sector. According to a National Housing Council (NHC) research, there is a shortage of capacity and expertise in South Africa's housing sector, making it difficult for the government to properly implement housing regulations (Nyakala et al., 2021). As a result, housing delivery has been delayed, making it more difficult for low-income people to receive affordable housing. Finally, there is a scarcity of data and study on South Africa's home finance sector. There is a dearth of data on informal sources of housing finance, which makes understanding the entire scope of the housing crisis and developing effective policy responses difficult.

The housing financing system in South Africa confronts various interconnected problems, according to the currently available literature, which must all be addressed at once. There is a lack of land for housing development, access to affordable housing for low-income households, a coherent and effective housing policy, a formalised and accessible housing finance system, a functioning mortgage market, a secondary market for mortgages, data and research on the housing finance system, and more. These issues are rooted in the institutional setting, making it difficult to get formal finance and inexpensive housing. By utilising an institutional economics framework to comprehend the shortcomings of South Africa's housing finance system and its ongoing housing difficulties, this study will add to the body of existing literature. The history of the home financing system is covered in the following section.

7.3 Housing Finance Evolution

Ancient civilisations are where the origins of house finance may be found, as people employed various financial methods to buy or construct homes (Blackwell & Kohl, 2017). The idea of home finance has seen tremendous development in the contemporary era and has been essential to the growth of the world economy. Many economic, social, and political issues have influenced the development of housing finance, which has seen substantial changes over time (Blackwell & Koln, 2017). Using personal resources to buy a home was one of the first housing financing methods. Mortgage loans, initially provided by lone lenders and later by banks and other financial institutions, were frequently used to augment this.

7.3.1 Evolution of Housing Finance in the African Context

Various political, economic, and social issues have influenced housing finance throughout Africa's history, making it a complicated and diversified topic (Mabin, 2020). However, a few important elements in African housing financing might be noted as illustrated in Fig. 7.1.

7.3.2 Evolution of Housing Finance in the South African Context

Many economic, social, and political factors have influenced the development of housing financing in South Africa, and these changes have been considerable over time. Savings and credit groups and rotating savings and credit associations, among others, have historically dominated the informal and unregulated home financing sector in South Africa (Kajimo-Shakantu & Evans, 2006). These systems frequently rely on social networks and neighbourhood loans. They have greatly helped low-income families and small enterprises in South Africa by financing housing (Schirmeister & Nadler, 1996). The development of housing finance in South Africa has been influenced by many political, economic, and social variables (UN Habitat, 2008). These are a few significant occasions and trends as shown in Fig. 7.2.

However, with the formation of housing finance institutions and the growth of the mortgage markets in recent decades, formal housing finance systems have also started to emerge in South Africa (SARB, 2000). Government initiatives and legislation, such as housing finance reforms and the creation of mortgage guarantee schemes have frequently supported these formal institutions. The expansion of microfinance and microcredit programmes, which offer small loans and other financial services to low-income households and small enterprises, has been a key development in

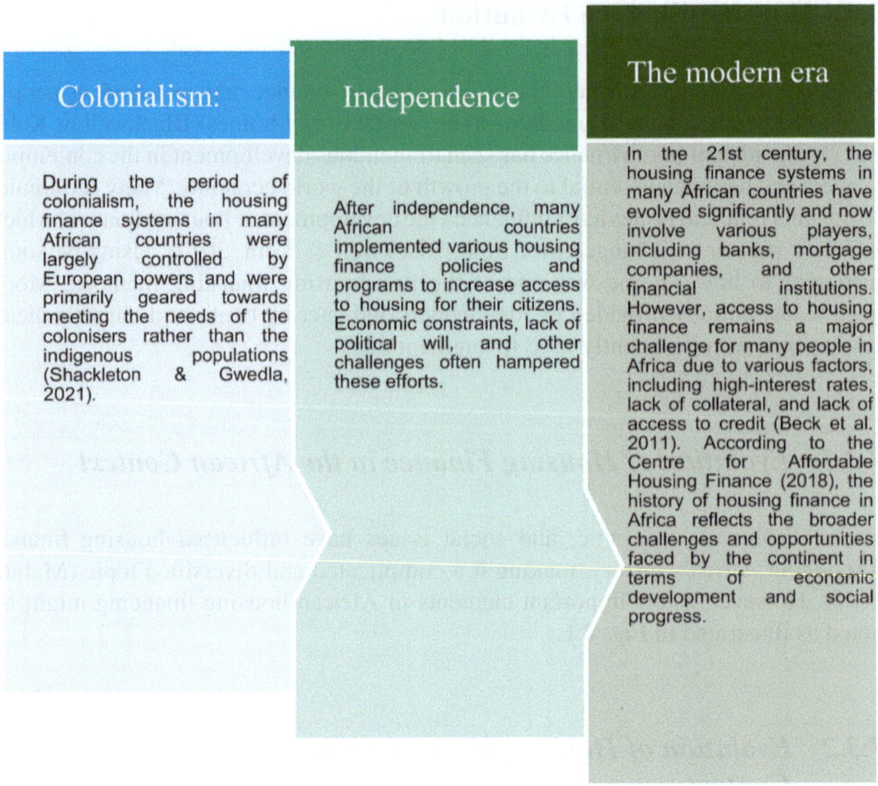

Fig. 7.1 Evolution of housing finance in the African context in three stages. *Source* Authors (2023)

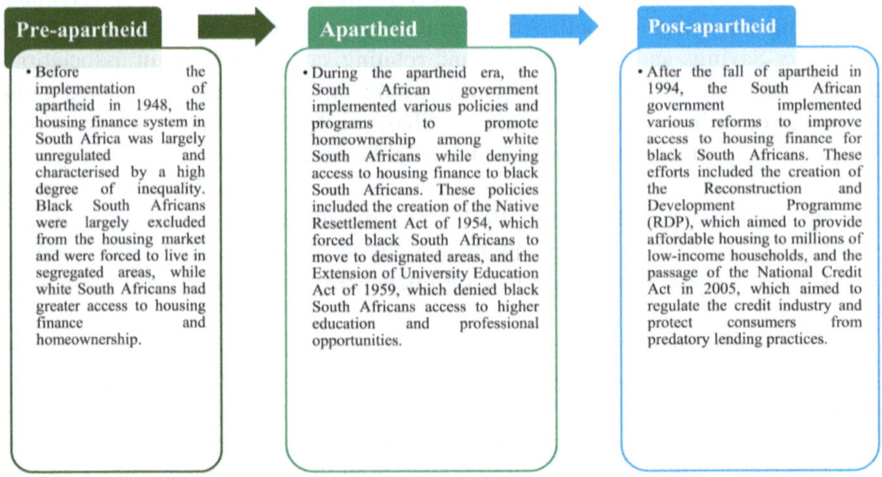

Fig. 7.2 Evolution of housing finance in the South African context. *Source* Authors (2023)

the evolution of housing finance in South Africa (van Rooyen et al., 2012). Many individuals who may not have been able to secure loans from conventional financial institutions now have greater access to house financing because of these programmes. The emergence of Islamic finance and adopting alternative financing methods like crowdfunding and peer-to-peer lending are recent changes in South African home finance.

Many noteworthy occurrences in recent years, such as the COVID-19 pandemic and the world financial crisis of 2007–2008, have also affected the home financing sector. These occurrences have changed the terms and availability of mortgage loans and increased industry regulation. The legacy of racial discrimination and inequality in South Africa's history of housing financing has long-lasting affected the housing market and the country's overall economy. Nevertheless, economic, social, and political variables have influenced how housing finance has developed in South Africa. It has continued to change in response to shifting market conditions and community demands.

7.4 Housing Finance in the South African Economy

Housing finance is crucial to the South African economy because it helps people and families get the money they need to buy or rent a home (Doling et al., 2013). Several possibilities exist for financing housing, including both formal and informal institutions. Mortgage loans from banks and other financial organisations and government-backed programmes, such as those provided by the National Housing Finance Corporation (NHFC), are all part of South Africa's formal housing financing systems (UN-Habitat, 2010). To eligible South African citizens and permanent residents, the NHFC offers long-term, low-interest mortgage loans and other types of housing financing, such as grants and bond financing (Department of Human Settlements SA, 2022).

Savings and credit clubs, rotating savings and credit societies, and unregulated mortgage markets are examples of informal housing finance systems in South Africa. These methods frequently rely on social networks and community-based lending, which have greatly helped low-income families and small enterprises finance their housing needs. Alternative financing possibilities for housing in South Africa include peer-to-peer lending and crowdfunding and microfinance and microcredit programmes, which offer small loans and other financial services to low-income households and small enterprises. Ultimately, there are many different ways for people and families to get the money they need to buy or rent a property in the modern South African housing finance market, which is broad and multifaceted.

7.4.1 Types of Housing Finance Systems in South Africa

There are a variety of different types of housing finance products that are available to individuals and families who are looking to purchase or rent a home. These products include:

Mortgages: A mortgage is a loan taken out to buy a home, with the asset being used as security for the loan. Mortgages can have fixed-rate interest rates, which remain constant during the term of the loan, or adjustable-rate interest rates, which fluctuate over time. Banks, other financial organisations, and government-backed initiatives like those provided by the National Housing Finance Corporation all offer mortgages (NHFC).

Home equity loans: A loan backed by the equity in a homeowner's house is known as a home equity loan. With this kind of loan, homeowners can borrow money using the value of their property as security. Banks and other financial organisations offer home equity loans.

Rent-to-own agreements: In a rent-to-own contract, the tenant rents the property for a predetermined amount of time with the opportunity to buy it after the rental term. Private landlords or businesses specialising in this kind of finance may offer rent-to-own agreements.

Lease-purchase agreements: In a lease-purchase agreement, a tenant rents a property for a predetermined amount of time with the option to buy the property at any point throughout the renting period. Private landlords or businesses specialising in this financing may offer lease-purchase arrangements.

These housing finance options depend on one's financial situation and the type of property one wants to purchase or rent.

7.4.2 South African Policies on Housing Finance

Since the end of apartheid in 1994, South Africa has had considerable housing problems, with a sizable segment of the population still residing in informal settlements or without suitable homes (Socio Economic Rights Institute of South Africa (SERI) (2018). The government of South Africa has implemented several policies, programmes, and initiatives to address these housing challenges and improve the living conditions of low-income earners since 1994. Figure 7.3 shows a timeline for some of the South African policies on housing finance.

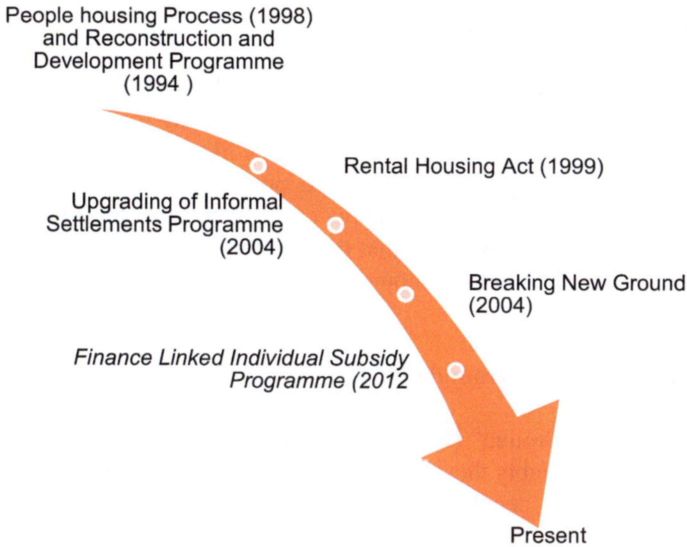

Fig. 7.3 Timeline for some of the South African policies on housing finance. *Source* Authors (2023)

7.4.2.1 Reconstruction and Development Programme (1994)

The "Reconstruction and Development Programme" (RDP), which was established in 1994, intends to provide low-income earners with homes and other services. It is regarded as South Africa's greatest social housing initiative (Corder, 1997). More than 3 million homes have been created due to the RDP, giving millions of people who were previously homeless or living in squatter camps a place to call home (Mabuya & Scholes, 2020). However, the programme has encountered issues emanating from poor implementation of the programme including insecure tenure, poor quality, lack of upkeep, and delays in housing construction.

7.4.2.2 People's Housing Process (1998)

In 1998, the "People's Housing Process" (PHP) programme got underway. The PHP has enabled communities to construct their homes actively but has encountered certain difficulties (Gumbo & Onatu, 2015). One of the primary issues is that not all low-income earners can access the PHP; it is only available to specific communities. The South African government has implemented other steps in addition to the above-listed policies to broaden access to housing financing and enhance the living conditions of low-income earners.

7.4.2.3 Rental Housing Act (1999)

With different incentives for developers and landlords, the "Rental Housing" programme seeks to enhance the supply of rental housing for low- and moderate-income earners. The programme offers incentives like tax exemptions, subsidies, and low-interest loans to promote the construction of rental housing. Although the policy has successfully increased the supply of rental housing, there have been some difficulties. One of the key issues is that the developed rental housing is frequently out of reach for people with low incomes due to the high rental costs.

7.4.2.4 Breaking New Ground (2004)

The "Breaking New Ground" (BNG) policy also seeks to provide homes for low-income individuals, notably those residing in informal settlements. The BNG policy is focused on enabling people to construct their homes while providing serviced sites with essential infrastructure, such as water and sanitation (Parikh et al., 2020). The BNG strategy was designed to help low-income individuals, particularly those residing in unofficial settlements, find accommodation. The goal of the strategy is to give those who reside in informal settlements and lack access to basic infrastructure and services permission to construct their own homes on serviced lands (Amin & Cirolia, 2018). The BNG programme has successfully supplied serviced sites with essential amenities like water and sewage so that people can construct their homes, but it has also encountered significant difficulties. The BNG policy's slow implementation, which results in many people still residing in informal settlements without suitable accommodation, is one of the major problems (Weimann et al., 2020).

7.4.2.5 Upgrading of Informal Settlements Programme (2004)

The "Upgrading of Informal Settlements" policy seeks to upgrade informal settlements by providing essential services and infrastructure. The programme intent recognition and formalisation of tenure rights, provision of basic services and community empowerment and social economic integration. Many people living in informal settlements have seen an improvement in their living conditions thanks to the programme, but there have also been significant difficulties (Del Mistro & Hensher, 2009). One of the major issues is that the strategy has not been implemented as rapidly as anticipated, leaving many people to continue living in squalid conditions in informal settlements.

7.4.2.6 Finance Linked Individual Subsidy Plan

The "Finance Linked Individual Subsidy Plan" (FLISP), offers subsidies to low to medium-income earners to help them acquire financing for the purchase of a home, is one of the important initiatives (CAHF, 2019). The financial aid is given to eligible people and is connected to the bond financing received from a participating financial institution. Low to medium-income individuals can receive financial aid from FLISP to help them secure mortgage financing. Several low- to medium-income individuals have benefited from FLISP's assistance in obtaining housing financing and becoming homeowners. However, the programme has faced some challenges. One of the primary issues is that only people who meet specified income and contribution requirements can participate in the FLISP. In addition to that beneficiary assessment and market inconsistencies as another major challenge. Sometimes people on the lower income range of the GAP market are outsmarted or booted out by those at the upper range.

7.4.3 Measures for Accessible Housing Finance in South Africa

The South African government established initiatives like the NHFC to make house finance more accessible to low-income people. The NHFC provides low-cost housing finance to low-income individuals and persons living in informal settlements, making buying or building a home easier. In addition, there is the "Human Settlement Development Grant" (HSDG) initiative that provides funds for housing and related infrastructure development in low-income neighbourhoods. The HSDG aims to assist low-income communities in upgrading their homes and infrastructure and improving their living conditions. Another programme that provides rental accommodation to low-income individuals, particularly those who cannot afford market-related rental charges, is the Social Housing Programme.

Also, the South African government has undertaken various steps to enhance the living conditions of persons living in informal settlements. Among these is the "Integrated National Electrification Project" (INEP), which aims to give those living in informal settlements access to electricity. The INEP attempts to enhance people's living conditions in informal settlements by providing access to electricity and other essential amenities. The government also has a programme known as the "Pro-poor housing policy," which tries to balance the poor's requirements and market forces to offer cheap housing for the poor (Pernegger, 2021). Since 1994, the South African government has developed many policies, initiatives, and measures to enhance access to housing finance and improve low-income earners' living conditions. Notwithstanding the implementation of these policies and programmes, South Africa continues to confront substantial problems in tackling the housing issue, including a lack of affordable housing, insufficient money, a lack of execution, and limited community participation. Furthermore, rapid urbanisation and population growth have placed additional strain on the country's housing resources, resulting in a persistent backlog of housing demand.

7.5 Results

The housing finance system in South Africa has various flaws. Despite efforts to solve South Africa's continuing housing difficulties, many South Africans struggle to find affordable accommodation. One key result is that the majority of South Africans cannot afford the expense of a standard mortgage. Because a small percentage of the population earns enough to qualify for a mortgage, forcing many rely on alternate housing financing, such as renting or living on the streets. Furthermore, the current housing finance structure favours the elite and middle classes, with little support for low-income individuals. This is mirrored in the fact that most housing finance is provided by commercial banks, which are more likely to lend to higher-income borrowers with good credit scores.

Another significant concern is the lack of affordable housing options. Most new housing developments are aimed at the high-end market, leaving low-income earners with few affordable options. This is aggravated further by the high cost of land and building materials, making it difficult for developers to provide cheap housing. Furthermore, the current housing finance system is not effectively reaching low-income people because of a lack of cheap mortgage products, education, and awareness about available housing options. Due to a shortage of affordable housing options, informal housing finance, such as stokvels and informal savings clubs, has increased. While they provide temporary relief, they are not long-term solutions. Existing government policies and programmes tackling the housing crisis have had mixed results. For example, the BNG policy, which aims to provide low-cost housing for low-income people, has been hampered by a lack of financing and inefficiencies in the bureaucracy. Similarly, the "Rental Housing" scheme, which promised to offer low-income earners rental homes, has been plagued by a shortage of cheap housing options and large rent arrears.

South Africa is experiencing a severe housing crisis, with many low-income earners unable to have affordable homes. According to 2011 census figures, over 2.1 million households, or 18% of all households in South Africa, reside in informal settlements or backyard houses (Marutlulle, 2021). Furthermore, according to a 2019 National Department of Human Settlements report, South Africa's housing backlog is anticipated to be around 2.2 million. The housing finance system in South Africa has also faced significant challenges. According to National Credit Regulator data, the percentage of South Africans in arrears on their home loans has risen in recent years, peaking at 14.3% in 2019. This indicates that many low-income earners are struggling to keep up with their mortgage payments and may be at risk of losing their homes.

Furthermore, data from the Reserve Bank of South Africa reveal a drop in the rate of home loan approvals in recent years. The rate of home loan approvals in 2020 was at its lowest in over a decade, showing that many low-income workers are having trouble obtaining housing finance. South Africa's housing crisis is a major issue, with many low-income earners living in informal settlements without suitable accommodation. South Africa's housing finance system has faced severe issues with an increasing number of persons in arrears on their home loans and declining home loan approvals. These statistics indicate that the South African government's housing policies and programmes have not successfully tackled the housing problem and provided appropriate home finance to low-income earners.

The findings of this study highlight several severe flaws in South Africa's housing finance system. While attempts have been made to solve the country's continuing housing challenges, data reveal that access to affordable housing remains a major concern for many South Africans. According to the 2019 National Department of Human Settlements report, more must be done to assist low-income earners and tackle the main causes of informal settlements to ensure that all South Africans have access to adequate, affordable housing.

7.6 Discussion

South Africa's housing finance system falls short of meeting the demands of the economically disadvantaged. Despite several government initiatives and laws to alleviate the housing crisis, a substantial portion of the population continues to lack access to affordable housing (Bah et al., 2018). This is due to various causes, notably high housing costs, a shortage of credit, and a lack of political will to address the issue. The high housing cost is one of the primary causes behind the inadequacies of South Africa's housing financing system (Pillay, 2006). Low-income households find it difficult to afford houses due to the high cost of land, building materials, and labour. This is exacerbated by a shortage of cheap credit, making it difficult for low-income households to fund the purchase of a home. As a result, many low-income households are compelled to rely on unregulated and informal credit sources, which are frequently associated with exorbitant interest rates and lending practices.

The government's dedication to affordable housing has also been insufficient, with a lack of funds and political will to address the problem. The private sector has likewise fallen short in meeting the demands of the impoverished. Due to the high risk and lack of collateral, financiers like banks have hesitated to lend to low-income households. This has resulted in low-income households not accessing the loans they need to purchase a house. In recent years, South Africa's housing finance sector has faced several issues, including high levels of family debt, low levels of home ownership, and a scarcity of affordable housing (Pillay, 2006). Numerous diverse options to address these challenges and improving South Africa's home finance system could be considered. South Africa's insufficient housing finance system has had numerous detrimental implications for the poor. Urbanisation and the rise of informal settlements have resulted from a lack of affordable housing. The result has negatively influenced the poor's health and well-being and the country's general economic progress. There are numerous approaches to increasing affordable housing available for the poor, which will be discussed in the next section.

7.7 Recommendations

The chapter emphasises the importance of a comprehensive strategy to tackling the housing issue, covering not only housing finance but also affordable housing and the underlying social and economic reasons that contribute to the housing crisis. To cater to the poor, the chapter advises that the government explore enacting laws and programmes encouraging affordable housing developments, such as subsidies for low-income earners. In as much as there are a couple in existence already, they should be reformed. Furthermore, the government should explore enacting policies and programmes that boost access to affordable home finance choices, such as expanding the reach of existing government housing finance schemes and developing new programmes that offer low-cost mortgages to low-income individuals. Furthermore, the research urges the government to invest in financial education initiatives to boost low-income earners' awareness of housing finance possibilities. The government should invest in affordable housing development by creating new housing units or offering subsidies or tax breaks to private developers who build cheap housing. This will increase the number of affordable housing options, increasing their accessibility to low-income individuals.

The General Assembly, in its resolution, called on UN-Habitat to take "urgent steps to guarantee a stronger mobilisation of financial resources at all levels, to boost the implementation of the Habitat Agenda, particularly in developing countries. The South African government should promote widespread access to adequate housing financing, so boosting the availability of affordable homes and establishing an enabling environment for long-term growth that will attract investment. Housing financing systems do not always meet the diverse demands of vast portions of the population, particularly those from disadvantaged groups living in poverty and low-income individuals. The government and other interested parties will need to address the core causes of the housing issue and take more broad and realistic methods to solve it. This could involve expanding the supply of affordable housing, restructuring the housing finance system to make it more accessible and equitable, and tackling underlying issues like income inequality and inadequate infrastructure. Cooperation and partnerships with the commercial sector, civic society, and community-based organisations will also be critical in developing and implementing realistic solutions.

A comprehensive and long-term approach is required to address the issue of insufficient housing finance for the poor in South Africa. The government should prioritise the supply of affordable housing by offering subsidies to low-income households. Banks and other financial institutions should also be encouraged to lend to low-income households through risk-mitigation tools such as guarantees.

7.8 Conclusion

The chapter thoroughly examines the shortcomings of South Africa's housing finance system and the country's ongoing housing challenges. South Africa's housing finance system is inadequate since it does not serve the poor. The housing sector in the country is characterised by a significant scarcity of affordable housing, with a considerable part of the population unable to afford decent housing because of financial limitations. The high cost of housing and a lack of affordable credit are significant impediments to homeownership for low-income families. The government's housing programmes have failed to meet the requirements of the underprivileged. Despite implementing numerous housing finance measures, most low-income households lack access to affordable housing. A substantial proportion of the population is still excluded from the formal housing market, perpetuating inequity and impeding socio-economic growth. The government's emphasis on social housing has also failed to meet the requirements of the poor, owing to a lack of funds and political will to address the issue. The private sector has likewise fallen short in meeting the demands of the impoverished. Due to the high risk and lack of collateral, banks and other financial institutions have been hesitant to lend to low-income households. As a result, low-income households have been forced to rely on unregulated and informal credit sources, which are frequently associated with exorbitant interest rates and predatory lending practices.

South Africa's insufficient housing finance system has had numerous detrimental implications for the poor. Overcrowding and the rise of informal settlements have resulted from a lack of affordable housing. This has negatively influenced the poor's health and well-being, as well as the country's general economic progress. A comprehensive and long-term approach is required to address the issue of insufficient housing finance for the poor in South Africa. The government should prioritise the supply of affordable housing by offering subsidies and tax breaks to low-income households. Banks and other financial institutions should be encouraged to lend to low-income households by offering guarantees and other risk-mitigation tools. The chapter's suggestions provide a road map for the government to follow to improve South Africa's housing crisis and provide adequate and affordable homes for all. A long-term solution will be accomplished by working together to guarantee all South Africans access to affordable housing.

References

Amin, A., & Cirolia, L. R. (2018). Politics/matter: Governing Cape Town's informal settlements. *Urban Studies*, *55*(2), 274–295. https://www.jstor.org/stable/26428439

Bah, E. M., Faye, I., & Geh, Z. F. (2018). Housing finance in Africa. In F. Afeikhena (Ed.), *Housing market dynamics in Africa* (pp. 43–63). Palgrave Macmillan. https://doi.org/10.1057/978-1-137-59792-2_3

Blackwell, T., & Kohl, S. (2017). *Varieties of housing finance in historical perspective: The impact of mortgage finance systems on urban structures and homeownership* (pp. 1–42). MPIfG Discussion Paper, No. 17/2. Max Planck Institute for the Study of Societies. https://hdl.handle.net/11858/00-001M-0000-002C-6801-F

Brevoort, K. P., Grimm, P., & Kambara, M. (2016). Credit invisibles and the unscored. *Cityscape, 18*(2), 9–34. http://www.jstor.org/stable/26328254

Centre for Affordable Housing Finance (CAHF). (2016). *The role of mortgage liquidity facilities in housing finance: Lessons learned from Egypt, Tanzania, Nigeria and Malaysia.*

Corder, C. K. (1997). The reconstruction and development programme: Success or failure? *Social Indicators Research, 41*(1/3), 183–203. http://www.jstor.org/stable/27522262

DBSA Integrated Annual Report. (2022). Retrieved February 28, 2023, from https://www.dbsa.org/EN/About-Us/Documents/DBSA-Integrated-Annual-Report-2022.pdf

Del Mistro, R., & Hensher, D. A. (2009). Upgrading informal settlements in South Africa: Policy, rhetoric and what residents really value. *Housing Studies, 24*(3), 333–354. https://doi.org/10.1080/02673030902794709

Department of Human Settlements SA. (2022). *Facilitating the development of integrated human settlements.* https://www.gov.za/about-government/contact-directory/human-settlements/facilitating-development-integrated-human-settlements

Doling, J., Vandenberg, P., & Tolentino, J. (2013). Housing and housing finance—A review of the links to economic development and poverty reduction. *Asian Development Bank Economics Working Paper Series No. 362.* Available at SSRN: https://ssrn.com/abstract=2309099 or https://doi.org/10.2139/ssrn

Ganiyu, B. O., Fapohunda, J. A., & Haldenwang, R. (2017). Sustainable housing financing model to reduce South Africa housing deficit. *International Journal of Housing Markets and Analysis, 10*, 410–430. https://doi.org/10.1108/IJHMA-07-2016-0051

Gilbert, A., Mabin, A., McCarthy, M., & Watson, V. (2011). Low-income rental housing: Are South African cities different? *Environment and Urbanisation, 9*, 133–148. https://doi.org/10.1177/095624789700900111

Gumbo, T., & Onatu, G. (2015). Interrogating South Africa's People's housing process: Towards comprehensive collaborative and empowering aided self-help housing approaches. *Journal of the Korean Housing Association, 26*(6), 27–34.

Kajimo-Shakantu, K., & Evans, K. (2006). The role of banks in the provision of low-income housing finance in South Africa: Can they play a different role? *International Journal of Strategic Property Management, 10*(1), 23–38. https://doi.org/10.1080/1648715X.2006.9637542

Kuddus, M. A., Tynan, E., & McBryde, E. (2020). Urbanisation: A problem for the rich and the poor? *Public Health Reviews, 41*, 1. https://doi.org/10.1186/s40985-019-0116-0

Mabin, A. (2020). A century of South African housing acts 1920–2020. *Urban Forum, 31*, 453–472. https://doi.org/10.1007/s12132-020-09411-7

Mabuya, B., & Scholes, M. (2020). The three little houses: A comparative study of indoor and ambient temperatures in three low-cost housing types in Gauteng and Mpumalanga, South Africa. *International Journal of Environmental Research and Public Health, 17*(10), 3524. https://doi.org/10.3390/ijerph17103524

Marutlulle, N. K. (2021). A critical analysis of housing inadequacy in South Africa and its ramifications. *Africa's Public Service Delivery and Performance Review, 9*(1), Article a372. https://doi.org/10.4102/apsdpr.v9i1.372

Mbandlwa, Z. (2021). Challenges of the low-cost houses in South Africa. *Social Psychology of Education, 58*(2), 6753–6766.

Nyakala, S., Ramoroka, D., & Ramdass, K. (2021). Factors influencing the quality of low-income housing in Polokwane municipality South Africa. *Acta Structilia, 28*(2), 23–52. https://doi.org/10.18820/24150487/as28i2.2

Office of the United Nations High Commissioner for Human Rights (OHCHR). (n.d.). *Special Rapporteur on the Right to Adequate housing.* https://www.ohchr.org/en/special-procedures/sr-housing/human-right-adequate-housing

Parikh, P., Bisaga, I., Loggia, C., Georgiadou, M. C., & Ojo-Aromokudu, J. (2020). Barriers and opportunities for participatory environmental upgrading: Case study of Havelock informal settlement, Durban. *City and Environment Interactions, 5,* Article 100041. https://doi.org/10.1016/j.cacint.2020.100041

Pernegger, L. (2021). Effects of the state's informal practices on organisational capability and social inclusion: Three cases of city governance in Johannesburg. *Urban Studies, 58*(6), 1193–1210. https://doi.org/10.1177/0042098020910111

Pillay, A. (2006). *Financing low-income housing in South Africa: Borrower experiences and perceptions of banks.* Habitat International.

Reserve Bank of South Africa. (2022). *Annual report 2021/22.* https://www.resbank.co.za/content/dam/sarb/publications/reports/annual-reports/2022/Annual-Report-2021-22.pdf

Scheba, A., & Turok, I. (2020). Informal rental housing in the South: Dynamic but neglected. *Environment and Urbanisation, 32*(1), 109–132. https://doi.org/10.1177/0956247819895958

Schirmeister, R., & Nadler, M. (1996). Promoting linkages between formal and informal finance in an African Country: A case study. *African Review of Money Finance and Banking, 1/2,* 135–154. http://www.jstor.org/stable/23027372

Schirmer, S. (2017). Property rights, institutional change and development in South Africa. *Acta Commercii, 17*(1), 1–9. https://doi.org/10.4102/ac.v17i1.358

Socio Economic Rights Institute of South Africa (SERI). (2018). *Informal Settlements and Human Rights in South Africa: Submission to the United Nations Special Rapporteur on adequate housing as a component of the right to an adequate standard of living.* https://www.ohchr.org/sites/default/files/Documents/Issues/Housing/InformalSettlements/SERI.pdf

Trusler, R. K. (2009). *Is 'Breaking New Ground' as a comprehensive plan for housing delivery, a realistic solution to the housing problem faced in South Africa?*

UN-Habitat. (2010). *Housing finance system in South Africa.*

United Nations, Department of Economic and Social Affairs, Population Division. (2019). *World Urbanization Prospects: The 2018 Revision – Highlights (ST/ESA/SER.A/421).* United Nations. https://population.un.org/wup/assets/WUP2018-Highlights.pdf

United Nations Human Settlements Programme (UN-Habitat). (2022). *World Cities Report 2022: Envisaging the Future of Cities.* Nairobi: UN-Habitat. https://unhabitat.org/world-cities-report-2022

Van Rooyen, C., Stewart, R., & De Wet, T. (2012). The impact of microfinance in Sub-Saharan Africa: A systematic review of the evidence. *World Development, 40*(11), 2249–2262. https://doi.org/10.1016/j.worlddev.2012.03.012

Weimann, A., Kabane, N., Jooste, T., Hawkridge, A., Smit, W., & Oni, T. (2020). Health through human settlements: Investigating policymakers' perceptions of human settlement action for population health improvement in urban South Africa. *Habitat International, 103,* Article 102203. https://doi.org/10.1016/j.habitatint.2020.102203

Williams, D. S., Máñez Costa, M., Sutherland, C., Celliers, L., & Scheffran, J. (2019). Vulnerability of informal settlements in the context of rapid urbanisation and climate change. *Environment and Urbanisation, 31*(1), 157–176. https://doi.org/10.1177/0956247818819694

Fortune Mangara holds a qualification in Rural and Urban Planning from the University of Zimbabwe and earned an M.Sc. and a Ph.D. in Urban and Regional Planning from the North-West University. Presently, he is a postdoctoral research fellow affiliated with the African Centre for Disaster Studies at the School of Geo and Spatial Sciences at the North-West University. Dr. Mangara's research focuses on spatial and urban resilience, disaster risk studies, disaster risk governance, urban disaster risk, social and economic geography, urban policy, urban governance, transport planning and mobility, and spatial transformations in African cities.

Hangwelani Hope Magidimisha-Chipungu is a professor in town and regional planning at the University of KwaZulu-Natal. Her record of service spans but not limited to the South African Research Chairs Initiative (SARChI) for Inclusive Cities, National Planning Commission (RSA), Alternate Chairperson Scientific for the South African Council for Planners (SACPLAN), Committee Member for the International Society of City and Regional Planners (ISOCARP), City Planning Commission for eThekwini Municipality, advisory committee of Cooperative Governance and Traditional Affairs (COGTA-KZN), member of KwaZulu-Natal Tribunal and up to late a Board member of SACPLAN, Researcher at the Human Sciences Research Council (HSRC) etc.

Chapter 8
The Search for Housing Finance in Post-2000 Contested Housing Schemes in Harare, Zimbabwe

Lovemore Chipungu and Hangwelani Hope Magidimisha-Chipungu

Abstract The Global Report on Human Settlements (1996) emphases that finance is the most critical factor which affects housing production. It cuts across the housing production process, i.e. from the time a piece of land is acquired to the time individual housing units are sold. It is also a key factor in the refurbishment of old housing stock. Unfortunately, evidence from around the world shows that financial institutions are more willing to give housing loans to high and middle income than to the low-income groups. The argument behind such a scenario is that low-income households have limited capacity to service their loans. In fact, they are viewed as a liability and financial institutions do not want to risk their resources. This chapter revisits the issue of housing finance by analysing it through the window of low-income housing that emerged in post-2000 in Harare, Zimbabwe. At the height of the fast-track land reform programme, the urban poor managed to access large swaths of land for housing purposes. This was free land on which they embarked on housing production through housing cooperative systems. Emerging evidence shows that while land was easily accessible, funding for all professional services proved to be a nightmare. As a package of services, urban housing is tied to huge capital outlay to finance professional services such as land survey, town planning, engineering and architectural services. While the government was nowhere to offer such financial support, private financial institutions were not ready to release funds for housing development whose land was contested by commercial farmers and driven by political whims. Hence, the urban poor were forced to dig deep into their coffers in search of alternative funding means which came through their meagre savings. Hence, the progress and quality of housing that emerged during this period shows how attrition between formal and informal housing finance mechanisms can either stall or enhance housing production.

L. Chipungu (✉) · H. H. Magidimisha-Chipungu
School of the Built Environment and Development Studies, Howard College, University of KwaZulu-Natal, Glenwood, Durban, Republic of South Africa
e-mail: Chipungu@ukzn.ac.za

H. H. Magidimisha-Chipungu
e-mail: MAGIDIMISHAH@ukzn.ac.za

L. Chipungu et al. (eds.), *Contemporary Models for Pro-Poor Housing in Urban Sub-Saharan Africa*, https://doi.org/10.1007/978-3-032-05489-0_8

Keywords Low-income housing · Housing finance · Fast track land reform · Land · Housing cooperatives

8.1 Introduction

The turn of the millennium in 2000 is a historical landmark in Zimbabwe's sovereignty and economic emancipation. It will go down memory lane as the year when the Zimbabwean government embarked on the fast-track land reform programme (FTLRP) whose intention was to acquire large scale commercial farms for distribution to the landless indigenous people. Through the FTLRP, productive large scale commercial farms were either invaded by landless people or designated by the government for redistribution to those in need. The dynamics of the land reform programme became a pivotal centre of change which was felt beyond the agrarian sector, as they impacted on every facet of the economy. In the ensuing contestations, urban housing equally became a benefactor as the urban poor invaded peri-urban farms which witnessed rapid transformation into housing through cooperatives. While houses that emerged were spurned by some, largely arising out of their political affiliation to the ruling party (Zimbabwe African National Unity—Patriotic Front—ZANU-PF), they prevailed in being one of the greatest innovative interventions of the decade which benefitted the urban poor around in the country.

It is from this perspective that this chapter revisits the decade 2000–2012 to interrogate this contested housing market with the intention of analysing housing finance systems that emerged to support housing production. The FTLRP was instrumental in releasing land from commercial farmers thereby enabling the government to redistribute it to local authorities and housing cooperatives at almost no cost (Chipungu & Magidimisha, 2020: 211). Although land was acquired for almost free, how did the urban poor manage to build houses given the economic challenges that prevailed in the period in question? These sentiments are captured by Chaeruka and Munzwa (2009: 20) who argued that:

> It is not the cost of land that prevents poor people from developing, but the associated transactions which invariably involve professionals that charge high fees in the housing development process.

This situation was complicated by the fact that most financial institutions were not willing to provide housing finance on land which was contested; especially where households did not have documentary proof of ownership.

In trying to explore the dynamics around housing finance in the period in question, this chapter uses the case study of Harare which witnessed massive land invasions and the formation of housing cooperatives. Such developments were expected given that housing provision in the years leading to the fast-track land reform programme has drastically fallen while the price of land for housing development was beyond the reach of the urban poor.

8.2 The Emergence of Contested Housing Schemes in Harare—An Overview

The onset of the FTLRP in July 2000 and the emerging subsequent housing schemes around the country and Harare in particular were a response to socio-economic pressure that was building up over the year. Chipungu and Magidimisha (2020) contend that housing under the FTLRP emerged as an appendage of the agrarian programme which was meant to redistribute land for agricultural purposes and to resettle landless households. This view is contested by some authorities such as Marongwe (2003) who argue that the Zimbabwean government had already conceived the idea of the urban component of the land reform programme prior to 2000 with focus on peri-urban settlements. This was witnessed in the Inception Phase Framework Plan of 1998 where the government emphasised the need to develop strategies to monitor urban growth in line with demand for land within and outside urban areas. On paper, this was a noble policy intervention which would create peri-urban areas as zones of transition where households would maximise the benefits of both the urban and rural environments.

Unfortunately, this proposed Framework Plan was never implemented since it was over-shadowed by the onset of the FTLRP in 2000. As Moyo (2001: 125) observes, the FTLRP

> … …was a free-for-all situation where violence, disorder and chaos reigned as the landless people occupied land.

It should be noted that this was the first time in the history of housing development in Harare (and Zimbabwe at large) where urban housing was developed outside the existing central and local government regulations. Indeed, it was unprecedented in the history of urban development in Zimbabwe where historically, urban development was guided by the conventional development process. As informality became the norm due '*to free land for all*', housing schemes for poor urban households emerged as they took advantage of the availability of land to build housing.

In a bid to unravel the forces at work, the period under observation stretches from 2000 to 2012 but divided into two phases. The first phase is what Chipungu and Magidimisha (2020) term '*the unregulated environment*' which began in 2000 and lasted till 2004. This period is associated with the onset of the FTLRP in 2000. These were the formative years of the FTLRP and were characterised by massive land acquisition and development of contested housing schemes. Most housing cooperatives which emerged during this period in question were associated with the government in power (ZANU-PF)—hence earned the name '*Dzimba dzechibakera*' *(housing acquired through violence)*. This was not a misnomer since it befitted the prevailing situation at the time which was violent, chaotic and disorderly in nature. The '*unregulated environment phase*' was followed by '*taming the unregulated environment*' phase that stretched from 2005 to 2012 (Chipungu & Magidimisha, 2020). This phase was associated with the return to normalcy under the Government of National Unity (GNU—under ZANU-PF and the Movement for Democratic Change—MDC).

This period was a stage for reorganisation and institutional development as the GNU put in place measures to stabilise the explosive dynamics that emerged in the first phase. From a housing perspective, both conventional and unconventional housing schemes among the urban poor became a dual package in the housing production system. With the user initiative taking centre stage in the housing production process, housing finance became a sticking point among poor urban households despite having access to land.

However, it is interesting to note that during the '*unregulated environment*' phase, mixed housing schemes emerged which had attributes of both formal and informal attributes. This was indispensable given that urban Zimbabweans had always shunned the idea of informality as they used to refer them as '*Kuma squatter*' (or simply squatter settlements). The situation was further aggravated by the government that never '*romantised poverty*' which manifested itself in the form of informal housing—the wrath of *Operation Murambatsvina/Restore Order (of 2005)* bears testimony to the then government's policy position on informality (see Chipungu & Adebayo, 2013). It is therefore interesting to note that most houses that emerged at the height of the FTLRP exhibited all attributes of formality in terms of being guided by residential layouts on acquired land with housing structures guided by building plans and built of proper materials. Emerging structures were built of durable materials even at a time when the situation was not economically viable due to the prevailing social-political dynamics—hence the question: How did poor urban households mobilise housing finance at the height of the FTLRP in order to build durable housing? The response to this question is explored through the case study of Ushewokunze Housing Scheme (UHS) which is one of the first contested housing cooperation to emerge at the onset of the FTLRP.

8.3 Housing Finance for the Urban Poor—A Conceptual Framework

Housing finance (HF) is more than just a mechanism for accessing funds to acquire or maintain housing; it is a cornerstone of socio-economic development and individual well-being. At its core, HF determines the capacity of nations to ensure adequate shelter, a basic human need intricately tied to health, productivity and social stability. As King (2009: 3) pertinently puts it, housing finance is the:

> … …money we use to build and maintain the nation's housing stock, enabling the production and consumption of housing through rents, mortgages, and repayments.

However, despite its centrality, HF remains one of the most complex and multifaceted sectors in global development.

The International Union for Housing Finance (IUHF) underscores this complexity by highlighting that a single definition cannot encapsulate HF due to its contextual nature, which varies significantly across countries and regions. Factors such as legal frameworks, economic structures, cultural norms and political systems play critical roles in shaping how HF operates (IUHF, n.d.). In developed nations, housing finance systems are typically characterised by well-regulated mortgage markets and expansive credit availability. In contrast, Chiquier and Lea (2009) reveal that developing countries often face systemic barriers, including weak regulatory environments, limited access to formal credit and persistent economic inequality, all of which hinder the equitable distribution of housing resources.

Housing finance (HF) occupies a critical space at the nexus of economic development, social equity and urban planning. To conceptualise HF effectively, it is imperative to navigate its theoretical foundations and contextual nuances. Scholars such as Chiquier and Lea (2009) highlight its complexity, describing HF as a multi-sectoral, dynamic, and context-driven system. This characterisation underscores the interplay between economic, legal, cultural and political forces that shape the availability, accessibility and functionality of housing finance systems. HF is not merely a financial mechanism; it is a socio-political construct that reflects broader systemic inequalities. For instance, in developing countries, the lack of robust legal frameworks and institutional capacities often exacerbates barriers to effective housing finance, leaving low-income households disproportionately excluded. Such systemic limitations contrast sharply with developed nations, where institutionalised HF systems are supported by stable economic and regulatory environments.

King (2009) provides a complementary perspective by emphasising HF's role in facilitating housing production, maintenance, and affordability. This perspective brings into focus the dual nature of HF: It operates as both a market-driven and a social equity tool. Mortgage systems and rental mechanisms, for example, illustrate how HF channels capital flows to meet housing demand while simultaneously serving as a barometer of economic inclusivity. However, these mechanisms are not universally accessible, especially in contexts where income inequality and informal employment prevail, such as in many developing economies. The inability to leverage HF for social mobility in such settings reveals a critical gap between its theoretical promise and practical limitations. The overreliance on conventional mortgage products, with stringent credit requirements and high interest rates, further alienates vulnerable populations, making affordable housing an elusive goal.

Boleat's (1985) insights deepen this critique by framing HF as a system for channelling investor funds to homebuyers. While this perspective highlights the financial intermediation that underpins housing markets, it also reveals the inequities embedded within global HF systems. In developed nations, financial institutions efficiently allocate funds, enabling broad access to homeownership. Conversely, in developing countries, HF systems are frequently marred by underdeveloped capital markets and insufficient financial infrastructure. This disparity reflects a structural inequity that transcends mere economic capability—it is also a reflection of policy priorities and governance deficits. Boleat's (1985) analysis compels a re-evaluation of how HF systems can be adapted to bridge these divides, particularly by leveraging innovative approaches such as microfinance and community-driven models like housing cooperatives.

Critically, the conceptualisation of HF as a universal solution is inherently flawed. Its functionality is deeply contingent on contextual variables that differ significantly across regions. For instance, while HF in developed nations thrives on long-term credit availability and institutional stability, developing countries often struggle with macroeconomic volatility, currency risks and inadequate property rights frameworks. The application of HF in such divergent settings cannot be one-size-fits-all; rather, it necessitates localised adaptations that account for socio-economic realities. This dichotomy between theory and practice underscores the need for an integrated approach to HF, one that reconciles its financial, legal and social dimensions.

Housing finance in developing countries is marked by a combination of structural barriers and innovative solutions that seek to address the specific challenges of housing accessibility and affordability. The conceptualisation of housing finance in these regions must consider not only the economic and institutional limitations but also the geo-political and socio-cultural contexts that influence financial systems. One of the most significant challenges in developing countries is the absence of robust regulatory frameworks and legal systems to underpin housing finance systems. A weak regulatory environment often results in market inefficiencies, such as speculative land markets, high levels of informality in housing provision and inadequate protection for lenders and borrowers. These issues are exacerbated by fragmented governance structures, which fail to harmonise urban planning, land tenure security and financial regulation (Chiquier & Lea, 2009). For instance, in Sub-Saharan Africa, insecure land tenure restricts formal financial institutions from using land as collateral, thereby limiting access to mortgages.

Traditional mortgage systems are largely inaccessible to low-income groups due to high interest rates and stringent eligibility requirements. Financial institutions often perceive these groups as high-risk, citing unstable incomes and insufficient credit histories as barriers to lending. This results in the exclusion of a majority of urban populations from formal housing finance markets. King (2009) notes that this financial exclusion leads to the proliferation of informal housing markets, which are typically associated with poor living conditions and vulnerability to environmental risks. Additionally, macroeconomic instability in many developing countries further exacerbates these challenges, with inflation eroding the purchasing power of potential homeowners and increasing the cost of borrowing.

The limited availability of long-term financing options poses another critical challenge. Developing economies often lack deep capital markets, which restrict the flow of long-term funds necessary for housing finance. This is compounded by high dependency on short-term deposits, which are inherently ill-suited for financing long-term housing needs (Boleat, 1985). In many cases, local financial institutions are unable or unwilling to extend long-term credit due to liquidity constraints and regulatory limitations.

Despite these challenges, innovative approaches have emerged to address the housing finance gap in developing countries. Microfinance institutions (MFIs) have played a transformative role in bridging the gap for low-income households. By offering small, incremental loans tailored to the needs of informal sector workers, MFIs circumvent the rigidities of traditional mortgage systems. These institutions not only provide financial assistance but also build the creditworthiness of borrowers, enabling them to eventually access more formal financing options (Chiquier & Lea, 2009). For example, the Grameen Bank in Bangladesh has successfully implemented microfinance models that empower low-income families to improve their housing conditions incrementally.

International development agencies have also become pivotal actors in supporting housing finance programmes in developing countries. Organisations such as the World Bank and UN-Habitat provide technical assistance and financial resources to governments and financial institutions, fostering innovation and capacity-building. These agencies often act as guarantors, mitigating the risks associated with lending to underserved populations. For instance, the World Bank's affordable housing finance programmes in East Africa have introduced blended finance models that combine public and private resources, creating sustainable pathways for housing finance development. It is this blended approach which has been at work at the peak of the FTLRP largely propelled by the socio economic and political dynamics as witnessed in the case study of UHS.

8.4 Research Methodology

The FTLRP was implemented at a time when the country's political-economic landscape was at cross-roads. From a political perspective, the ruling party, ZANU-PF was facing an unprecedented challenge from the newly formed political party—MDC. The ruling party's strongholds were being tested especially in urban areas where the MDC was getting a lot of support. Coincidentally, it is in urban areas where most people were feeling the wrath of the ailing economy which was punctuated by spiralling prices of commodities, unemployment and poor service delivery (among other factors). It is under such an environment that the FTLRP emerged with contested housing being an appendage of the agrarian land reform programme. Amid such political tensions and economic uncertainty, collecting data for this research was not an easy task. It had to be designed to suit the dynamics of the time in the search for the truth which was only peculiar to the context.

Given such a complex environment, a holistic research approach was undertaken using both quantitative and qualitative techniques. Among such research tools that were used to collect data from this case study included household surveys, interview guides and observations. The main research tool was the household survey where structured questionnaires were administered through face-to-face interviews on a sample size of 100 households that was randomly selected out of 580 housing plots that were built or undergoing construction under the housing cooperative. The questionnaire comprised of questions that revolved around the construction process with financial issues being the main focus. These questions were directed to household heads who (it was assumed) held all responsibilities in terms of construction and management of the housing development process for the family. It must be emphasised that UHS is a consortium of housing cooperatives since there are different areas under its fold.

These household surveys were supported by unstructured interviews with key informants from the Harare City Council, the central government and committee members of the housing cooperatives. Discussions with the city council were mainly held with the deputy director and a town planner while discussions from the central government were held with three town planning officers from the Department of Physical Planning. These were purposely selected based on their knowledge about housing and spatial development in Harare. Finally, the cooperative committee for UHS was also contacted where some members volunteered information about the housing cooperative in response to questions in the interview guide.

Observations were also conducted in the area with focus on the type and level of developments. During observation, a template was used to record information on the typology of infrastructure and the level of housing developments. Infrastructure covered included roads, electricity, sewer reticulation and water provision. This information was meant to complement other data gathering tools.

However, it must be emphasised that given the sensitivity of activities at the time of the research, data collection was undertaken under strict observation since all outsiders were accommodated with suspicion. For instance, observations and household surveys were mainly undertaken under the observation of cooperative members. This serves to show the political tensions that were prevailing and which in turn labelled any activity not only suspicious, but also political. Fortunately, this did not affect the quality of data for this research given that most of the questions were purely technical and did not insinuate any political connotations.

8.5 Ushewokunze Housing Scheme

The onset of the FTLRP under the command of war veterans also witnessed the invasion of peri-urban land for housing development. Ushewokunze Housing Scheme (UHS) emerged in November 2000 as part of a consortium of 16 housing cooperatives which was supposed to be develop 5,211 housing plots. This area was divided into three phases with UHS being allocated two phases—an area earmarked for developing 1,711 housing plots. The area covered by this research is limited to 2000 residential plots around the proposed pre-independence residential area called New Cerney. Despite being one of the contested housing schemes, UHS is located in proximity to Glen Norah and Highfield high-density residential areas with high accessibility to the Harare-Masvingo highway.

According to the housing cooperative leadership, the site was strategically chosen for its easy connectivity to sewer reticulation, water and electricity given that it is close to existing residential areas. More so, this housing scheme fitted into the Harare City Council's (HCC) urban development plan as proposed in the Harare Combination Master Plan (HCMP) which designated part of this area for housing development.

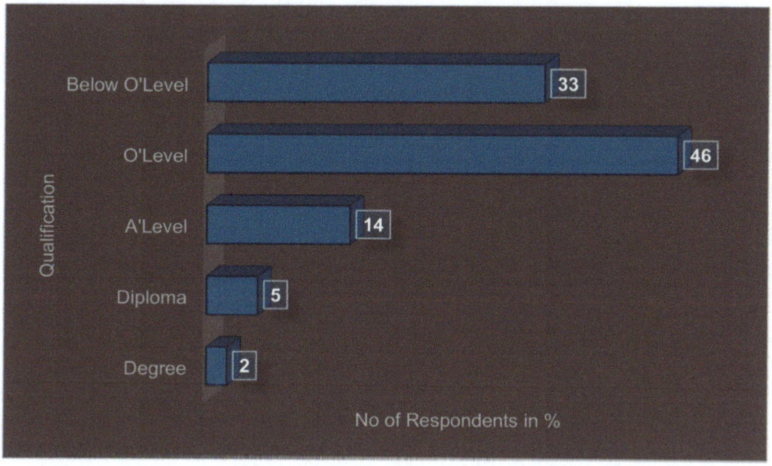

Fig. 8.1 Educational status of household heads. *Source* Fieldwork (2010)

The HCC's Saturday Retreat Local Development Plan No. 50 (SRLDP-50) clearly outlines the proposed vision for development of this area.

Ushewokunze Housing Scheme, like any other housing cooperative was born out of people's will in search of affordable housing. These were not mere opportunists, but households whose chances of securing a house in Harare were almost impossible largely due to their low salaries and the unavailability of serviced land to facilitate self-help housing. Most of these were literate household heads whose educational status would ensure them a secure job under normal circumstances (see Fig. 8.1).

As shown in Fig. 8.1, only 33% of the respondents did not have minimum educational qualifications. In addition, most of the households were also family-oriented with 90% of them being married as depicted in Fig. 8.2.

In essence, these were households that were desperately searching for decent and secure housing—hence, the FTLRP presented such an opportunity for them to initiate housing development.

8.5.1 Funding for Infrastructure Development

As already alluded to above, the proposed development of UHS was set out within the anticipated vision of the housing cooperative. It is important to note that while it is widely believed that the onset of the FTLRP allowed for invasion of land without payment, there was sanity among such anarchy. Once land was identified and invaded by prospective home owners under a particular cooperative, it was sold by the central government at intrinsic value of US\$7/m^2. Hence, housing cooperatives mobilised finance through their monthly contributions for land acquisition, infrastructure and housing development (among other financial commitments).

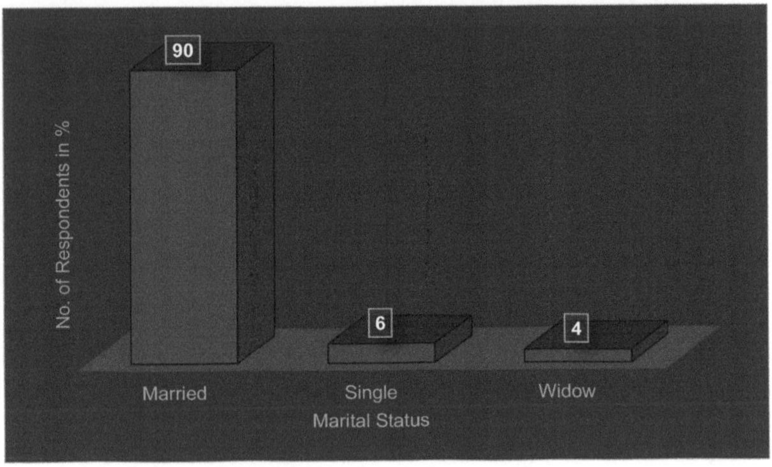

Fig. 8.2 Marital status of respondents. *Source* Fieldwork (2010)

Allocation of land parcels for individuals was not haphazard, but planned and managed by the housing cooperative's technical leadership. A town planning consultant was engaged to prepare a residential layout. This became the basis for the allocation of residential plots that ranged in size from 400 to 2000 m^2. What made this residential layout more ideal was that it fitted into an existing proposed residential layout—New Carney Township, that was planned on the same site by the colonial government but was abandoned in the late 1970s due to the intensification of the war of liberation. So, from a town planning perspective, it was an ideal site strategically located in proximity to existing off-site infrastructure such as Crowborough Sewer Works, Harare-Masvingo Highway, Jeffrey-Montons water works and the sub-station electricity. The technicality of the existing off-site infrastructure in terms of its capacity and connectivity can only be understood from the perspective of the municipality and the central government, but suffice to say that there were challenges (see Chipungu & Magidimisha, 2020; Harare Combination Master Plan, 1988). Therefore, the site was not outside and above the mandates of beneficiaries and the housing cooperative, as one of the town planners from Harare City Council stated:

> Among the conditions which the HCC places on all prospective housing cooperatives are the need to produce evidence of their capacity to fund infrastructural development. However, because of the overriding need for housing and political interference, most housing cooperatives construct houses without fulfilling this requirement.

However, beyond the mandates of off-site infrastructure, beneficiaries of the housing scheme were supposed to mobilise resources to provide both onsite infrastructure and connect to existing off-site infrastructure. On-site infrastructure came in the form of access roads, electricity, sewer and water reticulation. Funding to finance such developments came in two forms—(a) Firstly it came from the cooperative's savings account where each household contributed a monthly amount. However, this money was not enough to fund such capital developments. Only few gravel roads were constructed during the initial stages of land acquisition. (b) Secondary, the cooperative called for special levies that were directly meant for a particular type of development or infrastructure. Unfortunately, such contributions came in drips and drops—a situation which made it difficult to have meaningful impact on proposed infrastructure developments. It must be borne in mind that apart from mobilising funds to provide for immediate access to infrastructure within the neighbourhood and to individual plots, further funds were required to connect the neighbourhood to the main trunk sewer, water main and electricity power stations. More so, such funds were not only required for construction of infrastructure to connect to the existing one, but also for upgrading where the capacity was inadequate to meet the expected demand of the new housing scheme. In the case of UHS, the two 11 kW powerlines required upgrading to meet local demand as per the Zimbabwe Electricity Supply Authority (ZESA).

At the time of the survey, the level of infrastructure in the housing scheme was very low with gravel roads in place, no electricity; save for a few individuals who had access to generators and solar panels. On the other hand, standpipes for water found on site had very low pressure thereby forcing households to drill boreholes or dig shallow wells while pit latrines were dug for sanitation purposes. Given such ad hoc means of mobilising funds, it is not surprising that only 68% of the respondents had access to water-borne sanitation facilities.

8.5.2 Social/Public Infrastructure

The provision of social infrastructure (which comes in the form of schools, health facilities and shops) is a share responsibility. As per Circular No. 70 of 2004, the local authority (such as the HCC, investors and the local community) has a responsibility to provide these services. Field observations showed that these services were either inadequate or missing altogether. This is not surprising given that households and other stakeholders' focus was on houses (as in the superstructures) as opposed to housing (which includes houses and the entire neighbourhood). As such, this left a gulping gap in Ushewokunze Housing Scheme (UHS) whose local populations now depend on adjacent residential areas for public facilities. Some sites earmarked for these facilities as per the provisions of the residential layout have been overtaken by 'illegal' developments where they are now being used for make-shift schools and shops. Given the overriding immediate needs of houses and physical infrastructure, social infrastructure will take a long time to develop.

8.5.3 Funds for Housing Construction

The invasion of land for housing was largely driven by the burning desire to build houses. This sets a completely deferent tone in the level of commitment to building houses as opposed to housing. Hence from the onset, households were more than willing to invest into temporary shelter (such as wooden cabins) which they erected on site. This was a strategic move that enabled them to stay on their properties while enabling savings and investment in the much-anticipated housing scheme. For this reason, financial resources were mobilised from households' own savings, friends, relatives and financial institutions to fund the construction of houses as shown in Fig. 8.3.

While no tangibles savings were mentioned about their savings, respondents were quick to say that:

> … … under this economic environment, your bank account is not an indicator of your savings- … … … … instead, the number of bricks you lay on your wall are appropriate indicators of your savings … … ….

Although 90% of the respondents were gainfully employed (as shown in Fig. 8.4), only 4% obtained housing finance from the bank.

More so, although most of the respondents were gainfully employed, there are a number of factors that militated against them when it comes to formal housing finance. Formal housing finance mechanisms are driven by conventions which must be fulfilled by those in need of financial support. To be eligible, households are supposed to be gainfully employed with collateral security in place as shown in Fig. 8.5 (see summary of requirements in Fig. 8.5).

In as much as 84% of the households submitted their building plans to the local authority for approval, there were still other sticking issues that denied them access to

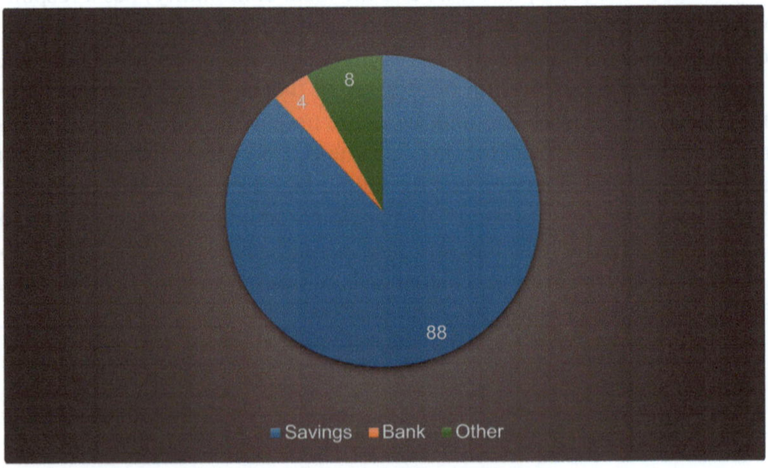

Fig. 8.3 Sources of housing finance. *Source* Fieldwork (2010)

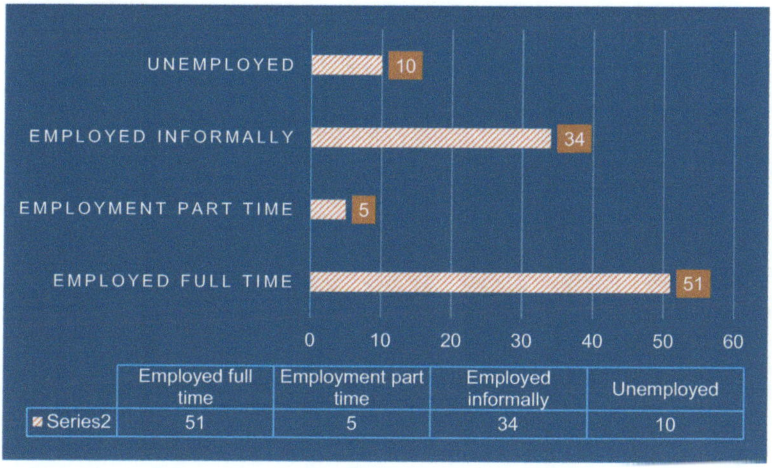

Fig. 8.4 Employment status of households. *Source* Fieldwork (2010)

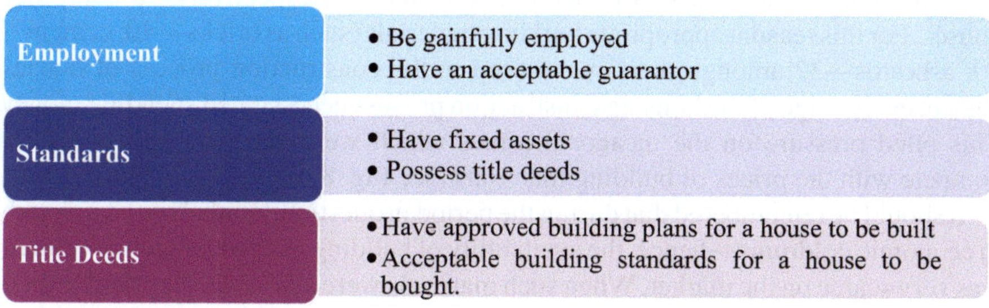

Fig. 8.5 Conditions to obtain a bank loan. *Source* Authors from various sources (2010)

housing finance from banking institutions. Most households did not have title deeds given that they were also first home seekers. The situation was complicated by the process of land acquisition, which apart from being contested, was under the umbrella of cooperatives. In addition, most households were in the low-income segment which meant that their income was low and could not afford monthly premiums. Above all, 33% of the respondents were informally employed—thus making it further difficult to submit evidence of employment. Amid all these negativities, housing finance from employers was nowhere to be seen.

As already alluded to above, the need to build decent housing pushed households beyond their financial limits to secure appropriate materials and services. Through household surveys, it was established that 84% of the respondents submitted their building plans to the local authority for approval. More so, 66% solicited services of the resident building inspectors to check and certify their houses at every appropriate stage. This was made possible because most households enlisted the services

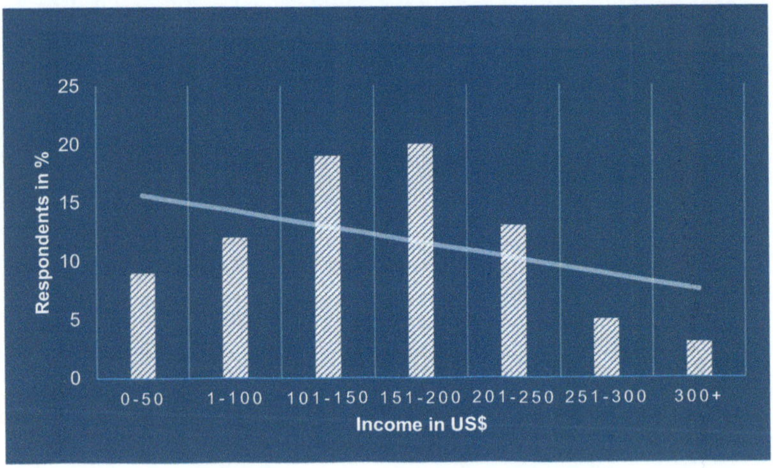

Fig. 8.6 Monthly income of respondents. *Source* Fieldwork (2010)

of professional builders and contractors (82% and 2%, respectively) to build their houses. For this reason, appropriate building materials (such as bricks—40; cement—40; asbestos—32 among others) were used in the construction process of houses. Hence, every stage of the housing construction process necessitated spending money. This piled pressure on the meagre incomes which were very low and could not compete with the prices of building materials (see Fig. 8.6).

It should be remembered that during the period in question, Zimbabwe was already in economic doldrums—hence, the availability of building materials (such as cement) was very scarce on the market. When such materials were available, they were expensive. For instance, a 50 kg bag of cement costed between US$12 and US$18, without factoring in transport and labour costs. This was extremely high bearing in mind that the majority of households earned less than US$200 per month, yet they pegged their dreams for decent housing very high as shown by the type of the building materials they used for construction. This in turn forced some households to engage innovative ideas such as moulding cement bricks onsite while some even built kilns for burning farm bricks.

As shown in Fig. 8.7, a variety of materials were used for building houses. While asbestos and bricks were commonly used, tiles were perceived as a luxury among households. On the other hand, timber, apart from being used for constructing roof trusses, was also mainly used for constructing temporary structures.

Observations from the field show that the construction process was not an easy fit. If anything, it required innovation and creativity to attain the best out of the financial resources that were invested. This was confirmed by 75% of the respondents who cited lack of financial support as the main hindrance to completing their houses. Given their close proximity from the central business district coupled with the high cost of materials, most households lamented the extra transportation cost that most delivery tracks added to the overall cost. Under such circumstances, they had to be

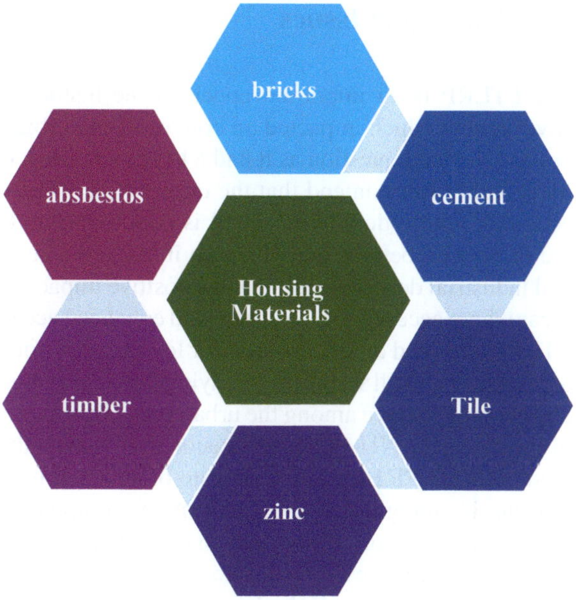

Fig. 8.7 Materials used for housing. *Source* Fieldwork (2010)

innovative in order to survive. It is therefore not surprising that some moulded bricks and dug building sand on their plots for construction and for sale. Indeed, the heavy construction costs tempered with their moral worthiness—hence, some ended up harvesting and selling both river sand and building sand illegally. Proceeds obtained from these sales were invested into housing construction—thus enabling almost 60% of the houses to be nearing completion by 2010.

Due to the short comings arising from both the housing environment and the houses themselves, households had mixed feelings about their housing scheme. It was encouraging to note that 62% of the households were happy with their houses, with only 14% who felt otherwise. Part of this positive perception stemmed from the fact that they were now owners of houses in the city where they could access other opportunities. Ownership for them entailed being in control of their construction process and the environment in which they were staying. However, 24% of the respondents expressed disappointment at the slow rate of the construction process and lamented the poor state of their housing environment especially the low level of infrastructure. Despite these negativities, housing development is continuing gradually in line with the principles self-help housing.

8.6 Emerging Issues

The FTLRP is an important epoch in the history Zimbabwe's land and housing environment which impacted on both rural and urban environments. Its impact went beyond the land question as it had a knock-on effect on other facets of the economy. Most authorities contend that the onset of the FTLRP marked the down turn of the economy whose downward spiral is continuing up to this day. Nevertheless, there are those who believe that although it impacted negatively on commercial farming and industrial development, it had a positive impact on low-income housing in urban areas where accessibility to urban land enhanced access to housing development. The FTLRP provided an opportunity for low-income households to access land in urban areas. One can still weigh-in by saying that it provided an opportunity to unlock land and housing finance among the urban poor. This narrative fits very well in the context of the ruling ZANU-PF Party's vision that 'land is the economy and the economy is land' (ZANU-PF Election Manifesto of 2000). It is against this argument that urban housing is emerging as a resilient major component of the urban economy which have endured the turbulent years ushered in by dynamics of the FTLRP.

However, at stake in housing development among the urban poor is not only land—but also housing finance which is critical in making sure that emerging housing development succeeded. While housing finance allows for the production and consumption of housing (King, 2009), it is a complex phenomenon and process especially when applied in a contested low-income housing environment like in Zimbabwe. Chiquier and Lea (2009) support this observation by further arguing that in its application, housing finance brings together complex and multi-sectoral issues such as the legal environment, economic make-up, the regulatory environment and the political system. All these factors were observable in the Zimbabwean context during the time in question. Hence to get insight into the operation of housing finance during the first decade of the FTLRP, one has to interrogate the dynamics associated with contested housing, the policy environment, the nature of the housing production system especially as it evolved around ownership and cooperative housing schemes.

8.6.1 Unpacking the Concept of Contested Housing Schemes

The concept of contested housing schemes is not new to the housing discourse. An insight into the history of urban development shows a myriad of examples which are labelled either as simply illegal and or informal housing schemes (Castells, 1983; Magidimisha & Chipungu, 2011). There are no universal standards for this 'label' suffice to say that they emerge in situations that are defined by a particular context especially when such housing fails to meet the expected aspirations of various stakeholders such as the central government, municipalities and the general populace. But more so, the concept of contestation operates within an understanding of global constitutional structures or norms. Wiener (2014) clearly captures the concept of contestation when he argued that it is associated with an objection in the implementation of the norm—which in essence means that it is a violation of compliance. Since norms are seen as a good prescription for rules, they form the basis for legitimacy of shared values and adopted regulatory mechanisms (Reder et al., 2020). Indeed, when these arguments are mirrored against the FTLRP, its moral validity comes into question especially during the '*unregulated environment*' phase.

The contested housing schemes were born out of an environment that was chaotic, disorderly and violent—with urban development emerging on designated agricultural land (in some cases). While there was genuine cause given the demand for housing especially among the poor urban beneficiaries, this does not change the moral campus of contestations that are associated with these housing schemes. In the case of UHS, most beneficiaries did not have any legal documents (such as title deeds) as proof of ownership for their plots and houses. This explains why financial institutions were not willing to provide housing finance to some beneficiaries simply because of the legitimacy of the land acquisition process and lack of collateral security. Housing finance is sensitive to socio-economic dynamics—hence it shuns areas of high risk as witnessed in the case of the FTLRP (Fuchs, 2018). The FTLRP ushered in a completely new housing market which was not familiar, let alone tested. The greatest challenge revolved around land and house ownership, which presented weak contractual frameworks and weak legal recourse in realising collateral value in the case of default by the borrower. The fact that only 4% of beneficiaries managed to get financial support from banks simply shows the high level of uncertainty which eclipsed the financial institutions about this new housing market. This in turn affected the construction progress of houses due to lack of timely housing finance while investment into infrastructure completely stalled.

8.6.2 Fitting the Housing Policy Context

Contested housing schemes emerged at a time of political turmoil pitying the ruling ZANU. PF and MDC—a situation which overshadowed some prevailing policies. While there were challenges in the acquisition of land and the eventual implementation of the FTLRP, the ruling principles at work in the development of housing schemes were in line with the government's strategies. An overview of the Zimbabwean government's approach to housing outlined in the Transitional National Development Plan (1982/83–1984/85) emphasised aided self-help and formation of housing cooperatives as key strategies for low-income urban housing schemes. Over the years, this strategy was very instrumental in creating new housing stock for the urban poor. Through this strategy, the government's role was limited to the acquisition and servicing of land while beneficiaries were supposed to mobilise resources for building the superstructure. Where resources permitted, the government provided core houses which beneficiaries would eventually expand at their own pace governed by their financial abilities. It is this enabling environment that has greatly contributed towards people's entrepreneurship to invest into housing.

Therefore, housing schemes that emerged during the FTLRP were governed by the same policy guidelines outlined above. These are driven by liberal convictions which argue that land acquisition and the provision of infrastructure are the most expensive components of the housing production process which push housing beyond the affordable limits of the urban poor. In a bid to reduce these costs, the government encouraged the formation of housing cooperatives as governed by the Cooperatives Act of 1990 which enabled the formation of the Zimbabwe National Association of Housing Cooperatives (ZNAHC) under which UHS falls. Land for registered housing cooperatives was sold at intrinsic value—thus reducing the financial burdens on beneficiaries. Further emphasis on this housing strategy came through the Short-Term Economic Recovery Programme (STERP) of 2009 which called on pension funds and insurance companies to participate. Despite such intervention measures meant to cut down land and infrastructural costs, beneficiaries still struggled to provide infrastructure. The situation was aggravated by the government's stance that it was the responsibility of households to connect onsite infrastructure to offsite infrastructure. Housing infrastructure represents up to 70% of total housing costs—yet its absence affects the construction of houses and impacts on the legality of housing schemes (Kihato & Horn, 2023). The dilemma of contested housing schemes perpetuates partly because of the absence of both public and private institutional finance in the housing production process. Hence, the label of contested housing schemes becomes a reality under such circumstances.

8.6.3 Home Ownership—A Befitting Magnet in Capital Mobilisation

The government's housing strategy, that revolves around self-help and home-ownership, has not only created an enabling environment for households and institutions to invest into housing; but also reconfigured people's mindsets to be self-reliant. As such, households do not look forward to get housing from the government, but to get land and at times infrastructural support. Hence, the perception of housing is not grounded in the finished product but on a piece of land that can be built gradually. This is further guaranteed by home ownership, which has become a general guarantor even under circumstances where there is no legal paperwork such as title deeds. For most households, the easiest route to own a house is through access to land and then to build gradually. Indeed, as Glickman's (1994) saying goes—'the best investment on Earth is earth'. This is equally summed up by the Zimbabwean government's vision that '*land is the economy and the economy is the land*'. This in turn explains the massive invasions of land which was witnessed during the FTLRP.

This acceptance of land as the best investment has also changed the perception of households, to not only value land, but also to perceive housing a construction process. The availability of housing finance is in turn, tied to people's attitudes, perceptions and beliefs that housing is not only a physical construct, but a gradual process as understood through site and service schemes. This enabling environment has been created by the government through policy interventions even at a local level where site and services are strictly governed by development control mechanisms through inspections undertaken by town planning officers and building inspectors. Hence, adherence to these regulations has seen households saving money to build decent houses irrespective of how the land was acquired. In essence, the mobilisation of housing finance is not found in generous income, but is hidden in people's conscience to achieve the best. Before UHS met its wrath of OM in 2005, it was the envy of many as they marvelled houses that were built through people's sweat equity. Therefore, poor performance of housing finance in UHS after 2005 is not a true reflection of people's ability, but rather evidence of fear and reluctance to invest largely arising out OM.

In essence, housing finance is not only tied to the urban poor's passionate need to invest their little savings, but is also sensitive to political dynamics that dictate both opportunities and adversities which affect them. Indeed, the political dynamics of the FTLRP shot UHS to fame—yet the same forces, as observed through OM, are responsible for its demise.

8.6.4 The Gradual Approach—An Ideal Saving and Investment Strategy

The emergence of contested housing at the height of the FTLRP is reminiscent of the '*whole sector development approach*' proposed by the World Bank (1983–1993) whose emphasis was on the retreat from the welfare approach in order to avail funding for municipal infrastructure in order to ensure adequate supply of land for housing and urban development in general. In the Zimbabwean context, the government was able to acquire land through the controversial land policy where it vowed only to pay land owners for capital developments—not for the soil (or land). Though marred with controversy, land was eventually acquired with housing cooperatives buying it at '*intrinsic value*'. Hence, housing cooperatives became the main vehicle through which finance was raised for buying land, connecting to off-site infrastructure, servicing the site and eventually for building houses. As the official from the HCC puts it,

> The vetting of cooperatives was based on their financial ability to fund major infrastructural developments for their sites.

However, this was never realised in the case of UHS where financial management problems pertaining to the cooperative committee were raised by 17% of the respondents. Among other factors, the cooperative abrogated on its mandate to service the site and construct houses for its members. This administrative quandary further explains why the residential layout was not approved and most houses being built without recourse to technical approval. Indeed, the financial woes of UHS underpin its failure to fulfil its mandate. In 2016, the NewsDay reported that the cooperative was continuously being rocked by political divisions, land grabbing and financial mismanagement by cooperative leaders. Similarly, the Herald (2016) documented how teachers under the same UHS (Waterfalls) were disgruntled and wanted transparency about how the management committee used over US$1 million of their finances. This situation is complicated by Section 85 (3) of the Cooperative Act (24:05) of 1996 which states that:

> No member shall have claim to share in the reserve fund … … … … the reserve fund and any general reserves shall be invisible.

Caught in the dilemma of the cooperative's administrative under-performance, households could not seek alternative funding avenues other than to dig deep into their coffers to finance the construction of houses. The situation, though by default for some, became an ideal saving and investment operandi since it enabled them to build their houses gradually while relaying solely on themselves. This is not only innovative, but also a viable option that guarantees ownership of a decent house. Geothert (2016) summaries this housing finance option by clearly noting that:

> … … … …. this pay-as-you-go process is a key way by which families succeed. However, the process takes a long time, with a clear burden on the families. Safety concerns of proper construction and lack of appropriate skills are challenges to the family-builder… … … ….

Despite these challenges, the pay-as-you-go is not only a cornerstone of contested housing schemes in Zimbabwe, but have since become a reality of the urban poor who see no light in their governments' struggle to provide decent housing with supporting infrastructure.

8.7 Concluding Remarks

It can therefore be argued that contested housing schemes that emerged at the height of the FTLRP were not simply ad hoc type of developments. Instead, they were underpinned by ideological frameworks that were governed by cooperative principles. Fanned by political dynamics of the time, they took advantage of the system to build dream homes for households in the low-income bracket. The availability of land, made possible through the government's agrarian land reform program, spurred their ambitions as land was the most difficult component of the housing development process to access by ordinary people, municipalities and even central government. Contested housing schemes fitted in the government's policy frameworks of housing production systems especially for people in the low-income brackets despite the fact that they contravened some of the existing laws which in turn attracted the government's wrath. Emerging at the peak of contested socio-political-economic cross-roads, conflicts were bound to occur and selective application of the law was common as UHS later witnessed through OM.

However, while housing cooperatives offer promise, they are not without their limitations. Challenges such as mismanagement, insufficient regulatory oversight and limited access to funding often hinder their scalability. Policymakers must, therefore, adopt a hybrid approach that integrates the efficiency of formal banking systems with the inclusivity of cooperative models (Sazama, 2000). Some authorities, such as (Durand-Lasserve & Selod, 2009) advise that tailored housing finance systems must also account for non-monetary factors, such as land tenure security, as these can significantly influence housing affordability and investment incentives in developing countries. Indeed, put together, poor cooperative financial management coupled with lack of access to formal funding systems (among other factors) clearly account for the continual woes UHS is experiencing. Above all, this case study serves as reminder to the nature of formal housing finance mechanisms that they are sensitive to socio-economic-political dynamics and shun away from such controversies.

References

Birchall, J. (2010). *People-centred businesses: Co-operatives, mutuals and the idea of membership.*

Boleat, M. (1985). *National housing finance systems—A comparative study.* Croom Helm.

Castells, M. (1983). *The city and the grassroots. A cross-cultural theory of urban social movements.*

Chaeruka, J., & Munzwa, K. (2009). *Low-cost housing development in Zimbabwe: An assessment of the effectiveness of the regulatory framework.* UN-Habitat and GoZ final report.

Chipungu, L., & Adebayo, A. A. (2013). The policy-planning divide: An evaluation of housing production in the aftermath of operation Murambatsvina in Zimbabwe. *Journal of Housing and the Built Environment, 28*, 381–396.

Chipungu, L., & Magidimisha, H. H. (2020). *Housing in the aftermath of the fast-track land reform programme in Zimbabwe.* Routledge.

Chiquier, L., & Lea, M. (2009). *Housing finance policy in emerging markets.* World Bank Publications.

Durand-Lasserve, A., & Selod, H. (2009). The formalization of urban land tenure in developing countries. In *Urban land markets: Improving land management for successful urbanization* (pp. 101–132). Springer.

Fuchs, C. (2018). Authoritarian capitalism, authoritarian movements and authoritarian communication. *Media, Culture & Society, 40*(5), 779–791.

Geothert, R. (2016). *Incremental housing: The new site and services.* KB CAHF. 11th December 2016. UN. This video is from the UN-Habitat Global Urban Lecture series.

Glickman, M. L. (1994). *Modigliani-Miller on capital structure: A post-Keynesian critique.* University of London.

Harare Combination Master Plan. (1988). *Harare combination master plan.* Harare Combination Master Plan Preparation Authority, Department of Works, Harare City Council.

International Union for Housing Finance (IUHF). (n.d.). *What is housing finance?* [Online]. https://www.housingfinance.org/about-us/what-is-the-iuhf/

Kihato, M., & Horn, P. (2023). Microfinance for housing in Africa: Achievements and challenges. *African Urban Futures, 12*(4), 56–72.

King, P. (2009). *Understanding housing finance—Meeting needs and making choices* (2nd ed.). Routledge.

Magidimisha, H. H., & Chipungu, L. (2011). Unconventional housing provision: Reflections on health aspects: A case study of Zimbabwe. *Journal of Housing and the Built Environment, 26*(4). https://doi.org/10.1007/s10901-011-9234-9

Marongwe, N. (2003). The fast track resettlement and urban development nexus the case for Harare. In *Symposium on Delivering Land and Securing Rural Livelihoods: Post-independence Land Reform and Resettlement in Zimbabwe, Mont Clair, Nyanga* (pp. 26–28).
Moyo, S. (2001). The land occupation movement and democratisation in Zimbabwe: Contradictions of neoliberalism. *Millennium, 30*(2), 311–330.
Reder, M., Filipovic, A., Finkelde, D., & Wallacher, J. (2020). *The concept of contestation of norms: An interview University of Hamburg & University of Cambridge Yearbook on Practical Philosophy in a Global Perspective (YPPGP-JPPGP) 2020, 4.* 2 Final Draft. 12 February 2020.
Sazama, G. W. (2000). Lessons from the history of affordable housing cooperatives in the United States: A case study in American affordable housing policy. *American Journal of Economics and Sociology, 59*(4), 573–608.
The Herald. (2016, September 15). *Teachers lose $1 million in housing scandal.*
The NewsDay. (2016, September 15). *Violence rocks Ushewokunze housing co-operative.*
UN-Habitat. (2011b). *Sustainable housing for sustainable cities: A policy framework for developing countries.* United Nations Human Settlements Programme.
Wiener, A. (2014). *A theory of contestation.* Springer.

Lovemore Chipungu holds a Ph.D. in Town and Regional Planning, a Master's Degree in Rural and Urban Planning as well as a Bachelor of Science Degree (Honours) in Rural and Urban Planning. He is an active member of the Zimbabwe Institute of Regional and Urban Planners as well as a Corporate Member of the South African Planning Institute. He is currently an Associate Professor in the School of the Built Environment and Development Studies, at the University of KwaZulu-Natal. Professor Chipungu is both a practitioner and academic. In addition, he is an editorial board member of various journals and the Managing Editor of the Journal of Inclusive Cities and the Built Environment. He has published widely in peer-reviewed journals and books; and his areas of interest are housing policy, land-use planning and urban design. He is currently a co-co-principal investigator of a project on inclusive cities in Southern Africa.

Hangwelani Hope Magidimisha-Chipungu is a professor in town and regional planning at the University of KwaZulu-Natal. Her record of service spans but not limited to the South African Research Chairs Initiative (SARChI) for Inclusive Cities, National Planning Commission (RSA), Alternate Chairperson Scientific for the South African Council for Planners (SACPLAN), Committee Member for the International Society of City and Regional Planners (ISOCARP), City Planning Commission for eThekwini Municipality, advisory committee of Cooperative Governance and Traditional Affairs (COGTA-KZN), member of KwaZulu-Natal Tribunal and up to late a Board member of SACPLAN, Researcher at the Human Sciences Research Council (HSRC) etc.

Part III Balancing the Formal and Informal—Innovative Trends

The bottom line that should govern this understanding is that housing for urban poor households is a process that involves migration from unconventionality to conventionality, inasmuch as it involves migration from illegality to legality. This part of the book illustrates both formal and informal housing production systems that are innovative. They provide a platform for people to showcase their creativity in order to survive the challenges associated with housing and its associated environment. But more so, they illustrate the convergence of both formal and informal creativity as governments try to provide lasting solutions to perennial housing challenges. These are seen through the manipulation of housing resources, land, tenure systems and architectural issues. Hence this part tackles this issue of innovative engagements in housing from two perspectives. On one hand, it traces existing trends (as provided through case studies) among the urban poor households who use illegal means to achieve housing. This in essence is a creative way for them to afford housing given their socio-economic status. On the other hand, is the formal approach which is a top-down process initiated by governments and informed through policies. Through this legal conduit, innovative engagement takes various routes such as through developing unique strategies to access land for housing and through using smart architectural designs to build adequate housing. When these formal and informal innovative engagements are put together, they reveal a complex array of systems that can be factored into the urban environment to house the urban poor.

Chapter 9
Sport and Recreational Facilities as Catalysts for Development and Social Change in Low-income Human Settlements in Greater Kokstad Municipality, South Africa

Vincent Myeni

Abstract The creation of sustainable low-income human settlements is a challenging task in the South African housing provision practice. This is because sustainable human settlements can only be achieved if the principle of integrated development informs all human settlements projects especially in the post-apartheid South Africa. This may include the use of space in relation to the provision of basic services, strategic location of settlements in relations to access and availability of employment opportunities, the consideration of social and environmental aspects of the settlements and the provision of sport and recreational facilities. The non-provision of basic services has led to several challenges especially for the youth. The study emphasises the provision of sport and recreational facilities as forming part of basic needs for any community that intends to pursue the goal of sustainable development. Underpinned by qualitative design, the study unpacks mainly the social challenges that are experienced by residents who reside in settlements where the provision of sporting and recreational facilities is non-existent. Qualitative data was collected through structured interviews, questionnaires, focus group discussion and observation. Two municipal officials from Greater Kokstad municipality were identified through purposive sampling method. A sample of 42 residents was secured through systematic sampling. This sample was drawn from a population of about 420 households in the study area. The study found that due to non-provision of sport and recreational facilities, there is high level of crime, drug and alcohol abuse, obesity mostly among the youth and violence. To a certain degree, the prevalence of some of these challenges if not all could be mitigated if sport and recreational facilities were provided in all communities irrespective of their economic status and location. Therefore, this study argues that the provision of sporting facilities must form part of project packaging so that such facilities form part of project implementation.

V. Myeni (✉)
School of Social Sciences, University of KwaZulu Natal, Durban, South Africa
e-mail: myeniv@ukzn.ac.za

L. Chipungu et al. (eds.), *Contemporary Models for Pro-Poor Housing in Urban Sub-Saharan Africa*, https://doi.org/10.1007/978-3-032-05489-0_9

Keywords Sport · Recreational facilities · Low-income settlements · Livelihood enhancement · Sustainability

9.1 Introduction

In post-apartheid South Africa, disparities between advantaged and disadvantaged areas are discernible in the provision of sport and recreational facilities. Accordingly, the former Sport and Recreation Minister, Fikile Mbalula, stated that information from municipalities revealed that a significant number of communities did not have access to even basic sport and recreation facilities (Mbalula, 2010). Further, where such facilities existed, their location made it difficult for local people to access them. For example, some were constructed in a manner that made their usability and accessibility problematic for the locals, and some were not well maintained for optimal use. Thus, there is need to consider disadvantaged communities particularly in rapidly growing urban areas to provide them with facilities for sports.

The provision of sport and recreational facilities forms an important aspect of integrated human settlements (Department of Human Settlements, 2004). Regardless of the resident's level of income, people need some time to relax and get physically active, with the latter mostly applying to children and the youth across all different income groups, social classes and races in both rural and urban areas. The main benefits of sport and recreational facilities are that young people get preoccupied and their involvement in alcohol and illicit drugs and criminal activities is usually minimised (Sibeko, 2007). Accordingly, the norms and standard set by the Department of Sport and Recreation (Sports and Recreation South Africa, 2010) include the provision of recreational facilities, a right to be enjoyed by all communities, thus creating socially sustainable human settlements (Department of Sport & Recreation, 2010). The same principle was envisaged by the Department of Housing (now Department of Human Settlements in South Africa) in its Breaking New Ground (BNG) Policy document where it is said that the delivery of housing should appreciate the interrelated nature of sustainable human settlements which consists of social, economical, institutional and physical components, which depict a full housing development process with each of the elements impacting on all others during the implementation process and beyond (Department of Human Settlements, 2004).

The above BNG assertion became clear after it emerged that most low-income human settlement developments were found to be dysfunctional, too basic and not in good locations environmentally and economically (Charlton, 2003). There is adequate literature emphasising the procedure and importance of sporting facilities for middle and high income residents (Henderson, 2012). However, there is little emphasis on the provision of such facilities for low-income residents in South Africa. This study aimed at bridging that gap, especially as various government departments, such as the Department of Sport and Recreation and the Department of Human Settlements, have recently made the provision of such facilities a part of the requirements at the design stage of every human settlement development project. According to the

Breaking New Ground, (Department of Human Settlements, 2004) housing policy, the creation of sustainable human settlements through the provision of facilities where people can play, study, work and engage in income-generating activities has become a necessity. This study was conducted to accomplish the following three specific objectives:

i. To explore the importance of the inclusion of sport and recreational facilities at the designing stage of a housing project.
ii. To reflect on the importance of creating human settlements that enhance human development and social change.
iii. To examine the challenges faced by municipalities in the provision of sport and recreational facilities for low-income settlements.
iv. The chapter consists of the following sections: research methodology which was used to collect data, literature review, research findings, discussion, conclusion and recommendations.

9.2 Study Area and Research Methodology

This study was conducted in Greater Kokstad Municipality, south of KwaZulu-Natal province. The residents of the study area are predominantly middle-income and high-income residents, that purchased their land through mortgage bonds or leases (COGTA, n.d.). The condition of approval was completed by the municipality and was submitted to Cooperative Governance and Traditional Affairs (COGTA) with an application for the establishment of community of Extension 7. This study was specifically focused on Extension 7 of the area because it is predominantly occupied by low-income residents. The subdivision of land was done according to Section 33 of the Town Planning Ordinance No. 27 of 1949.

9.2.1 Research Methodology

The study adopted a qualitative approach of data collection, however, some findings of the study were analysed quantitatively. Structured interviews, questionnaires, focus-group discussions and observation were used as data collection instruments. A purposive sampling method was used to identify two municipal officials with whom structured interviews were conducted from Greater Kokstad Local Municipality. These officials were considered important, especially as they participated in the design of human settlements development and the provision of sport and recreational facilities in the municipality.

The study area is home to an estimated 420 households; hence, systematic sampling was used to identify study participants. The starting point was the random selection, with every 10th house forming part of the study. Therefore, a sample size of 42 households, which constituted 10% of the total population, was considered eligible for the study. A questionnaire was administered to all the residents who were identified as participants in the study. Focus-group discussion was also conducted with the coach and eight members of Master Eleven, a football team which was active in the study area. The team, which was established in August 2001, catered for players aged between 12 and 19 years. The eight members were randomly selected by the coach as the whole team had the same experience and knowledge of the area. The team's achievements included having some of its players participating in various sporting codes at both national and international levels. For instance, one of the team's former players had a chance to play for Wits University's football Club, as some of the players were able to join teams at the provincial level. Some of the team's former players had represented the team and South Africa at the Olympic Games.

Field observation was another data collection method which was used in this study. The researcher collected data about the number of playgrounds that were available in the area. A camera was used to collect visual information on the physical appearance of the study area. Data was analysed using thematic method of data analysis. Further, tables were used to present statistical information. Before the collection of empirical data, the research project was ethically approved by the University of KwaZulu-Natal through the Humanities and Social Sciences Research Ethics Committee (HSSREC).

9.3 Literature Review

Literature that focused on the aspects of the availability of sport infrastructure in South Africa and the importance it has on human development and social change was reviewed. This section provides an overview of the provision of sport and recreational facilities during the apartheid period in South Africa. Literature on the integration of sport facilities during settlements design has also been reviewed. The last section explores the benefits or contributions of sporting facilities towards human development and social change.

9.3.1 An Overview of the Provision of Recreational and Sport Facilities in South Africa

According to Apartheid Museum (2008), the apartheid government, for the longest time intentionally provided most facilities on the basis of race and location in South Africa. Racial differences determined the provision and non provision of essential facilities in the country in terms of the type of service that was to be provided, the quality level of the service and location. There was a surge and intensification in the involvement of sporting activities from the beginning of the 1940s, which was due to the industrialisation of South Africa's economy (Modisane et al., 2019). The surge was stimulated by the availability of attractive physical facilities like those of soccer, netball, rugby, swimming, tennis, etc., high disposable incomes, more leisure time, improved education and general living standards. Just before the dawn of democracy in South Africa, various organisations began to advocate for the establishment of a discrimination-free sporting system that would create equal opportunities for participation by all South Africans (ibid). The dawn of democracy ushered in drastic changes in policy and legislative frameworks that addressed all aspects of people's political, economic and social life, including human rights. The democratic government which came to power in 1994 began to exert influence, not only on the provision and utilisation of sporting facilities, but also on the outdoor recreation behaviour patterns of many South Africans, irrespective of their racial background (South Africans National Sport and Recreational Plan, 2011).

The Sport and Recreational Department in South Africa aims to improve the quality of life of all South Africans by encouraging everyone to participate in all sporting activities in the country. For instance, the introduction of the Siyadlala (we are playing) Mass Participation Programme in 2005 is the cradle of community sports in South Africa. This programme was launched to facilitate access to sport for many South Africans, especially those coming from previously disadvantaged communities. The main purpose of this initiative was to establish at least one sporting hub in every municipality across the country. Hence, the Department of Basic Education and the Department of Sport and Recreation signed a partnership agreement which was intended to revive school-based sporting activities and physical education programmes. The focus of the programme was to build capacity, provide sport equipment for use at events, sustenance of the programme by supporting local sports assistants and local leagues and the provision of logistical support to participants (Modisane et al., 2019).

9.3.2 Integrating Sport Facilities During Human Settlements Design

The provision of sport facilities during human settlement design and implementation should not be left to chance; instead, it should inform the layout of a settlement from the conceptualisation, planning and implementation phases of housing development projects. The South African Constitution (1996) provides that 'everyone has the right to have access to adequate housing'. Accordingly, sporting or recreational facilities form part of the components that must be delivered together with housing, so that the goal of creating sustainable human settlements is achieved. This objective is in accordance with General Comments 4 on Article 11(1) (1991) of the UN International Covenant on Economic, Social and Cultural Rights (ICESCR); thus, 'a settlement is regarded as providing adequate housing if it considers the aspects of the right to housing in line with a particular context which considers among other things legal security of tenure; availability of services, materials, facilities and infrastructure; affordability; habitability; accessibility; location and cultural adequacy'. This assertion encourages the implementation of the principle of integrated development, which is one of the mechanisms that enhance human settlements that are socially, economically and ecologically sustainable.

The concept of a neighbourhood unit, which is associated with a form of residential design, is traditional but contemporary in nature. Since its inception, the concept has since become a recurring theme for planners when designing plans for human settlements (Marais, 2015). According to Perry (1929), a neighbourhood unit is defined as a populated area that needs the support of an elementary school and other community facilities so that self-contained, socially cohesive communities are created. This concept is based on a principle of good residential design, which stipulates that the settlement layout design should discourage the location of essential facilities far from people's residence and that children should not be exposed to hazardous activities that endangers their lives. Therefore, good planning of a neighbourhood is representative of an everyday landscape which should support residents' physical, psychological and social well-being (Ibid). Neighbourhoods, by their nature, are regarded as physical entities, and if they are planned well, they are more likely to nurture healthier and more socially interactive people and communities. Furthermore, well planned neighbourhoods are more unlikely to experience social ills, environmental, physical and political challenges than those that are not. This necessitates the development of settlements which have a strong sense of attachment, belonging and identity that foster social interaction among residents (Perry, 1929). The concept of neighbourhood unit prioritises the provision of basic human needs such as educational and recreational facilities as well as places of worship and other facilities to empower community members (Meenakasi, 2011).

This concept is further supported by the principle of urban design which speaks to the manner in which planning should be done to promote the development of a conducive and habitable environment for all residents (Marais, 2015). Therefore, apart from creating a sense of place for residents, the concept bears connection between environmental sustainability and community members's satisfaction (Dewar & Utenbogaardt, 1995). According to Carmona (2001), the principle of urban design, which borders on the need to create a sustainable human settlements and self-sufficiency, is a necessity as Planners are required to respond to environmental issues, local autonomy, consultation and the participation of the local community members during settlement design stages. It is for this reason that urban design must focus on the uniqueness of a place based on the community's needs and those of the environment.

The concept of neighbourhood unit has, over the years, received criticism from other commentators in the field of planning. According to Meenakasi (2011), a community is more likely to be a grouping of people in accordance with their social classes. Studies further indicate that this concept is too idealistic to represent real life situations, as neighbourhoods are sometimes too vast and large to foster neighbourly relations. Although the neighbourhood concept has been subjected to strong criticism, it continues to influence the design and development of many cities and towns around the globe (Brody, 2009).

The White Paper on Sport and Recreation (2011) clearly stipulates that sport and recreation facilities must be provided with the objective of improving the health and well-being of all communities. This objective could be achieved through the restructuring of sporting and recreational facilities in South Africa. Like many other pieces of legislation on development, the White Paper on Sport and Recreation puts emphasis on local government to ensure that the objectives are implemented at grass-root level. This development must be in line with the provincial and national policies so that sport and recreation in South Africa receives the necessary support. Accordingly, funding is to be provided to principal agencies, such as community-based organisations and clubs that are involved in sport and recreation. Municipalities are obligated to build and maintain the sport and recreational infrastructure in their areas of jurisdiction. The White Paper on Sport and Recreation further stipulates that all the three spheres of government (national, provincial and local) are to ensure that sport development networks are effectively and efficiently operational at community level. Hence, local municipalities have a duty to play a leadership role in encouraging sporting activities (White Paper on Sport and Recreation, 2011).

9.3.3 Sport as Catalyst for Human Development and Social Change

In 2015, a resolution which was passed by the United Nations (UN) General Assembly indicated the potential of sport in fostering tolerance and respect and the contribution it makes towards empowering women and young people, individuals and communities on the aspects of health, education and the social inclusion agenda. This assertion recognises sport as an important enabler of sustainable development in the sustainable development agenda (Dudfield, 2019).

According to Lyras and Welty (2011: 311), sport can exert a positive influence on public health, help socialise children, young and adults, foster the social inclusion of the disadvantaged members of society, functions as a vehicle for economic development of regions and states and foster intercultural exchange and conflict resolution. This does not only imply the importance of the provision of sport and recreational facilities, but also reveals the benefits which communities are more likely to enjoy through the provision of sporting facilities. It is public knowledge that physical activity positively contributes towards physical and psychological health as well as personal well-being of residents in any given community. Sport and recreation create a platform for individuals to showcase local talent and culture, enhancing social cohesion and encouraging a sense of community (Ware & Meredith, 2013). It has been established that:

> Sport can play a role in improving the lives of individuals, not only individuals…but whole communities. I am convinced that the time is right to build on that understanding, to encourage

> governments, development agencies and communities to think how sport can be included more systematically in the plans to help children, particularly those living in the midst of poverty, disease and conflict (Annan, 2004 Secretary-General of the United Nations).

Sport and recreation can also be seen as a mechanism for the improvement of learning and educational opportunities. It is counter-boredom for both sport participants and the community at large. It also reduces opportunities for anti-social behaviours such as crime, drug and alcohol abuse and violence. According to the United Nations International Children's Emergency Fund (UNICEF), sporting facilities and involvement in sport provide an opportunity for children of school-going age to freely express themselves through sport, art, music, movement, storytelling and indoor and outdoor activities (UNICEF, 2004).

According to UNICEF (2004), consistent physical activity strengthens and maintains the health of bones and strengthens muscles and joints. It also controls body weight, especially through reducing the chances of obesity and lowers blood pressure. Sport can directly or indirectly prevent chronic diseases. On the other hand, it is further said that passive participation in sport accounted for about 1.9 million deaths worldwide in the year 2000. In the same year, in America alone, lack of participation in sport costs the country more than $75 billion in terms of medical expenditure. On the other hand, active participation in sport helps to reduce symptoms of stress and depression in both adults and youth. A study conducted in the United States of America indicates that inactive children were more often depressed than their active counterparts. Another study found that high school athletes had less chances of committing suicide than those that were non-athlete (UNICEF, 2004). While the causes of crime might be complex and therefore very difficult to understand, involvement in sport can directly or indirectly yield positive results, by encouraging involvement in challenging adventures, using leisure time positively, developing coaches as role models and giving young people a sense of direction in times of despair. It has been found that sport could work better if it is aligned with comprehensive programmes that seek to address broader social and personal development issues at community level (Ibid). Sport can also be regarded as a powerful tool that fosters engagement in dialogue, especially during and post-conflict moments. People who are traumatised through involvement in conflicts can learn to incorporate the experience of pain, fear and loss into healing of emotional scars and the creation of a safe environment for all residents. Sport creates an enabling environment to dissolves all these conflicts, racial, cultural and political differences. For people who are involved in sport, interaction becomes a spontaneous habit that occurs without coercion and exploitation. Hence, social coercion becomes an ultimate achievement (Ibid).

A good example of the use of sport as a unifying factor manifests in former President Nelson Mandela's use of the Rugby World Cup to unite South Africa in 1995. Although rugby was not a priority sport for the former and late president of South Africa at that time, the event presented an opportunity for him to unite people of different racial and cultural background into one nation. Even though the remnants of apartheid were still discernible in political, economic and social spheres, former President Nelson Mandela still courageously decided to wear the Number 6 Springbok jersey which, by the way, was similar to that of the South African rugby team captain, Francois Pienaar (Marais, 2015). As much as nobody knew the outcome of the match, but President Nelson Mandela was fully aware of the significance of the tournament to both South Africans and the international community, as it was capable of uniting the country through playing for peace, simultaneously forging to the future irrespective of the blemished past the country had experienced (Cleary, 2013).

9.4 Research Findings

This section presents the findings of the study as per the objectives which were set to direct the collection of data from respondents. The findings represent the views of the residents, the local soccer team, the Professional Town Planner and the Youth Coordinator for Sports.

9.4.1 Age and Reason for Staying in the Area

Based on the questionnaires that were administered in the study area, about 47% of respondents belong to the youth category (18–35 years of age). About 63% of the respondents had stayed in the area for more than ten years, which suggests quite a long period. A quarter (25%) of the people who stayed in the area indicated that the area is quiet and peaceful, which explained why they chose to stay there. Furthermore, about 19% of the respondents said that they were staying in the area so that their children have easy access to educational facilities. About 16% said they continue to reside in the area because it was reasonably affordable to stay there. About 13% said that they were there because they wanted to own houses. Empirical findings indicates that the residents enjoyed staying in a peaceful place, as that contributes towards stability and sustainability of the settlement. A stable and sustained settlement comes with educational and economic opportunities that allow people to rise above their poverty-stricken conditions. The fact that a high percentage (63%) of the respondents had stayed in the area for more than ten years is a clear indication that educational and economic related opportunities are more important than owning a decent house; hence, people choose to live in informal settlements in order to fulfil their self-development and economic aspirations.

This study found a high level of participation in sporting activities in the area in spite of the inadequate facilities to support the residents, resulting in young people sometimes using streets as their playgrounds. Young people clearly stated that they had access to sporting and recreational facilities only at school and neighbouring settlements.

The study found that children were most affected by the absence of sport and recreational facilities. Based on Table 9.1, about 60% (24) of young people who are supposed to play sports have no access to formal sporting facilities. Family premises and streets are the only options available to young people to play. This has two implications for families and children: House windows are more likely to be broken and could be costly to their families and neighbours to repair. Secondly, the lives of the children who play in the streets are in danger as the risk of being knocked down by cars is high. The provision of designated and safe facilities for children's outdoor activities could make the area safer and more orderly for all residents. The fact that young people indicated the use of home premises and streets as their playgrounds clearly indicates a dire need for sport facilities in the areas.

Another challenge is that those children who are not active (24%), but watch television and or visit friends are susceptible to dangers related to obesity which could lead to other health-related problems for the young generation. A fraction of the participants (16%) indicated that their children were involved in activities like karate, fishing, camping and tennis, but these sporting codes are not found in the study area. This means that children could not attend to these sports as often as they would like to because of financial constraints related to travelling. Those who indicated non-availability of children in the households, stated that their children were attending tertiary institutions, or they were outside Kokstad for various reasons.

Most residents in the study area indicated that there were no sporting and recreational facilities catering for the youth, middle-aged and senior citizens as depicted in Table 9.2. About 64% of the respondents said that there was nothing they could do relating to sporting and relaxation in the area. Instead, most of them watched sport on television and listened to radio. Again, inactive adults reported experiencing health-related challenges. Of the 42 respondents, 10 (24%) indicated that they hardly exercised by jogging, and rarely socialised with friends on social gatherings like braais and small parties which are sometimes held outside the study area. The remaining 12% said that they attended community gatherings and meetings which are taking place outside the study area. The respondents were asked about their knowledge regarding the negative impact that the lack of sport and recreational facilities had on people's lives in their area. This knowledge was important in studying whether people understand the importance of participating in sport activities. The interviewed residents associated the lack of sport facilities in the area with the escalating statistics on crime and drug abuse. This finding is corroborated by those of Oksiutycz

Table 9.1 Children's activities after school

Activities	Frequency	Percentage (%)
Stay at home, play soccer on the street or home yard, watch television, visit friends and play video games	25	60
Do my homework and watch television	10	24
Karate, church, fishing, camping and tennis	7	16
No children in the households	0	0
Total	42	100

Source Author (2023)

and Azionya (2022) who reported that a lack of recreational facilities in a given settlement opens opportunities for young people to engage in illegal activities such as crime, gangsterism and alcohol and substance abuse. The focus-group discussion held with the soccer team in the study area revealed that young people play soccer so that they can become professional soccer players. Their four-day (Tuesday–Friday) practice held in a week helps them improve their soccer skills, enable them to make a career out of soccer as depicted in Table 9.3. This was confirmed by the soccer team coach who stated that his passion for soccer enabled him to see the need for transferring skills and knowledge that he possessed to prepare the youth to participate in the game at professional level. The coach's passion was clear when it emerged that soccer players who stayed far from the practice area were being transported from their various places of residence by the coach's own vehicle.

The study found that about 83% of the respondents knew that sport was a very important component of a sustainable settlement. They were aware that the absence of sport and recreational facilities facilitates the development of anti-social behavioural patterns that shutter young people's prospects in life. A small fraction of the population 5 (12%) did not know that sport is also an important aspect of nation-building. About 2 (5%) of respondents chose to leave the question unanswered because they could not relate the importance of sport to community development. The respondents who were able to relate sport to good behavioural patterns were mostly those who were actively participating in certain sports disciplines like soccer and others. In articulating their lived experiences, they regarded themselves as champions of their own lives.

Table 9.2 Activities by youth and adults in the study area

Activities by youth and adults	Frequency	Percentage (%)
Due to lack of sport and recreational facilities, we watch sports and news on television	27	64
Jogging, doing braais, socialise with friends and family and attend functions outside our area (study area)	10	24
Attend community meetings, attend sporting activities outside our area (study area)	5	12
Total	42	100

Source Author (2023)

Table 9.3 Resident's views on the negative impact of lack of sport facilities

Responses	Frequency	Percentage (%)
Aware	35	83
Not aware	5	12
Did not respond	2	5
Total	42	100

Source Fieldwork (2023)

Table 9.4 Level of consultation on their need for sport facilities

Responses by residents	Frequency	Percentage (%)
There was no consultation	33	79
There was consultation	7	16
Not aware	2	5
Total	42	100

Source Fieldwork (2023)

For any development to be successful, community participation must form part of the developmental plans, approval and implementation processes. The residents of the study area had differing views, with 33 (79%) of them saying that the municipality did not involve them during the planning phase of the project, while only 16% of the residents indicated that the community was consulted by the municipality regarding the need for sport facilities in their settlement. A small percentage (5%) indicated that the municipality rejected the views of the community because the land on which development took place was privately owned. Another point which was raised is that sporting facilities were promised by the municipality before development took place, but a soccer field was later provided far from the study area. The findings in Table 9.4 show that community participation in community-based projects is still low. This is a serious violation of legislative framework and the Constitution of South Africa (1996) which emphasises the participation of people who get affected by projects so that community members contribute towards their development (Republic of South Africa, 1996).

According to the White Paper on Sport and Recreation (2011), municipalities are responsible for the provision and maintenance of sport and recreational facilities in their areas of jurisdiction. The challenge is that municipalities that are located in rural areas lack strong financial muscles and capacity which are needed to maintain such facilities. The reason for this is that rural municipalities do not have a strong base for rate collection and therefore rely on funding from government which, in most cases, is limited. The study went on to find out if residents in Extension Seven were willing to pay for such facilities should they be provided to them. As reflected in Table 9.5, the study found that about 33 (78%) of residents were willing to pay for the services and maintenance of sporting facilities. Most people who responded positively to the question were the same people who said they knew about the negative impact of lack of sporting facilities for the youth in the area.

Table 9.5 Residents' willingness to pay for sport facilities

Reponses from residents	Frequency	Percentage (%)
Yes, we are prepared to pay	33	78
No, we are not prepared to pay	5	12
Did not respond	4	10
Total	42	100

Source Fieldwork (2023)

Residents were willing to actively participate in solving the social ills prevailing in the area. Interestingly, the respondents 5 (12%) who said that their children were at tertiary institutions or outside the study area were the same residents who indicated that they were unwilling to pay for sporting facilities. Understandably so, this indicates that parents can only pay for sporting facilities if they are beneficial to their own children. Participants in the focus-group discussion revealed that they would prefer sporting facilities that would accommodate sports like soccer, tennis, swimming and netball for children, the youth and adults in their area. Most interestingly, the study found that there is a female soccer player who had represented South Africa in the Olympic Games who came from the study area. This female soccer player played for Banyana-Banyana (South African national female soccer team). Another player was a young man who played for Bidvest Wits, which is one of the soccer teams in the Premier Soccer League in South Africa. Evidence shows that many of the players from the study area reach provincial, national and international levels in sports. The results from the focus-group discussion revealed that no consultation of any sporting structure was held by the municipality in the area.

The issue of maintaining sporting facilities came into the spotlight as shown in Table 9.6. It was observed that the municipality does not do maintenance of the existing soccer playground if they have not asked them to. The community felt that the municipality did not take full responsibility of making sure that the playground was always in a good working condition for everyday use by the community.

The results in Table 9.6 indicate that most of the residents (59%) were prepared to establish a committee which was going to deal with the maintenance of the sport facilities in the area because the municipality does not show any interest in doing their job. Again, the people who gave this suggestion were the ones who understood the importance of sports to young people and children. About 7% of the residents indicated that fundraising for such maintenance could assist the community in maintaining the facilities. This is so because they saw a dire need to keep their children away from the street. A fraction of the residents (14%) said that the maintenance of the facilities is the responsibility of the municipality, while others (10%) were prepared to pay a monthly fee and others 25 (59%) indicated that they were prepared

Table 9.6 Approaches for maintaining sporting facilities

Ideas for maintaining facilities	Frequency	Percentage (%)
Establish a committee to maintain and provide security	25	59
Fundraising and hold monthly meetings	3	7
It is municipality's responsibility	6	14
Pay a monthly fee	2	5
Actively involved in maintenance	4	10
No response	2	5
Total	42	100

Source Fieldwork (2023)

to actively participate in the maintenance of such facilities. In summary, about 81% of the residents who participated in the study were willing to play a positive role in ensuring that the facilities were maintained.

9.5 Discussion

9.5.1 The Inclusion of Sport and Recreational Facilities During Layout Design

Based on the objectives and findings of this research, sport should be considered a socio-cultural tool which must be used to reduce social tension and promote reconciliation and reconstruction especially in the post-conflict situations (Giulianotti, 2011), correctly so in the post-apartheid South Africa. Sport can also be used to attain different non-sporting objectives ranging from conflict resolution, gender equity, disease prevention, imparting social values in children, tackling poverty, fighting racism, combating violence or drug abuse and as a form of response to disaster (Coalter, 2010; Giulianotti, 2011; Kidd, 2008). According to the Council for Scientific and Industrial Research (CSIR, 2000), for all settlements to be sustainable, the site for the construction of sport and recreational facilities must be provided. These facilities must be accessible to all members of the community hence their location must be within a walking distance. Given the three spheres of government and their powers and functions, the onus is on municipalities to make sure that this objective is accomplished, irrespective of whether the development is privately, or state driven. Municipalities are responsible for providing final approvals of settlement layout plans before implementation takes place. The sustainability of human settlements should be the ultimate goal that needs to be enhanced through integrated developments. The Breaking New Grounds housing policy was introduced to create sustainable human settlements and to correct the imbalances that were created during the apartheid past. Therefore, layout designs must integrate environmental, social and economic factors to ensure sustainable development in all communities (Nixon, 1992).

Community participation can be defined as the involvement of affected and relevant stakeholders in any form of development that affects their lives, either negatively or positively (Duraiappah et al., 2005). This involves giving an opportunity to members of the community in terms of unlocking ideas and achievable solutions through engaging, promoting and supporting the effective implementation of projects which are aimed at improving their lives. Community participation has two fundamental and immediate benefits: It gives the community an opportunity to drive its own development agenda; secondly, it ensures community buy-in and protection of the development from vandalism (Schiavo, 2021). Given the fact that 79% of the respondents said that the community was not consulted during the design of the settlement at Extension 7, this clearly indicates that government projects are not doing enough to ensure that ideas from members of community are integrated into the decisions that affect them.

9.5.2 Anti-social Behaviour due to Lack of Sport Facilities

South Africa is currently experiencing a high level of crime and a high rate of inequality, a scenario which has been associated with the political and socio-economic transition from apartheid to democracy (Zungu, 2015). Most cities that experience high levels of urbanisation stand a good chance of experiencing higher levels of crime due to lack of adequate economic opportunities as compared to rural areas. So, there is high demand for services and facilities which can assist in keeping young people away from committing criminal activities (ISS, 2001). The increase in anti-social behaviour is related to generational conflicts and the lack of any prospect of decent living conditions (Scheidegger, 2015). In the South African context, this can be attributed to consequences of the discriminative past. For young men, the creation of sub-cultures helps them to attain social status and recognition, but it also generates solidarity and strong social bonds. There is a connection between protest, violence and anti-social behaviour and patterns of exclusion and lacking prospect of decent living conditions. This is so because such social ills are a few mechanisms left to be utilized by the impoverished and destitute masses that has nothing to lose if they were to use them. Making things even worse is the fact that South Africa is currently experiencing a high rate of unemployment which mostly affects young people. Given these challenges, young people find themselves in the state of hopelessness, and they ultimately resort to anti-social behaviours like crime, drug and alcohol abuse and violence. It is for these reasons that sport has become an alternative way of keeping the youth preoccupied. Using sport as a tool for political, economic and social or moral regeneration dates back to previous centuries (Darnell et al., 2019). For instance, sport was used in antiquity in ancient Greece during the Olympic Games. The Olympic truce was a symbol of peace; hence, the wars were stopped during the organisation of the antic Olympic Games. Recently, the South African rugby team won the World Cup, President Cyril Ramaphosa declared the 15th of December 2023 a public holiday for the whole country to unite and celebrate the national team's victory. This gesture indicates strong connections between sport and social transformation (Millington-Kidd, 2019).

9.5.3 Adequate Funding for Sporting Facilities and Payment for Services by Residents

The integration of the human settlement should be dealt with when the project is conceptualised. This is the phase when ideas about the end-product are brainstormed and discussed in full detail. This phase involves the identification of a need for the project, the problem that needs to be solved and opportunities which are to be created by the project (Clements & Gido, 2012). The identification of sectoral departments of the project is done when a plan for the project is developed (planning phase). In this case, such departments should involve the Department of Human Settlements, Education, Health and the Department of Sport and Recreation and other departments which will be invited depending on the concept the project intends to accomplish. The main reason for the identification of these departments is to explain the concept for development and make requests for financial support.

The Department of Human Settlements should be providing housing subsidies specifically for the development of houses. The Department of Sport and Recreation should provide funding for sport and recreational facilities and other departments would provide funding based on their executive powers and mandate. For big and resourceful municipalities like the Metros, such funding could be sourced from within due to various streams of income. The study found that about 88% of the residents in the study area were willing to pay for the services and maintenance of the sporting facilities. This is beside the fact that the construction, maintenance and management of such facilities are the responsibility of the municipality within which such facilities are located (White Paper on Sport and Recreation, 2011). Developers or local authorities must understand that recreational facilities are necessary for ensuring the social inclusivity of people from different ages, backgrounds, classes and races. In the same token, the participation of residents indirectly benefits the government in terms of achieving moral regeneration in communities and nation building initiatives. This corroborates the findings from the focus-group discussion where respondents clearly stated that besides playing for promotion to the professional level, players had other benefits like keeping their bodies healthy and fit and being away from their friends who are involved in anti-social behavioural activities.

9.5.4 Poor State of Service Delivery

There is a need to provide equitable access to expanded public facilities. The availability of appropriate educational, healthcare facilities, clean water, sanitation and sporting facilities, adequate transport to the poor communities are clear demonstration of developing sustainable human settlements. The substantial availability of these services can contribute to augmenting the productivity of poor communities and promoting their full participation in society (Sanbrook, 1982). This calls for the State to intervene in order to remove class biases in access to public services so that resources and services could be redistributed among powerful social classes and those who are still disadvantaged and vulnerable. This would also be responding to the principles of the basic-needs approach which, by its nature, has offered a valuable vision of an alternative society and international order which seeks to abolish poverty by reorganising production, distribution, consumption and political institutions (Sanbrook, 1982). For the State to provide basic services to all citizens, it requires that state officials that are efficient, honest and ideologically committed to serving the people irrespective of their social class. Part of the finding was that the layout plan was designed and hugely influenced by private consultants because they wanted to maximise profit; hence, their focus was on housing delivery and not basic infrastructure (Marais, 2015). One is then able to question the presence and influence of the municipal officials who are tasked with ensuring that government policies are properly implemented.

9.6 Conclusion and Recommendations

Based on the findings of the study, it can be concluded that the provision of sport and recreational facilities was not prioritised in the study area. Evidence given by the residents indicates that a site for the construction of a sport facility was made available during the settlement's design phase; however, the sporting facility was constructed in another area which is far from the settlement where the study was conducted. Given the plethora of the existing legislation (the Constitution of the Republic of South Africa, 1996; the Norms and Standards for Sport and Recreation Infrastructure Provision and Management, 2010; the White Paper on Sport and Recreation, 2011) guiding the provision of services and facilities in South Africa, the challenge of interpretation and implementation of legislation still exists. This is worsened by the fact that there is lack of clarity on who should take a leadership role in the provision of sporting facilities in communities, let alone the lack of clarity on who should fund such facilities. The lack of community involvement during the design and implementation of the project created a huge gap between deliverables of the project and what the community wanted. The study revealed that in the absence of sporting facilities, communities experience quite a number of social ills, such as high rate of criminal activities, alcohol and drug abuse, the increase in rape incidents and violence. These social challenges are a barrier to both the principles of sustainable development and the creation of sustainable human settlements, which the Department of Human Settlements and the Government intend to achieve in the post-apartheid South Africa. Therefore, the use of sport could be regarded as one of the vehicles that can hugely break the barriers to economic development, social inclusivity and sustainable development. Some high-profile politicians have publicly asserted that sport has an unmatched role to play in promoting understanding, healing wounds and mobilising support for a social cause (Kofi Annan, the former UN Secretary-General). On the other hand, former President Nelson Mandela clearly stated that:

> Sport has the power to change the world. It has the power to inspire. It has the power to unite people in a way that little else does. It speaks to youth in a language they understand. Sport can create hope where once there was only despair, it is more powerful than the government in breaking down racial barriers.

Given the aforementioned findings, the study therefore recommends that a strong evaluation and monitoring tool be introduced by the government, community concerned and civil society so that all community-based developmental projects, including those of providing sport facilities, are able to meet the needs of the local people. With all the challenges that South Africa is facing, including high rate of youth unemployment, inequality, drug abuse and increasing rate of crime, those in power in terms of developing plans and policies must ensure that programmes that are geared to develop and implement effective, coherent and sustainable policies leading to investment in physical infrastructure such as housing, sport, education to alleviate poverty and build a prosperous society are developed. These actions need to focus on delivering on the objectives of policy, through which the provision of opportunities to deliver, upgrade or maintain facilities that are meant to help communities live a better life. It is also recommended that more effort should be put into the building of community participation and partnerships between low-income groups, the private sector and the government.

References

Apartheid Museum. (2008). *The implementation of Apartheid* (online). Retrieved June 11, 2023, from http://www.apartheidmuseum.org/sites/default/files/downloads/Learners%20book%20Chapter3.pdf

Annan, K. Annan (the then Secretary-General of the United Nations) in United Nations International Children's Emergency Fund. (2004). *Sport and recreation and play. For the children, health, education, equality, protection.* Advanced Humanity.

Brody, J. (2009). *Constructing professional knowledge: The neighbourhood unit concept in the community builders handbook.*

Carmona, M. (2001). *Sustainable urban design—A possible agenda, in planning for sustainable future.* Spon Press.

Charlton, S. (2003). The integrated delivery of housing: A local government perspective from Durban. In P. Harrison, M. Huchzermeyer, & M. Mayekiso (Eds.), *Confronting fragmentation: Housing and urban development in a democratising society.* University of Cape Town.

Cleary, M. (2013) *Nelson Mandela seized the opportunity of Rugby World Cup 1995* (online). Retrieved July 27, 2023 from http://www.telegraph.uk/news/worldnews.nelson-mandela/10140763/Nelson-Mandela-seized-the-opportnity-of-the-Rugby-world-Cup-1995.html

Clements, J. P., & Gido, J. (2012). *Effective project management* (5th ed.). Joe Sabatino.

Coalter, F. (2010). The politics of sport-for-development. Limited focus programmes and broad-gauge problems? *International Review for the Sociology of Sport, 45*(3), 295–314.

Constitution of the Republic of South Africa. (1996). *Chapter 2: Bill of rights.*

CSIR. (2000). *Building and construction technology.*

Darnell, S. C., Field, R., & Kidd, B. (2019). *The history and politics of sport-for-development. Activists, ideologues and reformers.* Palgrave Macmillan.

Department of Human Settlements. (2004). *Breaking new ground: A comprehensive plan for the development of sustainable human settlements.*

Department of Sport and Recreation. (2010). *Norms and standards for sport and recreation infrastructure provision and management.* Vol. 1-Operation and management.

Department of Sport and Recreation. (2011). *White paper on sport and recreation.*

Dewar, D., & Utenbogaardt, R. S. (1995). *Creating vibrant urban places to live.* Rowman and Littlefield Publishers Incorporated.

Dudfield, O. (2019). SDP and sustainable development goals. In H. Collison, S. C. Darnell, R. Giulianotti, & D. P. Howe (Eds.), *Handbook of sport for development and peace* (pp. 116–127). Routledge.

Duraiappah, A. K., Roody, P., & Parry, J. (2005). *Have participatory approaches increased capabilities?* International Institute for Sustainable Development.

General Comment 4 on Article 11(1) (1991) Of the UN International Covenant on Economic, Social and Cultural Rights (ICESCR) (pp. 7–8).

Giulianotti, R. (2011). Sport, peace-making and conflict resolution: A contextual analysis and modelling of the sport, development and peace sector. *Ethnic and Racial Studies, 34*(2), 207–228.

Henderson, K. A. (2012). The contributions of leisure and active creation to health and wellbeing. *African Journal for Physical Health Education, Recreation and Dance*, 52–63.

Institute for Security Studies. (2001). *Crime in South Africa. A country and cities profile.*

Kidd, B. (2008). A new social movement: Sport for development and peace. *Sport in Society, 11*(4), 370–380.

Laverty, A. (2010). *Sports diplomacy and Apartheid South Africa* (Online). Retrieved April 26, 2023, from http://theafricanfile.com/politicshistory/sport-diplomacy-and-apartheid-south-africa

Lyras, A., & Welty, P. J. (2011). Integrating sport for development theory and praxis. *Sport Management Review, 14*(4), 311–326.

Marais, C. (2015). *A need for the provision of sport and recreational infrastructure to promote sustainable human settlements: A case study of Extension 7, Kokstad.* Unpublished masters dissertation, University of KwaZulu-Natal.

Mbalula, F. (2010). *Norms and standards for sport and recreation infrastructure provision and management.* Republic of South Africa.

Meenakasi, S. (2011). *Neighbourhood unit and its conceptualization in the contemporary urban context.* Institute of Town Planners.

Millington, R., & Kidd, B. (2019). The history of SDP. In H. Collison, S. C. Darnell, R. Giulianotti, & D. P. Howe (Eds.), *Handbook of sport for development and peace* (pp. 13–23). Routledge.

Modisane, N. M., Diale, A. J., & Moeti, K. B. (2019). Provision of recreational facilities by local government in South Africa. In *The 4th Annual International Conference on Public Administration and Development Alternatives. A Case of City of Tshwane*, 03–05 July 2019. Tshwane University of Technology.

Nixon, R. (1992). *Apartheid on the run: The South African sports boycott.* Transition issue.

Oksiutycz, A., & Azionya, C. M. (2022). Informal settlements: A manifestation of internal and cross-border migration. In P. Rugunanan & N. Xulu-Gama (Eds.), *Migration in Southern Africa.* IMISCOE Regional Reader. Springer Cham, eBooks.

Perry, C. A. (1929). The neighbourhood unit. In *Regional survey of New York and its environs* (Vol. 7). Committee on Regional Plan of New York and its Environs.

Sanbrook, R. (1982). *The politics of basic needs: Urban aspects of assaulting poverty in Africa.* Heinemann Educational Books Ltd.

Scheidegger, U. (2015). *Transformation from below? White Suburbia in the transformation of Apartheid South Africa to democracy.* Basel Southern Africa Studies 9, Basler Afrika Bibliographien.

Schiavo, R. (2021). What is true community engagement and why it matters (now more than ever). *Journal of Communication in Healthcare, 14*(2).

Sibeko, S. D. (2007). *The provision and utilization of recreation facilities for the youth at Ngwelezane Township, KwaZulu-Natal.* Unpublished dissertation submitted to the Centre for Recreation and Tourism at the University of Zululand.

United Nations International Children's Emergency Fund. (2004). *Sport and recreation and play. For the children, health, education, equality, protection.* Advanced Humanity.

Ware, V., & Meredith, V. (2013). *Supporting healthy communities through sport and recreation programmes: Resource sheet no. 26 produced for the closing the gap clearing house.*

Zungu, J. (2015). *Youth unemployment as an urbanization challenge in the creation of socially sustainable urban settlements: Case of Malukazi Area in eThekwini Municipality*. Unpublished dissertation at University of KwaZulu Natal, School of Built Environment and Development Studies.

Vincent Myeni is a Ph.D. candidate and a Lecturer in the School of Built Environment and Development Studies, University of KwaZulu Natal. He is based in the Discipline of Housing where he teaches Project Management modules. He holds a Master's degree in Housing, Post Graduate Diploma in Human Resources Management and Bachelor of Social Science. He has published on areas of housing and the environment, urban agriculture and housing, climate change and housing, sustainable human settlements development and the urban poor.

Chapter 10
Strategies for Sustainable Public Housing Programmes in Bauchi State, Nigeria

Sulaiman Muhammad Darazo and Aliyu M. Kawu

Abstract Urban residents, particularly government employees, are known to place high value on living in public housing and always welcome any process that can help achieve this goal. In pursuing this objective, Bauchi State Government, in North-eastern Nigeria, has been embarking on public housing provision for over four decades through different strategies. This study observed that the initial housing estates often contend with issues of inappropriate design and cumbersome modalities in accessing the products, thereby curtailing many of the envisioned success and continuity. Detailed surveys (physical, demographic and socioeconomic) and users' satisfaction assessments have shown that successive governments' attempt in engaging other stakeholders in the building industry was equally found to be inadequate and dotted with grave inconsistencies that also reneges on important aspects of infrastructure provision. From the beginning, the mass housing programmes have had modifications that inculcated participatory project execution, ownership transfer agreements and payment arrangements; however, in spite spanning through different shades of leaderships and democratic eras, products of the scheme still lack the expected public acceptance. Detailed assessment of the local social and environmental peculiarities before any wholesale adoption of a particular strategy studied through in this work is recommended. This is in addition to the fundamental exploration of related concepts of adequacy and quality in order to identify similarities and differences of issues paramount in developing procedures, regulations and theories in public housing provision and management.

Keywords Facilities · Government · Infrastructure · Mass housing · Sustainability · Urban management

---

S. M. Darazo
Physical Planning Directorate, Federal Polytechnic, Mobi, Nigeria

A. M. Kawu (✉)
Department of Urban and Regional Planning, Federal University of Technology, Minna, Nigeria
e-mail: aliyukawu@futminna.edu.ng

Centre for Human Settlements and Urban Development, Federal University of Technology, Minna, Nigeria

L. Chipungu et al. (eds.), *Contemporary Models for Pro-Poor Housing in Urban Sub-Saharan Africa*, https://doi.org/10.1007/978-3-032-05489-0_10

10.1 Introduction

Housing as a universal need has acquired diverse approach and policy instruments that recurrent issues and policy problems are global and inherently local (Buckley & Simet, 2016; Mahgoub & Khalfani, 2012; Van Dijk, 2008). In this regard, public housing is associated with descriptive terms such as social housing, state housing, state sponsored housing, welfare housing, non-profit housing, low-cost housing, affordable housing and mass housing (Ilesanmi, 2005; Jinadu, 2007; Kaitilla, 1993; Lux, 2005; Magutu, 1997; Van Vliet, 1990). Public housing is mainly characterized by the supply of housing by Government and Corporate bodies for public use. The concept of public housing in Nigeria dates back to the British colonial era when European Reservation Areas (now Government Reservation Areas, GRA) were created for the colonial masters and other top officials. From this initial beginning, Governments, through public-sector agencies such as the defunct Federal Ministry of Works and Housing (FMW&H), the Federal Housing Authority (FHA) including state housing corporations, are involved in the provision of housing; as practiced globally, for the poor, low-income and the vulnerable groups (Ilesanmi, 2005; Jinadu, 2007; Lux, 2005; Tibaijuka, 2009; Van Vliet, 1990).

In Nigeria, and many developing countries, public housing provision is characterized by complete control over the whole process of housing delivery by the Federal, State and Local Governments and their agencies in order to address housing shortages faced particularly in urban centres (Bolnick, 2018). Government is also a key player in public housing, in order to reach the low income at affordable level to ameliorate or improve the poor housing delivery system, boost the housing stock and thus enhance socioeconomic life of the people (Austen & Pauly, 2012; Boonyabancha & Kerr, 2015; Tibaijuka, 2009). Public housing provision is made up of two distinct activities of the production of housing units and the process of allocation of the housing units produced either directly through government effort—of her agencies, or, indirectly through the creation of enabling environments, such as provision of site and services schemes (Agbola, 1998; Holifield, 2012; Oladapo, 2002).

Studies have evaluated various aspects of public housing in many African and Asian countries including Nigeria (Ahmed, 2000; Oladapo, 2002; Trovalla & Trovalla, 2015). For example, it has been indicated that inadequate capacity of public housing agencies to deliver affordable housing was one of the key challenges of public housing in Nigeria (FGN, 1991, 2006; FMH&UD, 2006). However, this has placed less attention on products of public mass housing by examining residents' (particularly the target groups' preferences), accessibility to urban services as well as the underlying production and management frameworks through a study of the performance of the housing delivery strategy that were used. Further evaluation is also needed on residents' perception of the levels of adequacy of housing products and components in mass housing projects long after occupation. Post occupation evaluations have the overwhelming potentials of a chronological assessment of how users take advantage of the levels of satisfaction (or dissatisfaction) through the various elements of adaptations to various issues as they manifest (Adidi et al., 2021;

Ahmed, 2013; Zubairu, 2018). This has been known to aid methods of measuring the success and failure of housing programmes or plans (Matias et al., n.d.; Pinang, 2011; Potwarka & Wicker, 2021; Turnbull, 2017; Watson, 2014) and applying findings in the development of new public housing programmes in the United States and other developed countries since 1970s (Kantrowitz & Nordhaus, 1980; Lux, 2005; Ukoha et al., 1997). In order to increase housing supply in Nigeria, there is the need to evaluate the different public housing programmes in Bauchi State to identify strategies that work best and under what conditions. This has been known to aid policy and institutional framework formation for public housing programmes and similar interventions at different levels.

10.2 Concept and Rational for Public Housing Evaluation

Housing urban population particularly in developing countries is faced with myriad of issues like; lack of effective coordination among housing agencies, the politicization of housing and housing programmes; increasing constraints in accessing housing inputs and cost of providing infrastructure (Darby, 2010; Jinadu, 2007; Sullivan, 2011; Yuen, 2007). These are also responsible for the uncontrolled hike in cost of public housing, uncontrolled growth of informal housing; increases in property prices that further put house ownership beyond many city residents. This has not only forced government interventions in the sector, it has also necessitated residential satisfaction assessment hitherto based on physical features of housing units (Mohit et al., 2010; Mukhija, 2004; Ukoha et al., 1997); but now encompassing the various housing delivery strategies employed to ascertain the most effective approach to achieve mass home ownership.

Generally, evaluation is seen as a study involving collecting, analysing, interpreting and reporting information on a thing, place, process or event (Oladapo, 2002; Ukoha & Beamish, 1997) used in answering questions about human activities, habits, policy interventions, programmes and projects (Onibukun, 1985, 1990; Okofulure, 1994). Evaluation of social programmes like mass housing usually take two main forms: formative and summative evaluations (Cattaneo et al., 2007; Ilesanmi, 2005; Rivai et al., 2023; Tapager et al., 2022; Yap et al., 2022; Zubairu, 2018). The formative evaluation also called process or progress evaluation, seek to gather information during the process of implementation of the programme in order to address questions of programme implementation and how this relates to the achieved objectives on one hand (Ilesanmi, 2005) and programme activities and clients' satisfaction with the services provided on the other hand; by explaining how, why and under what conditions programmes or projects work or fail to work. In contrast, summative evaluation seeks to gather information on the effectiveness of a programme after it has been implemented (Daigneault et al., 2012; Yoon & Park, 2015). It seeks to answer the question of the relationship between the goals of a programme and its outcomes, measures how a programme works, its effectiveness and ways of improvement (Ahmed, 2000; Awotona, 1990; Mustapha, 2002; Obeng-Odom, 2009).

10.3 Public Housing and Bauchi State Housing Policy

The creation of the old Bauchi State in 1976 coincided with the period of Nigeria's Third National Development Plan (1975–1980) that witnessed federal government's active involvement in public housing provision (FGN, 1991, 2006). Consequently, Bauchi State and other states of the federation benefited from the first ever government assisted self-help housing programme and were further mandated to build 4,000 housing units each (Ilesanmi, 2005; Mustapha, 2002).The State also participated in the implementation of the National Low-Cost Housing Scheme of the Fourth National Development Plan (1980–85) which was in pursuant to the National Housing Policy objective of providing affordable housing to the low-income earners through direct construction of housing by Federal Government of Nigeria, but did not record remarkable achievement across the country (Awotona, 1990; Mustapha, 2002; Onibukun, 1985). Locally, one of the notable housing programmes implemented in Bauchi State then was the National Site-and-Services Programme launched in 1986 by the Federal Military Government that attempted to provide about 20,000 serviced plots in 20 States of the federation, including Bauchi State.

Prior to 1999, Bauchi state did not have a housing policy of its own. However, the Committee on Bauchi State Economic Empowerment and Development Strategy (BASEEDS) formed in 2003 among others came out with a Development Policy on Housing for the State which, is to 'ensure that all interested people in Bauchi State own or have access to decent, safe and healthy housing accommodation at affordable cost by the year 2007'. The policy was based on the need to develop housing schemes that would ultimately create employment opportunities, generate wealth and provide shelter for the people. It will also improve on the urban landscape of the State and improve socio-economic development and environmental sustainability. These factors informed the Bauchi State's strategies for housing development that was generally out to achieve: the construction of additional 2,000 units of houses and completion of work on existing 1,000 units of houses in partnership with the Local Government Councils on need and affordability basis. The strategy is also into the facilitation of mortgage schemes to enable people take advantage of the owner-occupier housing schemes.

In order to encourage private participation and economic development, the State housing policy also involved the design and implementation of Bauchi State building code to streamline urban development in the state. The privatization of refuse management, and introduction of facilities maintenance fee at the different government housing estates, were among the new policy issues. The State policy also includes the provision of conducive environment for the facilitation of private housing development through access to credit and serviced land (Ilesanmi, 2005; Jinadu, 2007; Lux, 2005). The State's housing policy was among a number of issues that necessitated the adoption of the different approaches used in realizing the objectives of public housing in the State through the strategies of: Government Direct Construction of houses, Purchase of Built residential housing estates and lately Public–Private Partnership (PPP) programmes.

By relying on, and assessing the completed housing units, this study shows the need for re-assessment and further evaluated the outcomes in relation to the specific goal of public housing in Bauchi State. As a summative evaluation; similar to the works of Taher (2001) and Obeng-Odom (2009) on public housing in Egypt and Ghana, respectively; this study adopted the Goal-based model—an approach of the summative assessment, to examined the extent to which public housing in Bauchi State over the last forty years has achieved set goals and objectives based on post-occupancy evaluation of products of the strategies employed.

10.4 The Study Area

The Old Bauchi State, comprising the present Bauchi and Gombe States, came into being in 1976 with the present Gombe State carved out in 1991 leaving behind the present Bauchi State with a total land area of about 49,119 km^2, representing about 5.3% of Nigeria's total land mass and a population estimate of 6,537,314 people in 2016 (NPC, 2006). The present Bauchi State is on two distinct ecological zones of Sudan Savannah at the southern part and the Sahel Savannah (semi-desert) characterized by isolated stands of thorny shrubs and sandy soils covering the western and the northern parts (Ahmed, 2000; James et al., 2022; Mohammed et al., 2021). Bauchi State is tremendously rich in tourism potentials like the famous Yankari Game Reserve which is perhaps Nigeria's leading tourism attraction hosting several African wildlife on a land of about 2,224 km^2; and host to largely untapped mineral resources like cassiterite (tin ore), limestone, gypsum, antimony, iron ore, gold, marble, columbite and zinc and other precious stones (Ahmed, 2000; Hussaini, 2013; Kaltho, 1982).

10.4.1 Sampling and Limitations

Qualitative and quantitative data were used to achieve the objectives of this study by evaluating the nature and types of public houses, housing policies implemented, institutional frameworks and strategies employed in the programmes. Results of field observations, structured questionnaire administration and interviews were useful in the evaluation of the beneficiaries' satisfaction of the programme, and, the assessment of the agencies involved with additional use of structured interviews with officials of Bauchi State Ministry of Lands and Housing (MOL&H), Yankari Savings and Loans Limited (YS&L) and Bauchi Investment Corporation (BIC) Limited. While cluster sampling was useful in accessing the housing units on the basis of the strategy for construction; purposive sampling was used on the respondents in the different agencies involved in the planning and execution of the State-wide housing projects. The sampling frame consists of 1,123 completed and occupied housing units in the twelve (12) public housing estates developed between 1976 to 2016 (that is about

90.34% of the 1,243 planned housing units). This study is limited to the public housing programmes initiated by the different governments of Bauchi State between 1976 to 2016 through three key strategies—Direct Construction, Purchase of Built Houses and Public–Private Partnership (PPP). Using Two hundred and ninety-one (291) structured questionnaires as instrument for data collection and assessment; the evaluation was then based on the quality of housing products, and, the beneficiaries' satisfaction—based on the building strategies adopted in implementing the programme and the payment schedules employed.

10.5 Results

10.5.1 Constructed and Purchased Housing Units

The State-wide government financed public housing projects are spread across the different sub-regional headquarters of the state. However, these projects are more and concentrated in Bauchi town which is the state's political and economic capital. In the Bauchi town, the housing estates are widely distributed across the town, but with higher concentration around the main traditional settlements of the metropolis. Many of the housing units were acquired by Bauchi State Government. For example, the Tambari Housing Estate Phase I (THEP1) was purchased by Bauchi State Government in 2004 from Federal Housing Authority. The different housing structures in the estate were at different stages of completion when it was acquired. There are three (3) different categories of housing units in the estate—three bedrooms with court yard (85 units), three bedrooms without court yard (19 units) and two bedrooms of 180 units. In addition, the estates have a primary and secondary schools, and shops on a land of about 35 ha. Unlike others, Gubi Housing Estate (GHE) possessed, in addition to four bedrooms (96 units) and three bedrooms (113 units), primary and secondary schools, shops, a health clinic and a police outpost on a land of about 32 ha (Fig. 10.1 and Table 10.1).

The State Low-Cost Estate (SLCE) was a Federal Government housing programme initiation with collaboration with state governments in 1979 with two hundred and thirty houses (230) constructed along Ahmadu Bello Way on about 12.60 ha of land. While the Water Board Quarters (WBQ) was built mainly for the Staff of the State Water Board comprising eighteen (18) housing units on a 4.4 ha land; the Nursing Quarters Housing Estate (NQHE) was built purposely to house the nurses employed by the state in its sixty (60) units houses on a 5.33 ha of land. Works Quarters (WQ) was built purposely to house the Staff of the State Ministry of works and Transport in 1980 on a land of about 4.12 ha. Raji Quarters (RQ) has three bedrooms of (30) housing units and two bedrooms (30 housing units) and covers a total land area of about 7.42 ha.

Bello Kaliel Housing Estate (BKHE) was also purchased from an individual after completing the houses in the year 1982. The four bedrooms of 9 housing units

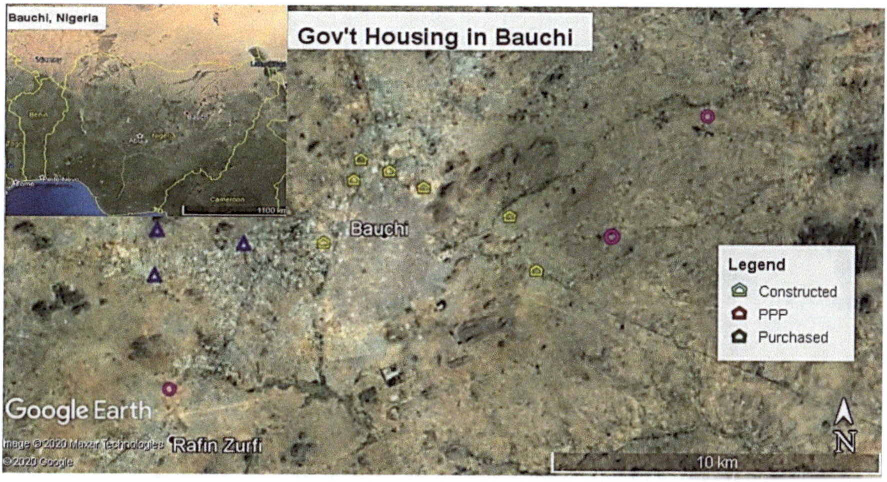

Fig. 10.1 Government housing estates in Bauchi. *Source* Google Earth (2020)

and three bedrooms of 31 housing units sits on a total land of 13.11 ha. Tambari Housing Estate Phase II (THEPII) was also bought over—and the two bedrooms of 120 housing units were constructed through direct construction strategy on a land of about 19.6 ha. Adamu Jumba Housing Estate (AJHE) was bought over by the state government in 1982 and comprise of three bedrooms of 42 housing units within a land coverage of 10.01 ha (Fig. 10.2).

10.5.2 Public–Private Partnership and Public Housing Agencies in Bauchi State

Prior to the year 2008, Bauchi State housing estates were either constructed directly or purchased, but in the Public–Private Partnership (PPP) arrangement, the government engaged ten (10) Private Estate Developers (PEDs) with financial backing from the Federal Mortgage Bank of Nigeria. The State Government also established and involved Yankari Savings and Loans Limited; Bauchi Investment Corporation Limited (BICL); and Bauchi State Ministry of Lands and Housing (MOL&H) to strengthen and support this particular housing programme. By the arrangement, the government in partnership with private developers would construct about 5,000 housing units in Bauchi. The State government then signed about ten (10) memoranda of understanding with the participating private estate developers who were allocated sites for the construction of the houses. This led to the acquisition of land and other properties for the execution of the PPP housing projects. However, compensation was only paid for three (3) out of the sixteen (16) schemes. This was against the Memorandum of Understanding (MoU) for the payment of 10% equity to the private

Table 10.1 Public housing estates in Bauchi

Housing estates	Year built	Built by	Bedrooms/units				Achievement rate			Total land (ha)	Facility and services			
			1	2	3	4	Total built	Total planned	% achieved		Water	Shops	Schools	Health clinic
ATBHE	1977	MW&T			96		96	100	96	8.72				
IBHE	1977	MW&T		1	2	60	63	63	100	21.72		1	1	1
SLHE	1979	MW&T	59	171			230	230	100	12.6				1
WBQ	1980	MW&T		18			18	18	100	4.4				
AJHE	1982	Individual			42		42	42	100	10.01				
NQHE	1982	MW&T		60			60	63	95.2381	5.33				
BKHE	1982	Individual			31	9	40	42	95.2381	13.11				
WQ	1980	MW&T		21			21	21	100	4.12				
RQ	1999	MW&T		30	30		60	60	100	7.42				
THEP1	2004	FHA		180	104		284	1286	22.084	35		1	1	
GHE	2004	FHA			113	96	209	1000	20.9	32		1	1	1
THEP2	2010	FHA & BIC		40	80		120	1038	11.5607	19.6				
PPP	2008	PEDs					0	5000	0	421.1				
Total			59	521	498	165	1243	8963	13.8681	586.4				

Source Field Surveys (2017)

Fig. 10.2 A typical house at Adamu Jumba Housing Estate, Bauchi. *Source* Field Surveys (2017)

developer. Hence, the Private Developers could not compensate land owners after formal acquisition; and eventually the PPP strategy could not yield the expected results.

10.5.3 Post Occupation Evaluations

10.5.3.1 The Beneficiaries of the Housing Units

Post Occupancy Evaluations (POEs) show that over half (51.2%) of the respondents were living in the purchased housing estate; and the remaining 48.8% in the government constructed housing estates. About half (44.57%) of the beneficiaries are between 46 to 59 years of age, with the least proportion of beneficiaries belonging to the younger generation (18–30 years old) that constitutes only 12.28%. Government establishments provided employment to average of 63.45% of the residents. While about 14.43% are self-employed, 4.91% are found to be retired civil servants. Overall, active civil servants are found to be the main residents of the housing estates, while retired civil servants—who are also the least beneficiaries are only found in Gubi Hosing Estate (GHE), Ibrahim Bako Housing Estate (IBHE) and the Water Board Quarters (WBQ). See Fig. 10.3.

With about one-fifths (8.24%) earning between N41,000 to N50,000 and almost half (47.36%) earning above N51,000; majority of the respondents can be said to be above average income earners as only 3.5% are in the low-income group—that earn less than N18,000—which is the country's minimum wage. Almost all (over 83%) of the beneficiaries of the housing estates initially rated the dwelling units as good and appropriate as at the time of occupation. Presently, the dwellings are mainly (62%) owner-occupiers, while renters constitute about 38%; as about 30% of the beneficiaries have lived in the housing estates for over 15 years, only about a

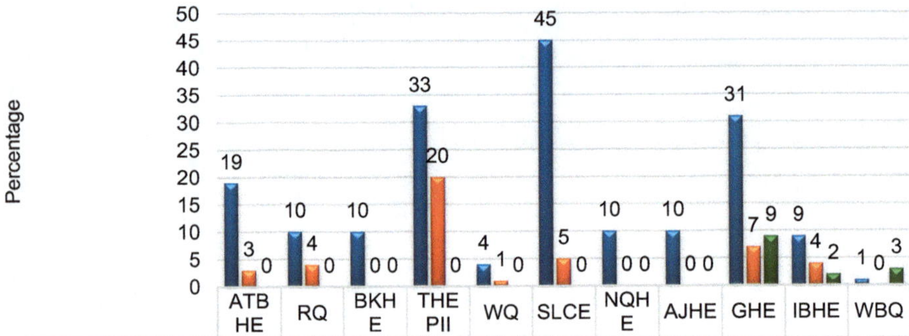

	ATB HE	RQ	BKH E	THE PII	WQ	SLCE	NQH E	AJHE	GHE	IBHE	WBQ
GOVT EMPLOYED	19	10	10	33	4	45	10	10	31	9	1
SELF EMPLOYED	3	4	0	20	1	5	0	0	7	4	0
PENSIONER	0	0	0	0	0	0	0	0	9	2	3

Housing Estates

Fig. 10.3 Employment status of residents of the housing estates. *Source* Field Surveys (2017)

quarter (22.8%) have occupied them between 1–5 years and only 12.63% between 11–15 years.

10.5.3.2 Housing Infrastructure Assessment

There are about three thousand, three hundred and twenty-seven (3,327) planned housing units (of one, two, three and four-bedroom units) but only 1,243 were completed. More so, out of those completed, only 1,123 of the housing units were occupied and 120 are not duly occupied (Table 10.1). The layouts of all the housing estates are on gridiron pattern to ensure easy accessibility, proper numbering and easy location and identification of the houses and the accompanying housing components in each street and unit. The roads in Gubi Housing Estates and Tambari Housing Estates are partly or fully tarred. However, none of the housing estates had walkways for pedestrian movement and virtually all lacked open spaces and landscaped green areas including recreational and sporting facilities, shopping facilities. However, health centres are in Ibrahim Bako, State Low-cost and Gubi Housing Estates; with only Gubi, Tambari and Ibrahim Bako Housing Estates having educational facilities (Fig. 10.4). There are now places of worship in all the housing estates—only two have them initially; and un-occupied shops at Tambari Housing Estate Phase I and Gubi Housing Estate (locally referred to as 'Gida Dubu'). Shopping facilities like make-shift and corner shops are now widely present in virtually all the roads and streets of the housing estates—the usual aftermath of the taken-over by beneficiaries in Owner-Occupier arrangements that often lack effective control building and land use development control in many cities in Nigeria.

Overall, only about 33.68% of the beneficiaries rated portable water availability has been fare (or intermittent), and less than half of that figure (14.73%) rated it

Fig. 10.4 A primary school at Gubi (Gida Dubu) Housing Estate. *Source* Field Surveys (2017)

as adequate (needing storage), while almost a quarter (23.85%) shows that it is inadequate (even stored water would be exhausted before the next supply). Similar assessment goes for electricity power supply in the estates. However, roads and linkages in the estates are well connected due to the gridiron layout patterns.

10.6 Implication of Findings

10.6.1 Performance of the Strategies

The PPP public housing strategy happens to be with the least target achievement percentage of zero, because many of those engaged were unable to go to the allocated sites to execute the projects due to varying reasons. While housing structures built in the 1970s and 80s were able to achieve their target number of houses, those commenced after year 2000 have virtually not achieved even a quarter of the target number of houses. Example of this is the THEP2 with only 120 (11.5%) completed out of the 1038 intended housing units. It is only those build in the 1970s/80s that achieved 100% or at least 95% and they are most acquired after been built by the Federal Government agency or individual or private investors at different capacities and time.

Although, over a thousand houses have been built and long occupied, the number of beneficiaries is still small compared to the State's work force of over 33,605 people or the individuals and families that are in need of public housing in the metropolis in particular and the State in general. This increasingly widening unmet needs gap could be due to a number of issues. First, are the complex and cumbersome modalities for securing loan from FMBN through the Primary Mortgage Institutions (PMIs). In 2007, government's pledge to construct over 5,000 housing units through PPP Scheme was completely derailed due to government's renege on the MoU by: not providing the 10% equity to the developers; non-provision of infrastructure on the sites, and, non-payment of compensation to the land owners. The PPP arrangement might have also become cumbersome to the government that is supposed to fund and monitor it. And, in such arrangement, it would be very difficult for the State Government—who is also a party to the arrangement, to objectively monitor and evaluate progress of a project that it has derailed in paying her own counterpart funding from the beginning.

Post occupation assessment revealed serious alterations and lack of effective control of land uses in the housing estates. For example, there is the continuous change of use of land as plots set aside for social facilities and services (like schools, clinic, police outpost, playground, etc.) in the estates were parcelled and subdivided into plots for the ever intruding commercial and residential uses as against what they were initially intended for. This brought in serious issues of incompatibility of land uses in addition to growing congestion, and ineffective land use control.

User satisfaction in regard to the needed and accompanying housing facilities was found to be low. The houses were also found to be quite expensive and beyond the reach of the low-income and even the mid-income earners, just as many also fail to meet the expectations and requirements of the target beneficiaries. For example, a one-bedroom house usually allocated to a family of six (6) persons is grossly inefficiently and can hardly bring the much-needed satisfaction for adequate shelter. High interest rates on loans from the commercial banks and the inability of the FMBN to provide the needed adequate mortgage coverage for the facilities have also contributed to heightened low satisfaction amidst compounded funding challenges for the State as in many instances across the nation.

The State lack elaborated and all-encompassing State-wide Housing Policy and functional State Housing Authority in charge of formulating housing policies, coordinating and supervision for effective implementation of public housing programmes and schemes. This gap has further widened the official bureaucratic bottle neck that have continued to hinder needed rapid provision of public housing in the different towns in the state where housing provision was embarked on. For example, there is little avenue to address the adopted criteria and or requirements for granting and accessing loans for housing from the National Housing Fund (NHF); although, it has been proven—by different bodies and agencies, not to favour the low-income earners. The absence of these highly important bodies and organs of government also means that there was little or no attention or consideration for other means of encouraging housing construction like; Site and Services Provision—where government provide basic infrastructure for the layouts and the beneficiaries build the structures—to their own taste, composition and with the required facilities affordable to them. There was no success recorded through PPP approach basically due to government's lack lustre performance in that aspect of the State-wide public housing provision. This further shows the consequences of not having any dedicated agency to effectively handle that otherwise robust aspect of housing delivery system.

10.7 Conclusion and Recommendations

In order to achieve impactful and widespread success in the State's provided shelter, the capital-intensive nature of mass housing projects requires the availability of a large pool of long-term savings and adequate funding through secured budgetary allocation from government. Although, the institutional capacity of the Primary Mortgage Institutions (PMIs) is said to be grossly inadequate, they have enhanced access to building structures by facilitating loans to qualified National Housing Funds (NHF) contributors who are the main targets and beneficiaries of the mass housing schemes. This study has shown that the adoption of different housing delivery strategies by different government agencies in housing provision can result in the provision of varied adequate housing which will in turn lead to improved quality of life and livelihood activities of the beneficiaries and residents of public housing in the study area and beyond.

Bauchi State Government had provided housing mostly for the middle- and high-income earners of the State's civil service; and was also able to realize proceeds from the endeavour. However, the housing estates generally lacked healthcare facilities, recreational and education facilities; that has negative implications on user satisfaction. As field records showed, only IBHE, GHE and THEPI have a number of the needed infrastructure and facilities, virtually all the remaining housing estates are lacking in one or all of these essentials for convenience, liveability, and increased quality of life. This scenario has made residents of the affected areas to further spend large amount to procure and maintain these essential amenities even while paying for the mortgage on the same property.

Beside been cumbersome, the PPP strategy is also least effective in screening out less committed investors—as only two out of the over ten of the investors actually went to the field (without completing a single structure). Virtually all of the participating investors were awaiting government funds in order to commence work. Already, different case studies have shown that PPP is known to be problematic in the area of housing provision (Lohri et al., 2014; Ndayako & Kawu, 2011; Rufai, 2010; Warsen et al., 2018). This has made the strategy less attractive and less satisfying for result-oriented schemes like that of the State-wide mass housing programme by Bauchi State government in Nigeria. This has necessitated the need for a renewed public partnership arrangements and subsequent engagement. A Public–Private-Partnership (PPP) scheme, but, with the institutionalized capacity in public housing to take full advantage of opportunities through collaborations that rely less on government for direct funding by attracting other investments outlays. Such approach would only allocate to the Government or the State what it has and can easily give—like land, policy instrument, manpower in areas of monitoring and evaluation, negotiations in terms of multiple partners that often cut across spatial jurisdictions.

With declining State revenues, the pressing need for housing residents in bourgeoning cities of developing countries will continue to attract attentions. Governments and other stakeholders can meet the demand for decent accommodation through the adoption of an efficient and result-oriented approach. From the three approaches studied, direct acquisition of already build houses (either by higher level government of private investor) has yielded more positive results than the other adopted approaches to public housing provisions. This has brought to the fore the fruitful search for an enduring public housing procurement system that can help in closing the widening housing deficit gap currently bedevilling the entire country and similar communities in the global south. This strategy possessed the tendency to attract only committed investors. Aside of been competitive, the investors would be automatically compelled to bring out durable products in good time in order to gain patronage. This can free government from other costly engagements involved in other strategies that are at the end less satisfactory. This is apart from given the government and urban authorities adequate opportunities of evaluating the project from inception.

The lacklustre performance of government and her agencies in the initiation and implementation of PPP arrangements and the self-built housing strategies has further highlighted the need to embrace direct purchase as the State's strategy for mass housing provision. This strategy has proven to be the only working approach that have yielded higher result in the last four decades, and it is still available with less involvements by the State and her agencies. It is relevant that government redirect investors on the other strategy to embrace this particular one.

Also, important, is the exploration of other alternatives or variants of the direct purchase strategy. With different approaches, varieties of the strategy can be established in highly sustainable manner. For example, based on local socioeconomic and environmental peculiarities, variations can be derived by either; inculcating the provision of site-and-services, reduction or subsidizing the housing products for wider patronage, instituting a contributory scheme that can be supported by government or employers of labour for the intending beneficiaries of the scheme, or a combination of these variants. Also important is the involvements of cooperative and labour organizations formed by the workers who are also the intending beneficiaries of these interventions. This will be a welcome idea for the social organizations as attempt by many of them in the past have met with little success—which is likely to be achieved if government were to engage them in an enduring partnership like mass housing provision for their members through the backing of the government.

Whatever the performance of any of the evaluated strategy, before any wholesale adoption, there is the need for detailed assessment of the different planning approaches available. This should go beyond plan approval and even the cross examination of the architectural designs of housing prototypes in public housing provision in the specific area to be adopted. This would also be in addition to the exploration of how related the concept of housing adequacy and housing quality are for identifying local similarities or differences. These requisite environmental, socio-cultural and political assessments can further help in developing theories in public housing provision that are not only workable and been result oriented, but, are engrained by local socio-cultural and environmental attributes.

Community Involvement in Developing State Housing Programmes and Policies is critical in the provision of mass housing in a growing metropolis like Bauchi.

Equally important is giving support to informal sector partners in housing development through people-oriented strategies that can also lead to the development of mechanisms to make credit available to small-scale stakeholders, provide houses that has higher user satisfaction, and help instil the needed bottom-up approach. Public agencies responsible for housing development should provide licenses and contracts to informal bona fide community associations for the execution of specific tasks in the construction of housing infrastructure and services.

References

Adidi, J. E., Alhassan, M. M., & Adekiya, O. A. (2021). Challenges of housing development in the federal capital city, Abuja. *Environmental Technology & Science Journal, 12*(2), 114–121. https://doi.org/10.4314/etsj.v12i2.11

Agbola, T. (1998). *The role of Non-Governmental Organizations (NGOs) in housing provision in Nigeria*. Ford Foundation.

Ahmed, A. (2000). *An evaluation of the spatial planning input in the Ningi development area of the Bauchi state integrated rural development programme (BASIRDP)*. Ph.D. (URP) thesis, Ahmadu Bello University Zaria. www.abu.edu.ng

Ahmed, A. (2013). Affordable housing supply as a sustainability strategy for the federal capital city, Abuja. In *"De-leveraging urban bias and imbalances in cities"—FCT annual planners day* (p. 14). Yar'Adua Centre, Central Area.

Austen, T., & Pauly, B. (2012). Homelessness outcome reporting normative framework: Systems-level evaluation of progress in ending homelessness. *Evaluation Review, 36*(1), 3–23. https://doi.org/10.1177/0193841X12439704

Awotona, A. (1990). Nigerian government participation in housing: 1970–1980. *Habitat International, 14*, 17–40.

Bolnick, J. (2018). Poverty no longer compounded daily: SDI's efforts to address the poverty penalty built into housing microfinance. *Environment and Urbanization, 30*(1), 141–154. https://doi.org/10.1177/0956247817752234

Boonyabancha, S., & Kerr, T. (2015). How urban poor community leaders define and measure poverty. *Environment and Urbanization, 27*(2), 637–656. https://doi.org/10.1177/0956247815600945

Buckley, R. M., & Simet, L. (2016). An agenda for Habitat III: Urban perestroika. *Environment and Urbanization, 28*(1), 64–76. https://doi.org/10.1177/0956247815622131

Cattaneo, M. D., et al. (2007). *Housing, health and happiness* (Vol. 14).

Daigneault, P.-M., Jacob, S., & Tremblay, J. (2012). Measuring stakeholder participation in evaluation: An empirical validation of the Participatory Evaluation Measurement Instrument (PEMI). *Evaluation Review, 36*(4), 243–271. https://doi.org/10.1177/0193841X12458103

Darby, S. (2010). Smart metering: What potential for householder engagement? *Building Research and Information, 38*(5), 442–457. https://doi.org/10.1080/09613218.2010.492660

FGN. (1991). *The national housing policy*. Federal Republic of Nigeria.

FGN. (2006). *The national housing policy*. Federal Republic of Nigeria.

FMH&UD. (2006). *Promoting adequate shelter and sustainable urban communities in Nigeria*. The Federal Ministry of Housing & Urban Development (FMH&UD), and Petral Digital Press.

Holifield, R. (2012). Environmental justice as recognition and participation in risk assessment: Negotiating and translating health risk at a superfund site in Indian Country. *Annals of the Association of American Geographers, 102*(3), 591–613. https://doi.org/10.1080/00045608.2011.641892

Hussaini, I. U. (2013). Building design practice and household energy use in urban centres in Nigeria: A case study of Bauchi Town. In S. Laryea & S. A. Agyepong (Eds.), *Proceedings of the West Africa Built Environment Research (WABER) Conference 2013*, Accra, Ghana, 12–14 August (pp. 251–261). www.waberconference.com

Ilesanmi, A. O. (2005). *An evaluation of selected public housing schemes of Lagos State Development and Property Corporation, Lagos, Nigeria.* Unpublished Ph.D. thesis, School of Postgraduate Studies, Obafemi Awolowo University (OAU).

James, S., et al. (2022). Potential deleterious effects of paracetamol dose regime used in Nigeria versus that of the United States of America. *Toxicology Reports (Elsevier B.V.), 9*(January), 1035–1044. https://doi.org/10.1016/j.toxrep.2022.04.025

Jinadu, A. M. (2007). *Understanding the basics of housing: A book of notes for students in tertiary institutions.* Jos University Press Ltd.

Kaltho, J. B. (1982). *Accommodation of change in an indigenous Hausa-Fulani settlements: A case study of Bauchi.* Unpublished M.Sc. (URP) thesis, Ahmadu Bello University.

Kaitilla, S. (1993). Satisfaction with public housing in Papua New Guinea: The case of West Taraka housing scheme. *Environment and Behaviour, 25*(4), 514–545.

Kantrowitz, M., & Nordhaus, R. (1980). The impact of post occupancy evaluation research: A case study. *Environment and Behavior, 12*(4), 508–519.

Lohri, C. R., Camenzind, E. J., & Zurbrügg, C. (2014). Financial sustainability in municipal solid waste management—Costs and revenues in Bahir Dar, Ethiopia. *Waste Management (Elsevier Ltd.), 34*(2), 542–552. https://doi.org/10.1016/j.wasman.2013.10.014

Lux, M. (2005). On housing satisfaction among Czech citizens. Sociologicky, Casopis-Czech. *Sociological Review, 41*(2), 227–252.

Magutu, J. (1997). An appraisal of Chaani low-income housing programme in Kenya. *Environment and Urbanization, 9*(2), 307–320.

Mahgoub, Y., & Khalfani, F. (2012). Sustainability of gated communities in developing countries. *Developing Country Studies, 2*(6), 53–64. www.iiste.org

Van Dijk, M. P. (2008). *Urban management and institutional change An integrated approach to achieving* (Vol. 16).

Mohammed, M., et al. (2021). Comparative analysis of environmental justice among urban neighbourhoods. *Journal of Inclusive Cities and Built Environment, 1*(2), 27–40. https://doi.org/10.54030/2788-564X/2021/v1i2a4

Mohit, M. A., Ibrahim, M., & Rashid, Y. R. (2010). Assessment of residential satisfaction in newly designed public low-cost housing in Kuala Lumpur, Malaysia. *Habitat International, 34*, 18–27.

Mukhija, V. (2004). The contradictions in enabling private developers of affordable housing a cautionary case from Ahmedabad India. *Urban Studies, 41*(11), 2231–2244.

Mustapha, I. (2002). Overview of housing and urban development programme since independence. *Journal of the Association of Housing Corporations of Nigeria, 1*(6), 28–30.

Ndayako, F. M., & Kawu, A. M. (2011). 'Beneficiaries' assessment of public private partnership in housing delivery in Minna, Nigeria. *CHSUD Journal—Centre for Human Settlements and Urban Development, Fed. Univ. of Tech, Minna, Nigeria, 2*(June), 41–54.

NPC. (2006). *2006 National population census figures.* National population commission. Minna, Nigeria.

Obeng-Odom, F. (2009). Has the habitat for humanity housing scheme achieved its goal? A Ghanaian case study. *Journal of Housing and the Built Environment, 24*, 67–84.

Okofulure, J. O. (1994). An assessment of government housing programme in Nigeria. In *Housing data collection and analysis in Nigeria, A National Workshop*, Ibadan, 4–5 May.

Oladapo, M. A. (2002). Procurement systems and project organization model for low-cost housing. *Housing Today, 1*(6), 32–36.

Onibukun, P. (1985). Housing finance in Nigeria. In P. Onibukun (Ed.), *Housing in Nigeria.* Nigerian Institute of Social Economic Research (NISER).

Onibukun, P. (1990). A review of the Nigeria housing policy and programmes. In P. Onibukun (Ed.), *Housing in Nigeria.* Nigerian Institute of Social Economic Research (NISER).

Pinang, P. (2011). Measuring satisfaction with student housing facilities Nurul Ulyani Mohd Najib, Nor Aini Yusof and Zulkifli Osman School of Housing, Building and Planning. *American Journal of Engineering and Applied Sciences, 4*(1), 52–60.

Potwarka, L. R., & Wicker, P. (2021). Conditions under which trickle-down effects occur: A realist synthesis approach. *Sustainability, 13*(69). https://doi.org/10.3390/su13010069

Rivai, F. R., Rohman, M. A., & Sumantri, B. (2023). Assessment of social sustainability performance for residential building. *Sustainability: Science, Practice and Policy (Taylor & Francis), 19*(1), 33–45. https://doi.org/10.1080/15487733.2022.2153575

Rufai, H. L. (2010). *User satisfaction study of selected public private partnership projects in Abuja.* MSc (URP) Thesis, Ahmadu Bello University, Zaria - Nigeria.

Sullivan, T. O. (2011). Building sustainable housing markets: Lessons from a decade of changing demand and housing market renewal. *Housing Studies, 26*(1), 182–183. https://doi.org/10.1080/02673037.2011.530489

Tapager, I., Rose, K., & Vrangbæk, K. (2022). Social science and medicine exploring equity in accessing diabetes management treatment: A healthcare gap analysis. *Social Science & Medicine (Elsevier Ltd.), 292*(October 2021), 114550. https://doi.org/10.1016/j.socscimed.2021.114550

Tibaijuka, A. K. (2009). *Building prosperity: The centrality of housing in economic development.* Earthscan.

Trovalla, E., & Trovalla, U. (2015). Infrastructure as a divination tool: Whispers from the grids in a Nigerian city. *City, 19*(2–3), 332–343. https://doi.org/10.1080/13604813.2015.1018061

Turnbull, S. (2017). Democratizing the wealth of cities: Self-financing urban development. *Environment and Urbanization, 29*(3), 237–250. https://doi.org/10.1177/0956247816685985

Ukoha, O. M., & Beamish, J. O. (1997). Assessment of residents satisfaction with public housing in Abuja Nigeria. *Habitat International, 21*(4), 445–460.

Van Vliet, W. (Ed.). (1990). *International handbook of housing policies and practices.* Greenwood Press.

Warsen, R., et al. (2018). What makes public-private partnerships work? Survey research into the outcomes and the quality of cooperation in PPPs. *Public Management Review, 20*(8), 1165–1185. https://doi.org/10.1080/14719037.2018.1428415

Watson, V. (2014). Co-production and collaboration in planning—The difference. *Planning Theory and Practice, 15*(1), 62–76. https://doi.org/10.1080/14649357.2013.866266

Yap, W., Janssen, P., & Biljecki, F. (2022). Free and open source urbanism: Software for urban planning practice. *Computers Environment and Urban Systems (Elsevier Ltd.), 96*(April), Article 101825. https://doi.org/10.1016/j.compenvurbsys.2022.101825

Yoon, J., & Park, J. (2015). Comparative analysis of material criteria in neighborhood sustainability assessment tools and urban design guidelines: Cases of the UK, the US, Japan, and Korea. *Sustainability (Multidisciplinary Digital Publishing Institute), 7*(11), 14450–14487. https://doi.org/10.3390/su71114450

Yuen, B. (2007). Squatters no more: Singapore social housing a profile of Singapore poverty. *Global Urban Development, 3*(1), 1–22.

Zubairu, S. N. (2018). Post-occupancy evaluation: A necessary tool for architects, facilities managers and other professionals in the building industry. In *The 68th Inaugural Lecture of Federal University of Technology, Minna, Nigeria,* 4th October.

Sulaiman Muhammad Darazo has a Bachelor's Degree in Urban and Regional Planning (Honours). He worked for the Bauchi State Government Civil Service from May 2001 to early 2010 before joining The Federal Polytechnic, Mubi in Adamawa State, northeastern Nigeria where he is presently the Chief Town Planning Officer and Chief Campus Planner in the Institute's Physical Planning Directorate. He is a Corporate Member of the Nigerian Institute of Town Planners (NITP) and a Registered Town Planner with Town Planners' Registration Council of Nigeria (TOPREC).

Aliyu M. Kawu is Associate Professor, Department of Urban and Regional Planning, and The Director, Centre for Human Settlements and Urban Development (CHSUD), Federal University of Technology, Minna (FUT Minna), Nigeria. He attended Ahmadu Bello University, Zaria, Nigeria where he obtained B.Sc. (Geography), and both M.Sc. and Ph.D. in Urban and Regional Planning. He joined FUT Minna in September 2006 as a full-time academic staff with interest and publications in the fields of urban management and infrastructure finance; regional development; urban groups and self-help organizations; and, African urbanization. He is a member of the Association of Nigerian Geographers (ANG), Nigerian Institute of Town Planners (NITP), and a registered town planner and teacher with the Town Planners' Registration Council of Nigeria (TOPREC) and Teachers' Registration Council of Nigeria (TRCN) respectively. He is currently the Editor-in-Chief of the Journal of Human Settlements Research and Development—a bi-annual journal published by the CHSUD, FUT Minna.

Chapter 11
Urban Gatekeepers' Perception and Challenges of Planning Implementation in North-Central Nigeria

Samuel Medayese, Hangwelani Hope Magidimisha-Chipungu, Edmore Mutsaa, Ummulkhair Hussaini, and Lovemore Chipungu

Abstract The challenges of infrastructure provision and delivery have been a major point of discourse across the global planning space. African cities have also grappled with these infrastructure challenges which ranges from transport infrastructure, healthcare and housing infrastructure among others. Country like Nigeria in West Africa currently has a housing deficit of about 28 million housing units with respect to the over 200 million population. This housing infrastructure challenge is however more system in terms of the causative issues than physical. Most of the statutory agencies of government that are saddled with the responsibility of providing housing and housing infrastructure have oftentimes proposed mass housing programmes which are public private sector driven but which eventually fail as a result of failures on either the part of government to meet its own obligations or sharp practices on the part of the private sector to maximize profit. This research therefore employed a qualitative technique to appraise the overarching challenges of implementing housing and other planning implementation challenges in North-Central Nigeria. The chapter reviewed stakeholders' concerns in terms of the current challenges militating against public–private partnership in providing and ensuring planning implementation in the area, through a direct interview with some of the key stakeholders saddled with the

S. Medayese (✉) · U. Hussaini
Department of Urban & Regional Planning, School of Environment & Technology, Federal University of Technology, Minna, Nigeria
e-mail: medalandgroup@gmail.com; hussainiummulkhair01@gmail.com

U. Hussaini
e-mail: hussainiummulkhair01@gmail.com

H. H. Magidimisha-Chipungu · E. Mutsaa · L. Chipungu
School of the Built Environment and Development Studies, Howard College, University of KwaZulu-Natal, Glenwood, Durban, Republic of South Africa
e-mail: MAGIDIMISHAH@ukzn.ac.za

E. Mutsaa
e-mail: MutsaaE@ukzn.ac.za

L. Chipungu
e-mail: Chipungu@ukzn.ac.za

L. Chipungu et al. (eds.), *Contemporary Models for Pro-Poor Housing in Urban Sub-Saharan Africa*, https://doi.org/10.1007/978-3-032-05489-0_11

responsibilities of ensuring implementation of planning programmes in the region. The chapter identified poor coordination, absence of plans and poor political will on the part of government as the major factors limiting infrastructure provision, while, unearthing the challenge of sharp practices, poor monitoring and implementation and resistance to agreed memorandums of understanding (MoU) as the related issues and concerns for the private sector operators in meeting their obligations of planning infrastructure provision and especially housing. In view of the successes of public-–private partnership (PPP) in housing provision and the innovations which keeps emanating from these sectors in the developed world, it is therefore imperative to conclude that PPP activities in Nigeria especially the North-Central zone of the country requires a complete overhaul to reflect the challenges of housing deficit and planning opportunities in these zone of the country and promoting the values of adherence of to the integrity of the evolved undertakings for the proposed programme either in housing or other infrastructure provision.

Keywords Stakeholders · Housing · North-Central Nigeria · Housing deficit · Innovations

11.1 Introduction

Planning is a process through which people and organizations utilize proper management of resources and is a precise application of the process of decision-making. Any decision that has an influence on the future or be implemented in the future, is what is known as a planning decision. On the basis of this, planning is defined in different ways; planning is a process carried out to define objectives or results to be achieved; it also is one which is undertaken to ensure the achievement of results; planning is an activity that is done to decide on actions that will affect the future (Victoria et al., 2019). Planning for urban areas should however incorporate extensive adaptation and mitigation approaches for climate change, in a way that ensures a transverse reach to physical scales, jurisdictions and democratic timeframes. All of these activities are done with the aim to bring an improved quality of life for the citizens as the key performance result (Raven et al., 2018).

According to reports from the IMF (2013), infrastructure encompasses all the necessary services, facilities, equipment and devices required for the physical and mental well-being of both individuals and families. In a discussion on the 'Impact of Infrastructure on Economic Development', Kessides (2006) defines infrastructure as durable engineered structures, equipment, and facilities that not only play a role in economic production but also serve households. Crump (2011) characterizes infrastructure as goods and services that may not be directly productive on their own but are crucial for the smooth operation of a comprehensive economy. Infrastructure constitutes a wide array of facilities that enable a city to function efficiently. This encompasses the creation of physical structures and the provision of services by either public or private entities to enhance the effective operation of an urban centre, as pointed out by Nubi and Oyalowo (2010). It forms the foundational framework of physical systems within an urban area and can be described as the underpinning on which development flourishes, as highlighted by Udoudo and Udoidem (1999).

11.2 Planning and Infrastructure Intertwined

Urban planning has been profoundly ingrained in infrastructure and physical planning. An ontological base of this has remained to create urban places out of available spaces vide the introduction and use of infrastructure. Presently, a description of the connection between urban planning and infrastructure may be understood as 'numerous but nonstrategic and incomprehensive', although the bond between the two stays strong. This is partly due to the fact that the profession of planning has left vital roles in infrastructure, to other professions (Kalogeropoulos et al., 2010).

Infrastructure is primary to creating sustainable cities, and adequate planning is essential to creating sustainable infrastructure. At present, systems of infrastructure are usually planned for separately, isolated from one another and isolated from urban planning. This involves countless social, economic and ecological expenses, even bigger than when unplanned, and involves greater risks (Altshuler & Luberoff, 2004). Infrastructure plays an important role in shaping the spatial development of cities at a city wide and more local scale (Klein et al., 2018). Easy movement, convenience for pedestrians, feasibility and rates of public transport, as well as safety for pedestrians are done through the layout and design of local areas (Curtis, 2016).

Conceivably, the most-obvious connection is between planning and infrastructure is the link between transport, land use and city form. At the very centre of this relationship, is the relevance of accessibility for economic activities and development of residential areas, as well as its consequences for land values. An example is that economic processes that require high levels of accessibility, usually cluster around transportation paths and nodes of importance to their users (Boarnet and Haughwout, 2000). Water, electricity, sewerage and telecommunications systems play a role in shaping the spatial forms of cities, predominantly through huge secure investments and available capacity, both of which permit growth in precise areas (Klein et al., 2018).

11.3 Stakeholders in Urban Infrastructure Planning

Yang et al. (2014) explains that the public, civics and communities are referred to as stakeholders in the field of urban development, in various studies. Whatever the terminologies used, the key concepts of the studies pinpoint and explain the participation and interests of the groups or individuals that have a stake in projects regarding urban development. Stakeholders in urban development are individuals or societies who can influence or be influenced by the accomplishment of a project.

According to Stewart (2024), the planning of a city or an urban area is a progression which necessitates the participation of several stakeholders, including local government, businesses, community groups and citizens. The roles of different stakeholders include:

Local Government—This is the major stakeholder with regards to urban or city planning, responsible for developing and implementing long-term sustainable plans. These include land use plans, transportation planning, housing plans and plans for economic development, among other facets of urban life. Local governments also ensure that the developed plans are in tune with the needs and desires of the local community (Stewart, 2024).

Businesses—Businesses also have an important stake in the planning of a city. They are accountable with regards to economic resource provision required to support the city or urban area. Businesses are also responsible for job creation and service provision that profit the local community, while also providing input into the planning process by giving feedback on the plans created by local government (Stewart, 2024).

Community Groups—Community groups have a stake in urban or city planning due to the fact that they are in charge of providing input into the planning process and guaranteeing that the plans developed by local government, are in tune with the wishes of the local community. Community groups can also make available feedback on the plans formed by local government and can assist in ensuring that those plans are implemented in a way that will profit the local community (Stewart, 2024).

Citizens—Citizens are the final stakeholder in the process of planning an urban area, as they are in charge of providing final feedback on the developed and implemented to confirm whether or not they are beneficial to the local community (Stewart, 2024).

There is an unstable stakeholder engagement that explains the reality that democracy in projects of urban development, is more regularly rhetoric than realistic in practice. With the complicated situation of rapidly growing population, huge net migration, changes in global economy, climate change, energy and resource restrictions within a nation, policymakers are facing significant challenges in order to address various interests and objectives, visible among stakeholders (Major Cities Unit, 2010). Stakeholders should be given a seat at the decision-making table with regards to urban development projects, in order to resourcefully obtain a perfect understanding of stakeholders' concerns, and efficiently manage conflicts between stakeholders (Robinson, 2005).

11.4 Urban Infrastructure Planning and Plans

Urban planning is a field of profession that assists urban leaders in transforming a vision of sustainable development into reality by utilizing available space as an important resource for development, while engaging various stakeholders. Good urban planning plans towards medium- and long-term goals, using resources to achieve them. Proper urban planning ideally distributes economic development within a given urban area, to spread to extensive social purposes (UN-Habitat, 2013). If urban areas are meant to be resilient and sustainable in the future, they must improve physical and established capacities in order to respond to constant changes and uncertainties. Long-term strategies and commitments are required across democratic and political spheres, to establish functional urban areas (Raven et al., 2018).

The urban planning arrangement, measures and methods are effective for as long as they recognize the urban essentials. The mindful shift from a technique-based planning strategy to an approach that focuses on service delivery and quality of life, necessitates both effort and time. Urban planning constitutes 'values' and taking decisions cannot be left to technical specialists or assumed representative leaders. The requirement for taking planning from an opaque closed box to a more clarified transparent medium is imperative (Toutain & Srinivasan, 2006). Locations of facilities have a tendency to have a substantial influence on places of development, where certain densities may or may not occur sustainably and efficiently, over time. Consequently, their locality has to be carefully considered against a wide range of factors, given their impact on spatial growth within an urban area (Klein et al., 2018).

11.5 Challenges of Infrastructure Planning and Urban Development

11.5.1 Global Challenges

Rao et al. (2021) explain that global challenges of planning for urban infrastructure cover a range of problems that relate to urban residence, power, water, business premises, transport, airports, sewerage, railways, bridges, seaports, roads, tourism infrastructure, entertainment, solid waste management, healthcare, communication. A tremendous growth of IT, pharmaceutical industries and manufacturing, has caused the inadequate infrastructure in India to be consumed by them. The poor and downtrodden majorly bear the brunt of inadequate infrastructure. The 'well to do' among the urban populace, manage to get access to basic facilities and amenities on their own or by taking benefit of the urban as well as pro-rich bias of service provision in the towns and cities. It is for the poor to suffer.

Challenges in the United States of America as reported by Kalogeropoulos et al. (2010), follow a long history of urban development inhibited by entrepreneurs using government provided infrastructure. Majorly, American cities developed streetcars, telephones, gas light, telegraphs, electricity and water supply from private companies operating under municipal licences. More recently, utilities such as cable TV, mobile phones, computer networks and other also followed suit.

Benedict and McMahon (2012) give an insight to problems of urban planning in the United Kingdom and United States. An emerging concept known as 'Green Infrastructure' in the UK and the US is that of green infrastructure, which is 'a unified system of green space that preserves natural ecosystem values and functions and offers accompanying benefits to human populations'. Green infrastructure differs from orthodox approaches to open-space planning because it deals with not only conservation values but also actions as well as development of land, growth management and planning for physical infrastructure. This type of infrastructure has a great impact on urban form (sensitive grassland, wetland areas or coastlines), as it usually restricts urban development in these areas, resulting in either further urban sprawl or the need to densify developments. In certain cases, the necessity to preserve green infrastructure can result in increased physical infrastructure costs (Klein et al., 2018).

11.5.2 Regional Challenges (Africa and West Africa)

Infrastructure development faces a multitude of challenges in third-world countries. Typically, demand surpasses supply, and the availability of funds to expedite rapid provisioning is lacking. This creates a significant disparity between the provision of infrastructure and the actual demand, contributing to financial strain across various sectors of leadership. The prevailing political climate dissuades foreign investors, and the government often fails to establish appropriate priorities for infrastructure development. In many instances, proposed projects intended to achieve specific objectives culminate as nonfunctional endeavours, resulting in their abandonment (Oyedele, 2012).

The infrastructural frameworks inherited from colonial rule in Africa facilitated relatively robust economic growth from the early 1960s until the oil shocks of the 1970s. However, from that point until the mid-1990s, an extended period of economic downturn, coupled with increasing interest in regional trade and other economic pacts, prompted shifts in Africa's economic structure. The continuous, steady expansion in household demand for infrastructure, driven by rapid population growth and escalating urbanization rates, contributed to fuelling this demand independently of alterations in the economic production structure (Estache, 2010).

Across West Africa according to Aruofor (2015), infrastructure development challenges present disturbing similarities such as inadequate project preparation, unresolved project viability and bankability issues, deficient governance frameworks and insufficient funding. While comprehensive project planning is perceived as costly in the short term by project sponsors, the shortcuts often taken result consistently in costly project failures and improperly designed or executed capital ventures. Numerous West African nations have also struggled historically with navigating projects through public procurement processes. Capacity and expertise gaps within the public sector, coupled with perceived transparency and governance deficiencies, as well as protracted bureaucratic procedures, collectively decrease investor interest and risk tolerance. Ultimately, insufficient project oversight and the lax enforcement of performance contracts in West Africa exacerbate issues related to infrastructure quality.

11.5.2.1 Challenges in Nigeria

Jiboye (2011) argues that these issues raised by the uncontrolled urbanism in Nigeria include loss of Biodiversity, high population density, risk of epidemic outbreak, security and reduced access to social infrastructure. The lack of sufficient long-term planning for the advancement of both urban and rural areas stems partially from inadequate capacity, limited resources and the absence of up-to-date data required to facilitate effective planning (Jibril, 2019). The Nigerian Federal government has made repeated efforts to implement policies that aim to maximize population growth while also addressing the challenges posed by urbanization. These development policies encompass initiatives such as the National Economic Empowerment and Development Strategy (NEEDS), the Nigerian Vision 20:2020, the Transformation Agenda and the 7-Point Agenda (Oluwatayo et al., 2017). A report by Onwuamaeze (2022) explained that the Lagos Chamber of Commerce and Industry (LCCI) has identified inadequate infrastructure as Nigeria's most significant hindrance to socio-economic progress, second only to corruption.

Wisner (2025) stated that over the years, after corruption, substandard infrastructure has emerged as a substantial socio-economic challenge in Nigeria. Inadequate infrastructure stands as one of the obstacles to our economic development. The condition of our infrastructure not only hampers our nation's economic progress but also severely impacts our societal environment. It has adversely influenced the quality of life for Nigerians, thereby affecting our overall psychological well-being. The deficiency of essential infrastructure across various sectors has hindered the provision of effective services and contributed to a deficit in Nigeria's growth trajectory. Despite multiple attempts by successive governments to address the infrastructure gap, factors like lack of funds, subpar implementation, maintenance practices, corruption, weak fiscal policies and policy reversals have collectively thwarted progress in infrastructure development.

Nigeria's infrastructure deficit is estimated at $100 billion annually, a figure that far exceeds the country's yearly infrastructure budget in recent years. The African Development Bank suggests that a nation's infrastructure spending should constitute at least six percent of its Gross Domestic Product (GDP) to achieve reasonable sustainable development. Unfortunately, Nigeria's infrastructure spending over the past decade has fallen below five percent of GDP, considerably lower than the commitments made by other developing countries. This significant gap has made it challenging for the Nigerian government to adequately address the nation's pressing infrastructure needs (Wisner, 2025).

11.5.3 Challenges in North-Central Nigeria

Adamu et al. (2015) describe the North-Central Nigeria as one of the 6 geopolitical regions (or zones) in the country and comprised of six states; Benue, Kogi,

Kwara, Nasarawa, Niger and Plateau and the Federal Capital Territory (FCT). Nigerian government being the sole provider for housing and infrastructural development in Abuja because the private sector makes provision for only 5% of infrastructure. After the shift of the nation's capital from Lagos to Abuja in 1994, Abuja has seen a drastic rise in population till date. This increase brought pressure on the still infant and developing infrastructure of Abuja city; this is most notably in the area of housing (Alkali, 2005), giving rise to rapidly rising rents, overcrowding, upward leakage of units, shared households, growth of small-scale private sector housing in the outlying settlements of the city, and, in not a few cases, homelessness (Ikejiofor, 1998). This has overstretched the fiscal, technical and managerial capacities of the Federal Capital Development Authority (FCDA) to properly managing the usage of lands, provision of housing and basic infrastructure, maintenance of public services and keeping up with the incessant population growth (Ingiabuna & Uzobo, 2016).

There is no doubt that this rapid urbanization rate has resulted in various economic, cultural and environmental issues. The problems and challenges have created uncontrolled and unplanned cities resulting in millions of urban dwellers living in substandard dwellings mainly slums and shanty towns (Daramola & Ibem, 2010). The massive, unplanned development at the Nyanyan–Mararaba–Masaka growth corridor along the Abuja–Keffi road is a serious issue to contend with in North-Central Nigeria. Urban growth in the area is haphazard with low quality development and the attendance of socioeconomic and environmental problems (Junaid et al., 2022). The emerging spatial conurbation in the axis along the highway is witnessing a high rate of urban growth leading to sprawl development. Rapid rate of urbanization and uncontrolled population growth in the Greater Karu Urban Area (GKUA) adjacent the Federal Capital Territory (FCT) of Nigeria has witnessed accelerated land consumption and haphazard urban development in the recent years (Rikko et al., 2022).

A study by Kuma and Ighalo (2015), therefore concludes that the predominant source of access to residential land is through the informal land markets and developments on such lands have eluded government's effective planning and control. Hence, the pattern of access to land here has contributed significantly to the growing poor urban housing situation in the sampled North-Central States of Nigeria. Road infrastructure in north-central region of Nigeria does not only represent a substantial share of the economy, but also plays an essential role in supporting nearly all aspects of the Nigerian society's operation. According to Mudi and Lowe (2015) North-Central Nigeria has the largest road network in Nigeria being the seat of the Federal Government with an estimated 58,000 km of road network made up of 18,000 km, 15,000 km and 25,000 km for Federal, State and Local Government, connecting villages to cities, the distant with the near and the inter-land with the urban market. However, the nearly exclusive concentration of road infrastructure provision in the hands of the public sector, especially in north-central region of Nigeria, has led to the failures in the delivery of many road infrastructure facilities in the zone (Adamu et al., 2015).

Faced with the declining economic fortunes and dwindling revenue, most state government in the North-Central Nigeria increasingly found it very difficult to keep pace with the development and maintenance of road infrastructure. Hence, the perception of both the Federal and State Government that these economic infrastructures were social services affects the investment in the road infrastructure and consequently the effectiveness of their provision. Similarly, the traditional inefficiency associated with public sector monopolies in charge of the road infrastructure development affects the quality and reliability of services provided by the public sector in road infrastructure development (Adenikinju, 2003). The current state of road infrastructure in north-central region of Nigeria poses a significant problem; and the financing gap has proven to be the thorn in the flesh of the efforts to alleviate the road infrastructure problem in the zone. The Federal and State Government in the zone therefore took a bold step to embark on public–private partnership (PPP) initiatives gear towards tackling the current and impending challenges facing the development of road infrastructure in north-central region of Nigeria.

Energy infrastructures in North-Central Nigeria are inadequate and grid electricity is unable to meet suburban housing electricity demand. The alternative power-supply options proposed by government for the region require appropriation analysis for selection. The two major water resources management problems identified in North-Central Nigeria are: (1) inadequate funds for further agricultural, hydroelectric, navigation and industrial development; (2) poor data collection and banking (Okeola & Raheem, 2016).

11.6 Theory of Urban Infrastructure—World System Theory

The world-system theory is a macro sociological perspective that seeks to explain the dynamics of the 'capitalist world economy' as a 'total social system'. Its first major articulation, and classic example of this approach, is associated with Immanuel Wallerstein in 1974 and 1976. This is Wallerstein's landmark contribution to sociological and historical thought, and it triggered numerous reactions, and inspired many others to build on his ideas. Because of the main concepts and intellectual building blocks of world-system theory, it has had a major impact and perhaps its warmer reception in the developing world (Martínez-Vela, 2001).

Wallerstein's work developed at a time when the dominant approach to understanding development, modernization theory, was under attack from many fronts, and he followed suit. He himself acknowledges that his aim was to create an alternative explanation (Wallerstein, 2000). He aimed at achieving 'a clear conceptual break with theories of "modernization" and thus provide a new theoretical paradigm to guide our investigations of the emergence and development of capitalism, industrialism and national states' (Skocpol & Trimberger, 1977). In reacting to criticisms of modernization theory, Wallerstein outlined a research agenda with five major subjects: the functioning of the capitalist world-economy as a system, the how and why of its origins, its relations with non-capitalist structures in previous centuries, comparative study of alternative modes of production and the ongoing transition to socialism (Goldfrank, 2000; Wallerstein, 1979).

Chirot and Hall (1982) lay emphasis on three major intellectual building blocks of world-system theory, as conceived by Wallerstein: the Annales School, Marx and dependence theory. These building blocks are associated with Wallerstein's life experience and exposure to various issues, theories and situations. World-system theory is in many ways an adaptation of dependency theory. Wallerstein draws heavily from dependency theory, a neo-Marxist explanation of development processes, popular in the developing world. Dependency theory focuses on understanding the 'periphery' by looking at core-periphery relations, and it has flourished in peripheral regions like Latin America. It is from a dependency theory perspective that many contemporary critiques to global capitalism come from.

For Wallerstein, 'a world-system' is a social system, one that has boundaries, structures, member groups, rules of legitimation and coherence. Its life is made up of the conflicting forces which hold it together by tension and tear it apart as each group seeks eternally to remould it to its advantage (Wallerstein, 2000).

11.6.1 Linkage of Infrastructure Theory to Urban Infrastructure Provision

Wallerstein conceptualizes a 'world-system' as a 'world economy', which is interconnected through market dynamics rather than a centralized political authority. Within this framework, multiple regions rely on each other for essential resources such as food, fuel, and security. Additionally, various political entities vie for supremacy without the permanent establishment of a singular dominant centre (Goldfrank, 2000). Notably, Wallerstein's focus was directed away from American politics; instead, he concentrated on non-European regions. His attention was drawn to events unfolding across a geographical span stretching from India to Africa and beyond.

In his own first definition, Wallerstein (1993) said that a world-system is a 'multicultural territorial division of labour in which the production and exchange of basic goods and raw materials is necessary for the everyday life of its inhabitants'. This division of labour refers to the forces and relations of production of the world economy as a whole and it leads to the existence of two interdependent regions: core and periphery. These are geographically and culturally different, one focusing on labour-intensive, and the other on capital-intensive production. The core-periphery relationship is structural. Semi-peripheral states act as a buffer zone between core and periphery and has a mix of the kinds of activities and institutions that exist on them (Skocpol & Trimberger, 1977).

Among the most important structures of the current world-system is a power hierarchy between core and periphery, in which powerful and wealthy 'core' societies dominate and exploit weak and poor peripheral societies. Technology is a central factor in the positioning of a region in the core or the periphery. Advanced or developed countries are the core, and the less developed are in the periphery. Peripheral countries are structurally constrained to experience a kind of development that reproduces their subordinate status (Chase-Dunn & Grimes, 1995). The differential strength of the multiple states within the system is crucial to maintain the system as a whole, because strong states reinforce and increase the differential flow of surplus to the core zone (Skocpol & Trimberger, 1977).

This is what Wallerstein called unequal exchange, the systematic transfer of surplus from semi-proletarian sectors in the periphery to the high-technology, industrialized core. This leads to a process of capital accumulation at a global scale, and necessarily involves the appropriation and transformation of peripheral surplus. In the realm of global politics, several concepts hold significance within the world-system framework. According to Wallerstein's perspective, nation-states are dynamic factors, constituting essential components of the system. These states are leveraged by different class interests, particularly evident in core countries, where powerful classes utilize them to advance their objectives. Imperialism denotes the control exerted by potent core states over weaker peripheral regions. Hegemony refers to a temporary scenario in which a single core state surpasses its counterparts, maintaining equilibrium and enforcing unrestricted trade that aligns with its interests. Nevertheless, hegemony is transitory due to conflicts between classes and the widespread dissemination of technological advancements (Wallerstein, 2000).

In addition, a global struggle exists among classes. The present world-economy displays recurring cyclical patterns, forming the foundation for Wallerstein's classification of modern historical periods. Looking ahead, Wallerstein envisions a future stage where a socialist world government emerges. This alternative world-system is seen as the sole entity capable of sustaining elevated productivity and reshaping wealth distribution by amalgamating political and economic decision-making processes (Goldfrank, 2000).

11.6.2 Merits and Demerits of Urban Infrastructure Planning

Raven et al. (2018) affirm that urban planning and urban design play crucial roles as strategic platforms for effectively addressing both the mitigation and adaptation challenges posed by climate change. These disciplines offer a unique opportunity to go beyond the conventional responsibilities of professionals and policymakers, by incorporating insights from climate science, natural systems and urban structure. This is particularly relevant in the context of compact urban forms, enabling the creation of dynamic, appealing and healthy communities.

Traditionally, urban planning and design have centred on factors such as settlement arrangements, optimized land utilization, close proximity, community involvement, place identity, improved quality of life and urban vibrancy. However, their scope has now broadened to encompass additional principles such as resilience, comfort, efficient resource utilization and ecosystem services. The integration of these principles into urban policies aids in identifying and reinforcing specific measures, performance benchmarks and a more comprehensive set of indicators to gauge urban performance (Raven et al., 2018).

The development of urban infrastructure remains indispensable for holistic progress. Within the context of today's rapidly changing, competitive and globally interconnected economy, areas with robust infrastructure are poised to attract development and investments due to their capacity to provide necessary support (Rao & Hans, 2014). The influence of information and communication technologies (ICTs) on urban form however remains a subject of debate. While emerging trends like remote work are observable, predictions about the decline of cities and central locations have not materialized due to the persistence of agglomeration benefits and the significance of face-to-face interactions (Klein et al., 2018).

11.7 Methodology

Hermeneutics provides the intellectual foundation for phenomenological researchers that use qualitative research methods. Text, an image, written facts, or a discussion call for in-depth study and inquiry (Neuman, 2014). Interpretivists, or qualitative researchers, hold that the outside world can only be accessed indirectly, through the constructions that people have built of it, according to Ormston et al. (2014). As Stiles (2010) argues, interpretivism depends on the individual's ability to make sense of the information, supporting Stiles's claim. People's interpretations of the world are viewed as the primary source of the world's existence. Ardley (2008) further states that interpretivism considers the individual's experience and the linkages between human awareness and objects existing in the natural world.

When it comes to analysing organizational methods and characteristics in terms of rationalist approaches, interpretivism contends that the current study cannot hope to offer anything relevant to social reality (Dainty et al., 1998). There is a lack of consideration of sustainability and liveability performance and qualities in positivist research techniques, according to Dainty et al. (1998). As a result, these methodologies may limit the researcher to simple, 'unproblematic' observations or well-understood phenomena. A phenomenological method was taken into consideration by the researcher. While not philosophically anchored in interpretivism, this study was informed by the phenomenological notion of identifying significant elements of the organization that could disclose the intricacy surrounding their performance.

According to Dainty et al. (1998), reality is both complex and disputable. No two cities are exactly alike; each one is unique. Because the interplay between urban sustainability and liveability is diverse, this research needed to utilize an interpretive method to investigate the links and tensions between urban environmental liveability and the larger view of built environment sustainability. Word frequency queries and close thematic coding (using NVivo 10 with the aid of MS Excel to plot the charts) were used to analyse the transcribed interviews of the stakeholders in the selected North-Central cities of Nigeria. The criteria for the word frequency queries included: synonym word matches, the 1000 most frequently used words and two-letter minimum words.

11.8 Efforts by the Department/Ministry to Promote Infrastructure Provision

Several efforts have been put in place by the various sectors of the government ministries in the selected cities to address the challenges of infrastructure in North-Central Nigeria. These efforts put in place by the various ministries representing the different facets of governance included increased public participation, public education and awareness, continued infrastructure provision, and grants by the supra-national, national and subnational governments (federal, state and local governments). Other efforts included improved access to infrastructure, skill acquisition and leveraging existing donor agencies and partners.

The various stakeholders interacted with, noted that the first effort put in place to address the challenges related to infrastructure provision was developing a model for public participation that would reinforce the commitment of urban residents to utilizing infrastructure correctly and recommend appropriate and desired infrastructure that would be acceptable to all. This was noted in a response by one of the stakeholders from the Ministry of Works who posited that:

> The secret to a successful project in terms of implementation and utilisation is to get the right people to whom the project is targeted to recommend the project of their choice and get involved in ensuring appropriate implementation (Director of Planning, research and statistics Niger State Ministry of Works).

This statement implied that public participation was a critical effort that helped address the issues around poor infrastructure and development activities. As analysed, another measure to deal with the infrastructure deficit in North-Central Nigeria was public awareness and education. Most of the stakeholders interviewed opined that public infrastructure was often overstretched and overused because of use by too many people who did not even need such infrastructure and only desired to use it because it was available. An interviewee in the Ministry of Health in Niger State opined that:

> Our people have this sense of entitlement which often drives the rich and the poor to think that all must freely use public infrastructure…including health, as they were meant to cater to all (Project Manager, Niger State Ministry of Health).

This statement showed that even residents who were economically well to do still preferred to contest for the use of public infrastructure such as healthcare facilities and depriving the poor residents who are supposed to be the rightful beneficiaries of such infrastructure. As such, conducting public education and awareness would help

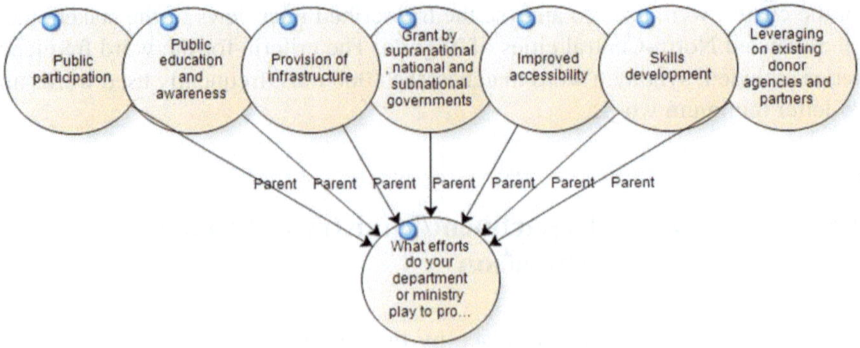

Fig. 11.1 Thematic analysis of the efforts by stakeholders to manage the demand for urban infrastructure development and provision in North-Central Nigeria. *Source* Author's Analysis (2021)

to better inform the people about the need to use alternative infrastructure instead of everyone trying to use the same centrally located infrastructure at the same time and over-stretching the facilities. Figure 11.1 intervention efforts by stakeholders in the region in order to manage and improve access to infrastructure.

The stakeholders interviewed also identified three critical efforts often employed to manage the challenges associated with the infrastructure deficit in North-Central Nigeria. These efforts, they opined, included the conscious provision of critical infrastructure, accessing various types of grants, and leveraging donor agencies. They believed that all of these efforts had definitely helped to improve urban residents' accessibility to infrastructure. They also noted that among the efforts made to address the infrastructure deficit in the region were the efforts towards improving skill acquisition. The stakeholders interviewed believed that all of these efforts made helped to address the challenges associated with the urban infrastructure deficits in North-Central Nigeria.

11.8.1 Availability of Development/Master Plans for the Cities

The use of plans as instruments for development has been emphasized over the years. This research also examined the availability of master plans for monitoring and guiding developments within the cities and as the cities expanded. The findings implied that a considerable percentage of the stakeholders and the state governments in North-Central Nigeria still lacked development plans, which were a critical tool for development (Comfort et al., 2019). Comfort et al. (2019) posited that among the many problems that hampered the implementation of community and social development projects in North Central Nigeria, there were corrupt development officials and community elites, and rural elites heading projects as an avenue for self-enrichment and political gains while ignoring the need for plan preparation and adherence to plans, and this often ended with negative outcomes for the residents.

As development took place on the urban peripheries, planners had to be aware of potential conflict between farmers and herdsmen, as conflict in these areas was problematic. Avoidance of conflict required meticulous planning in conjunction with consultation with those affected by further infrastructure provision. Adisa (2015) recommended the creation of three-tier conflict management committees to help reduce potential conflicts. Adisa (2015) recommended that these committees review the planners' extension plans periodically to ensure increased awareness and compliance with agreed upon stock routes and farming areas, so that crop farmers and herders could be accommodated and kept apart to ensure their harmonious coexistence while also meeting the needs of the growing urban population. Adisa (2015) suggested training of the workers building the infrastructure and clients on how to deal with conflict's psychological and social impacts.

11.8.2 Master Plan/Development Plan Updates

After establishing the availability of development plans in the selected cities in North-Central Nigeria, the research then further interrogated some of the stakeholders' claims of the availability of master plans for the developments in the selected cities by confirming the life spans of the master plans in place and when these plans had last been updated. During fieldwork, it was noted that 40% of the stakeholders interviewed noted that the plans available had not been updated since they were designed. An example was the case in Minna, where the General Manager of the Niger State Urban Development Board noted the following:

> Please let us be clear about this; Maslock developed the available masterplan in 1979, and this is the same plan that we have been using. Government has made several efforts to update this plan, but my view is that it is entirely outdated, and we deserve an utterly new master plan as Minna had extended beyond the 1979 limit when that plan was designed (General Manager, Niger State Urban Development Board, 2021).

The statement above was the submission by the representative of the planning board that was saddled with the responsibilities of physical planning and development in the city of Minna in Niger State. The One-Stop-Shop Land Bureau's Director-General in Lokoja in Kogi State made another profound statement when he claimed the following:

> The issue of a master plan is not applicable in Lokoja city space because we have made several efforts to commission people to design a master plan for the city, not just now but a long time ago, even when we still operated the urban board, but all these efforts ended in severe conflict (The Director-General, One-Stop-Shop Land Bureau Lokoja, Kogi State).

The Lafia State Urban Development Board's representative noted the following:

> Successive governments in this state, since its establishment in 1996, have not even given any thought to preparing a master plan as the state is so poor in its city capital and the significant

> revenue urban centres are those places like Karo and Nyanyan that are near Abuja.... So, they will instead use the money to provide infrastructure in those areas that invest in a master plan that may never be implementable (General Manager, Nassarawa State urban Development Board).

These three statements pointed out that the selected cities in North-Central Nigeria lacked the required development plans and blueprints that could help monitor and guide spatial development. This finding, therefore, implied that attaining urban liveability and sustainability in these cities would be very difficult if these attitudes towards spatial development plans were sustained. However, this also provided policymakers with an excellent opportunity to catch up with development requirements if efforts could be implemented to provide a broad-based inclusive plan for each city that guaranteed liveability for all.

Lloyd-Price et al. (2016) believed that public engagement opportunities should be considered in light of the actual environmental conditions and distinct political and cultural dynamics in place where such engagement was required. These five environmental issues were at the heart of their debate: (1) nature and biodiversity protection; (2) local pollution control; (3) green company image construction; (4) sustainable household practices and (5) participation in international conventions and treaties. Their context analysis examined civil actors' (potential) roles in each of these five environmental issues. Government policymakers often used a functional (objective) definition of liveability to guide their decisions. An analysis of crucial arguments in the literature explored the particular challenges that politicians, urban planners and developers faced in supplying the social (subjective) dimensions of liveability, linked to the community, social interaction and social cohesiveness. As a result of their findings, the authors said that there was a need to understand better how people constructed their sense of liveability in modern cities.

11.9 Discussion of Findings

Findings reveal that there is poor coordination on the part of the government in providing urban infrastructure. This implies that the government has failed to structure the affairs of the areas in a way that ensures quality delivery of infrastructure. Proper monitoring and review of infrastructural projects, as well as adequate implementation are conspicuously absent in the selected cities, which has caused challenges in proper infrastructure delivery. This is in agreement with the UN-Habitat (2016) that many local governments, particularly in small and medium-sized cities in developing countries, lack the organizational structures and authority to undertake infrastructure development initiatives. For example, they might lack properly staffed departments of finance, and the proper mechanisms and regulations for public–private investment are often not in place.

The absence of development plans and master plans are also reported in selected states. This implies that the government has not invested in infrastructure provision and delivery in North-Central Nigeria, which has led to an inability of planning departments to prepare plans that would create a systematic pattern for provision of infrastructure. The haphazard infrastructure development seen in these cities is a direct cause of this. This is the view of Chen et al. (2013) that the single investment of the government has always been the main limitation of the urban infrastructure. Broadening the financing channels and doing investment in the market are the effective ways to encourage operation of the urban infrastructure.

Woetzel and Pohl (2013) also have a similar opinion that robust national and municipal institutions serve as the cornerstone of a prosperous infrastructure programme. They cultivate the necessary skills to sustain initiatives, even as specific experts move in and out of administrative roles. Enhancements in governance, institutions, procedures and internal competencies are frequently overlooked, with officials often lamenting their financial constraints. However, directing attention towards these domains can significantly narrow the funding deficit, empowering cities to extract more substantial benefits from their investments.

Poor political will on the part of government has been found as one of the major factors limiting infrastructure provision in North-Central Nigeria. This implies that the government has poor or no determination in ensuring provision of infrastructure. Due to corruption, some feel there are no financial returns for them as political leaders during this process, while others do not want to start a project that is long-term in a short-term tenure, so as not to give the glory to their successors. This way of thinking has led to the deplorable state of infrastructure in the North-Central. This is in agreement with Aruofor (2015) that another inherent challenge in West African infrastructure development arises from the need to reconcile relatively short 'political life cycles' with the typically longer 'infrastructure life cycles'. Few politicians or decision-makers are willing to invest in infrastructure projects that might not yield results during their time in office. Furthermore, the risk of long-term projects being abandoned after a change in government further complicates the task of aligning political and infrastructure timelines.

With regards to Private Sector Participation, Private Sector's sharp practices are reported as a challenge associated with infrastructure delivery. This implies that the private sector finds 'short cuts' and 'easy ways out' when saddled with provision and delivering on infrastructure in these cities, because it is looking for every means possible to minimize cost therefore maximizing profit for itself. Without proper monitoring of the private sector's activities by the government, this cycle is bound to continue. Substandard infrastructure provision within the North-Central is a direct result of this cycle. In line with this, White and Wahba (2019) are of the view that securing private investment for municipal infrastructure in developing countries necessitates the presence of robust institutional, fiscal and regulatory frameworks, which are frequently lacking. The establishment of these frameworks often mandates policy adjustments and institutional reform, much of which falls outside the jurisdiction and authority of city governments.

Poor monitoring and implementation of infrastructure projects on the part of the private sector was found in the selected North-Central Cities. The implication is that negligent behaviour by the private sectors in these cities has led to the delivery of poor infrastructure in certain areas of the cities, and even non-provision in other areas.

Aarabi and Sarvari (2019) are of a similar view that challenges related to infrastructure development encompass deficiencies such as; limited management expertise and oversight within private firms when it comes to construction projects; contractors' inattentiveness to cost-reduction strategies and the persistence of traditional and less efficient construction methods; insufficient systematic collaboration between the public and private sectors, absence of innovative financial techniques and tools for infrastructure financing.

11.10 Lessons for the Developing Countries

Haphazard, substandard and inadequate infrastructural developments have risen due to inappropriate preparation, planning, funding, monitoring and implementation of development plans, as well as myriad of challenges caused by actions from both the public and private sectors in developing countries. These have directly affected the standard of living and quality of life of urban residents in third world countries. It is therefore necessary to learn better ways of adequately making provision for standard infrastructure in developing nations, some of which include the following:

i. Certainly, various methods can be employed by governments to fund infrastructure development in developing nations. These methods encompass tax increments, municipal bonds and other mechanisms for capturing land value.
ii. The rapid expansion of towns has direct implications for the ever-increasing demand for infrastructure. To ensure that infrastructure development keeps pace with urban growth, substantial efforts must be directed towards strengthening the municipality's financial foundations. Additionally, the municipality should possess the expertise to efficiently oversee infrastructure planning with the active involvement of the community.
iii. Inadequate infrastructure not only hampers economic growth but can also lead to its regression, resulting in higher unemployment rates, increased crime and urban deterioration. It has the potential to exacerbate urban tensions by widening gaps between different ethnic or income groups and between long-time residents and recent immigrants. Furthermore, it can contribute to a general sense of disillusionment that saps a city's vitality and vibrancy.
iv. Municipalities and urban utility organizations undoubtedly have a significant task ahead in enhancing their financial management practices, elevating the quality of their data, maximizing revenue generation, preparing projects that are attractive to potential investors and exploring avenues to enhance revenue security for private investments.
v. Cities in the process of development must adopt a more strategic approach to their infrastructure development initiatives. Elected officials should meticulously assess the relevance of the infrastructure they are constructing, the optimal utilization of existing infrastructure and the efficient execution of projects. They should also weigh the potential advantages that public–private partnerships

(PPPs) can offer. Cities that succeed in advancing their infrastructure projects during the face of current challenges will be those led by individuals who look beyond financial constraints to effectively serve their citizens with the available resources.

References

Aarabi, M., & Sarvari, H. (2019). Identifying barriers to private sector participation in urban construction projects. *Amirkabir Journal of Civil Engineering, 53*(5), 1705–1720.

Adamu, B., Tansey, K., & Ogutu, B. (2015). Using vegetation spectral indices to detect oil pollution in the Niger Delta. *Remote Sensing Letters, 6*(2), 145–154.

Adenikinju, A. F. (2003). Electric infrastructure failures in Nigeria: A survey-based analysis of the costs and adjustment responses. *Energy Policy, 31*(14), 1519–1530.

Adisa, O. (2015). Investigating determinants of catastrophic health spending among poorly insured elderly households in urban Nigeria. *International Journal for Equity in Health, 14*, 1–11.

Alkali, Z. (2005). *The descendants*. Tamaza.

Altshuler, A. A., & Luberoff, D. E. (2004). *Mega-projects: The changing politics of urban public investment.* Rowman & Littlefield.

Ardley, B. (2008). Articles of faith and mystic matrices: Marketing textbooks and the misrepresentation of reality. *Qualitative Market Research: An International Journal, 11*(4), 372–385.

Aruofor I. (2015). *Understanding West Africa's infrastructure potential.*

Benedict, M. A., & McMahon, E. T. (2012). *Green infrastructure: Linking landscapes and communities*. Island Press.

Boarnet, M. G., & Haughwout, A. F. (2000). Do highways matter? Evidence and policy implications of highways' influence on metropolitan development.

Chase-Dunn, C., & Grimes, P. (1995). World-systems analysis. *Annual Review of Sociology, 21*(1), 387–417.

Chen, Y., Li, F., & Wang, H. (2013). *Study on Urban Infrastructure Bottlenecks*. https://doi.org/10.1007/978-1-4471-4844-9_17

Chirot, D., & Hall, T. D. (1982). World-system theory. *Annual Review of Sociology, 8*(1), 81–106.

Comfort, P., Dos', T., Beckham, G. K., Stone, M. H., Guppy, S. N., & Haff, G. G. (2019). Standardization and methodological considerations for the isometric midthigh pull. *Strength & Conditioning Journal, 41*(2), 57–79.

Crump, K. S. (2011). Use of threshold and mode of action in risk assessment. *Critical Reviews in Toxicology, 41*(8), 637–650.

Curtis, C. (2016). Place-making. In *Handbook on transport and urban planning in the developed world* (pp. 321–337). Edward Elgar Publishing.

Dainty, D., Mior, S., & Bereznick, D. (1998). Validity and reliability of an isometric dynamometer as an evaluative teal in a rehabilitative clinic. *Journal of Sports Chiropractic & Rehabilitation, 12*(3), 109–117.

Daramola, A., & Ibem, E. O. (2010). Urban environmental problems in Nigeria: Implications for sustainable development. *Journal of Sustainable Development in Africa, 12*(1), 124–145.

Estache, A. (2010). Infrastructure finance in developing countries: An overview. *EIB Papers, 15*(2), 60–88.

Goldfrank, W. L. (2000). Paradigm Regained? The Rules of Wallerstein? World-system method. *Journal of World-Systems Research*, 150–195.

Habitat, U. N. (2013). *State of the world's cities: Prosperity of cities*. Routledge.

Ikejiofor, U. (1998). Access to land, development control and low-income housing in Abuja, Nigeria: Policy, politics and bureaucracy. *Planning Practice & Research, 13*(3), 299–309.

Ingiabuna, T. E., & Uzobo, E. (2016). Population and development in Nigeria: An assessment of the national policy on population and sustainable development. *International Journal of Development and Management Review, 11*(1), 80–102.

International Monetary Fund. Research Dept. (2013). World Economic Outlook, April 2013: Hopes, Realities, Risks.

Jiboye, A. D. (2011). Sustainable urbanization: Issues and challenges for effective urban governance in Nigeria. *Journal of Sustainable Development, 4*(6), 211.

Jibril, Y. M. (2019). *A meta-analytical review of the relationship between total quality management practices and the operational performance of manufacturing companies*. Doctoral dissertation, Sakarya Universitesi (Turkey).

Junaid, M., Siddiqui, J. A., Sadaf, M., Liu, S., & Wang, J. (2022). Enrichment and dissemination of bacterial pathogens by microplastics in the aquatic environment. *Science of the Total Environment, 830*, Article 154720.

Kalogeropoulos, A., Georgiopoulou, V., Psaty, B. M., Rodondi, N., Smith, A. L., Harrison, D. G.,... Health ABC Study Investigators. (2010). Inflammatory markers and incident heart failure risk in older adults: The Health ABC (Health, Aging, and Body Composition) study. *Journal of the American College of Cardiology, 55*(19), 2129–2137.

Kessides, C. (2006). *The urban transition in Sub-Saharan Africa: Implications for economic growth and poverty reduction*. Cities Alliance.

Klein, J., Araos, M., Karimo, A., Heikkinen, M., Ylä-Anttila, T., & Juhola, S. (2018). The role of the private sector and citizens in urban climate change adaptation: Evidence from a global assessment of large cities. *Global Environmental Change, 53*, 127–136.

Kuma, S. S., & Ighalo, J. I. (2015). Effects of access to land on housing delivery in the North central States of Nigeria. *ATBU Journal of Environmental Technology, 8*(1), 26–36.

Lloyd-Price, J., Abu-Ali, G., & Huttenhower, C. (2016). The healthy human microbiome. *Genome Medicine, 8*(1), 1–11.

Martínez-Vela, C. A. (2001). World systems theory. *Engineering System Division, 83*, 1–5.

Mudi, A., & Lowe, J. (2015). Public-private financed road infrastructure development in north-central region of Nigeria. *Journal of Management and Sustainability, 5*, 58.

Neuman, D. (2014). Qualitative research in educational communications and technology: A brief introduction to principles and procedures. *Journal of Computing in Higher Education, 26*(1), 69–86.

Nubi, T., & Oyalowo, B. (2010, March). Housing finance between social needs and economic realities: The dilemma of policy transfer under neo-liberalism. In *Comparative Housing Research: Approaches and Policy Challenges in a New International Era*, Delft, The Netherlands, March 24–25, 2010. Delft University of Technology, OTB Research Institute for Housing, Urban and Mobility Studies.

Okeola, O. G., & Raheem, K. A. A. (2016). Prioritization of water resources management problems in North central Nigeria using rapid impact assessment matrix (RIAM). *Journal of Water Resource and Protection, 8*(03), 345.

Oluwatayo, A., Opoko, A., & Ezema, I. (2017). Employee participation in decision-making in architectural firms. *Urbanism. Architecture. Constructions/Urbanism. Arhitectura. Constructii, 8*(2).

Ormston, R., Spencer, L., Barnard, M., & Snape, D. (2014). The foundations of qualitative research. *Qualitative Research Practice: A Guide for Social Science Students and Researchers, 2*(7), 52–55.

Oyedele, O. A. (2012, September). Public-private partnership (PPP) and infrastructure provision in Nigeria. In *AARCHES 2012 Biennial General Meeting and Annual Conference*.

Rao, N., Prakash, A., Hans, A., & Patel, A. (2021). Gender, climate change and the politics of vulnerability: An introduction. In *Engendering climate change* (pp. 1–16). Routledge India.

Raven, J., Wurie, H., & Witter, S. (2018). Health workers' experiences of coping with the Ebola epidemic in Sierra Leone's health system: A qualitative study. *BMC Health Services Research, 18*, 1–9.

Rikko, L. S., Laka, I. S., Pam, H. M., Namo, J. A., & Tupkop, N. N. (2022). Residential segregation and implications in Jos North local government area, Nigeria. *African Journal of Geographical Sciences, 3*(1).
Robinson, F. (2005). Results of a survey of performance deviations on major industrial construction projects in Alberta (1990–2003). In *Report to Construction Owner Association of Alberta.* Springer.
Skocpol, T., & Trimberger, E. K. (1977). Revolutions and the world-historical development of capitalism. *Berkeley Journal of Sociology*, 101–113.
Stewart, N. J. (2024). *America's dream reimagined: An exploration of extent of reparations enacted through reparative planning.* Master's thesis, Tufts University.
Stiles, W. B. (2010). Theory-building case studies as practice-based evidence. In *Developing and delivering practice-based evidence: A guide for the psychological therapies.*
ThankGod, N. C., & Onwuamaeze, A. (2022). Social Exclusion in La Greve Des Battu by Aminata Sow Fall: A Sociocritical Reading. *International Journal of Arts, Languages, Linguistics and Literary Studies, 11*(1).
Udoudo, F. P., & Udem, J. O. (1999). Urban infrastructure provision in Nigeria: A critique of the funding strategy. *State Government, 30*(15.8), 32–000.
UN Habitat, United Nations Human Settlements Programme. (2016). *Finance for city leaders handbook* (pp. 194–207). https://doi.org/10.18356/a4b3dafa-en
Unit, M. C. (2010). *State of Australian cities 2010.* Major Cities Unit, Infrastructure Australia.
Victoria, M., Zhu, K., Brown, T., Andresen, G. B., & Greiner, M. (2019). The role of storage technologies throughout the decarbonisation of the sector-coupled European energy system. *Energy Conversion and Management, 201*, Article 111977.
Wallerstein, I. (1993). *The modern world system.* Social Theory: The Multicultural and Classic Readings, 427.
Wallerstein, I. (1979). The Ottoman Empire and the capitalist world-economy: Some questions for research. *Review* (Fernand Braudel Center), 389–398.
Wallerstein, I. (2000). Globalization or the age of transition? A long-term view of the trajectory of the world-system. *International Sociology, 15*(2), 249–265.
Warren, J. A., Berner, J. E., & Curtis, T. (2005). Climate change and human health: Infrastructure impacts to small remote communities in the north. In *Impacts of global climate change.*
Wellmann, D. (2012). The legal framework of state land leases and concessions in the Lao PDR. IRDP, NUDP, GIZ.
White, R., & Wahba, S. (2019). Addressing constraints to private financing of urban (climate) infrastructure in developing countries. *International Journal of Urban Sustainable Development, 11*(3), 245–256. https://doi.org/10.1080/19463138.2018.1559970
Wisner, B. (2025). Disaster vulnerability. *Reflecting on Ben Wisner's contributions to scholarship and scholars.* The Benschrift, 81.
Woetzel, J., & Pohl, H. (2013). *Infrastructure: Doing more with less.* World Bank. http://hdl.handle.net/10986/17599 License: CC BY 3.0 IGO.
Yang, A. S., Wen, C. Y., Juan, Y. H., Su, Y. M., & Wu, J. H. (2014). Using the central ventilation shaft design within public buildings for natural aeration enhancement. *Applied Thermal Engineering, 70*(1), 219–230.

Samuel Medayese brings a great team knowledge in organization, Planning and Logistics gathered over the years as a Lecturer in Urban Planning at Federal University of Technology, Minna Nigeria, Practice in Spatial Planning, and hands-on skill in academic planning. Sam is a Spatial Planner in the true sense with all his degrees to Doctorate level in Urban and Regional Planning. He is a registered Town Planner in Nigeria with the Town Planners Registration Council. Sam has been a part of the SARChI Chair for Inclusive-Cities Hosted by University of KwaZulu Natal under the Chair of Prof. H. H. Magidimisha-Chipungu. Sam Obtained his Bachelor of Technology in Urban and Regional Planning from Federal University of Technology Minna, his Master of

Science in Urban and Regional Planning from University of Ibadan and Doctor of Philosophy from University of KwaZulu Natal, Durban South Africa. Sam is currently a research fellow in the SARChI Chair for Inclusive Cities Team in University of KwaZulu Natal having completed recently one Year Post-Doctoral research. Sam has authored and co-authored over 50 academic papers ranging from journals, book chapters, conference papers, and blog articles.

Hangwelani Hope Magidimisha-Chipungu is a professor in town and regional planning at the University of KwaZulu-Natal. Her record of service spans but not limited to the South African Research Chairs Initiative (SARChI) for Inclusive Cities, National Planning Commission (RSA), Alternate Chairperson Scientific for the South African Council for Planners (SACPLAN), Committee Member for the International Society of City and Regional Planners (ISOCARP), City Planning Commission for eThekwini Municipality, advisory committee of Cooperative Governance and Traditional Affairs (COGTA-KZN), member of KwaZulu-Natal Tribunal and up to late a Board member of SACPLAN, Researcher at the Human Sciences Research Council (HSRC) etc.

Edmore Mutsaa is a Ph.D. candidate and master's graduate in Town and Regional Planning from the University of KwaZulu-Natal. He is an experienced researcher in urban planning, inclusive smart cities and sustainable development, with publications in reputable journals. Edmore has lectured and supervised master's students and participated in various research projects. He is proficient in GIS and Remote Sensing and has nearly a decade of experience in applied research, lecturing, consultancy, and skills development. His accolades include multiple academic excellence awards, and professional awards.

Ummulkhair Hussaini is a registered Town Planner with the Town Planners Registration Council (TOPREC) in Nigeria. She is Assistant Lecturer in the Department of Urban and Regional Planning at the Federal University of Technology, Minna, where she has been teaching and supervising undergraduate dissertations since 2022. She currently is pursuing a PhD in Urban Governance and Poverty Study in the same department. Her interests include environmental management and sustainability.

Lovemore Chipungu holds a Ph.D. in Town and Regional Planning, a Master's Degree in Rural and Urban Planning as well as a Bachelor of Science Degree (Honours) in Rural and Urban Planning. He is an active member of the Zimbabwe Institute of Regional and Urban Planners as well as a Corporate Member of the South African Planning Institute. He is currently an Associate Professor in the School of the Built Environment and Development Studies, at the University of KwaZulu-Natal. Professor Chipungu is both a practitioner and academic. In addition, he is an editorial board member of various journals and the Managing Editor of the Journal of Inclusive Cities and the Built Environment. He has published widely in peer-reviewed journals and books; and his areas of interest are housing policy, land-use planning and urban design. He is currently a co–co-principal investigator on a project on inclusive cities in Southern Africa.

Chapter 12
Inclusionary Housing and Mixed-Use Development as a Strategy to House the Urban Poor: Proposal for a Selected Precinct of Bloemfontein, South Africa

Monaphathi Timothy Lehobo, Stuart Denoon-Stevens, and Johannes Bhanye

Abstract Housing is undoubtedly a huge problem for the urban poor, particularly in developing countries. Sub-Saharan Africa bears the brunt of this housing challenge. This chapter proposes inclusionary housing and mixed-use development to include poor people in housing and development and respond to the need to redress spatial inequalities and imbalances. This chapter is based on a selected precinct in the South of Bloemfontein a secondary city in South Africa wherein a methodical spatial analysis was undertaken with the view to make the most suitable proposal/ plan for the area. The objectives of the proposed plan are to (i) provide for housing opportunities for people of all income classes, (ii) provide for an inclusive, integrated and mixed-use neighbourhood, (iii) facilitate diversification of the local economy to contribute to the reduction of unemployment. The study employed a two-pronged approach involving the collection of both primary and secondary data. The first task was a comprehensive desktop study of various economic, statistical, and spatial data about Bloemfontein. The second task involved visiting the study site and conducting a thorough spatial analysis. Analytical and creative skills were applied to develop a suitable plan for the study area based on all the data gathered and scrutinised. The study reveals that there is a strong lack of housing and supporting infrastructure in the Mangaung Metropolitan Municipality (MMM). The proposed strategy recognises that providing housing alone is not enough to address the broader challenges facing low-income communities, such as access to jobs, education, healthcare, and

M. T. Lehobo (✉)
Department of Urban & Regional Planning, University of the Free State, Bloemfontein, South Africa
e-mail: timothylehobo@gmail.com

S. Denoon-Stevens
Nottingham Trent University, Nottingham, UK
e-mail: stuart.denoon-stevens@ntu.ac.uk

J. Bhanye
University of Cape Town, African Climate & Development Initiative, Cape Town, South Africa

L. Chipungu et al. (eds.), *Contemporary Models for Pro-Poor Housing in Urban Sub-Saharan Africa*, https://doi.org/10.1007/978-3-032-05489-0_12

public services. The study thus argues for and proposes inclusionary housing and mixed-use development as a suitable response to the housing and economic problem in Bloemfontein, South Africa.

Keywords Mixed-use development · Inclusionary housing · Urban poor · Spatial justice · Bloemfontein · South Africa

12.1 Introduction

Housing is a fundamental human right, yet in many cities around the world, the urban poor face numerous challenges in accessing safe and affordable housing (Gouverneur, 2015:1; Onatu & Baloyi, 2020:64; United Nations, 2014:1). Bloemfontein, South Africa, is no exception, as the city continues to grapple with a growing housing crisis, which has led to the proliferation of informal settlements and other substandard living conditions. In response to this crisis, various strategies have been proposed to address the housing needs of the urban poor, including inclusionary housing and mixed-use development (Klug et al., 2013: 668; Nzimande, 2020:3). Inclusionary housing is a policy approach that requires developers to set aside a certain percentage of new housing units for low-income residents, typically through a combination of rent control and subsidies (Calavita & Mallach, 2010:1; Thaden & Wang, 2017:2). Mixed-use development, on the other hand, involves the integration of different land uses, such as housing, retail, and office space, in the same building or complex (Bibri et al., 2020:10; Wardner, 2014:1). Proponents of these approaches argue that they can help address the housing needs of the urban poor by increasing the supply of affordable housing and promoting greater economic and social diversity (Calavita & Mallach, 2010:1; Nzimande, 2020:13).

It is the ultimate goal of spatial planners and human settlement practitioners to create people-oriented, sustainable, resilient, inclusive, equitable and liveable places for people from all walks of life. Good human settlements should be those that improve people's lives in social, economic, environmental and even political contexts. Planners want to transform abstract spaces into places where people can identify, socialise, feel a sense of community and safety, and be economically productive. Based on these convictions, this study set out to propose a spatial plan detailing how best to develop the assigned study area. The chapter proposes inclusionary housing and mixed-use development to include poor people in housing and development and respond to the need to redress spatial inequalities and imbalances. The chapter is based on a selected precinct in the South of Bloemfontein a secondary city in South Africa, wherein a methodical spatial analysis was undertaken to make the most suitable proposal/plan for the area. The proposed strategy seeks to tackle the housing crisis in Bloemfontein from a holistic perspective by promoting affordable and inclusive housing within sustainable urban development. In doing so, it recognises that providing housing alone is not enough to address the broader challenges facing low-income communities, such as access to jobs, education, healthcare, and public services.

The objectives of the proposed plan are to (i) provide housing opportunities for people of all income classes, (ii) provide for an inclusive, integrated and mixed-use neighbourhood, (iii) facilitate diversification of the local economy to contribute to the reduction of unemployment. Ultimately, the chapter aims to contribute to the ongoing debate on how best to address the housing needs of the urban poor in South Africa and beyond. The Literature review focuses on urban regeneration, inclusionary housing, gentrification, and policy and legislation review in the South African context. The next section presents the Area of study—Bloemfontein South Africa, followed by a presentation of the research methodology. The section that follows presents the spatial analysis focusing on the ownership of the land on site, national and municipal policy and legal documents, the socioeconomics of Bloemfontein and the larger MMM, the wider locality of the study site, and the mapping of statistical data as well as environmental data. This is followed by a section on the wider locality of the study area, which looks at the study area within its wider locality, demonstrating how the locality should impact the development proposals and final precinct plan. The section that follows presents the Mapping of Census Data through heat maps to inform the planning and development of the study site. The Synthesis and Spatial Plan section brings everything together, demonstrating how the discoveries inform the final proposed spatial plan. The final section concludes the chapter.

12.2 Literature Review

12.2.1 Urban Regeneration

Couch et al. (2011:3) draw attention to the complexities of urban regeneration, acknowledging the socio-political, economic and environmental aspects that would lead to the regeneration of an urban area. Urban regeneration (alternatively known as urban renewal) is, therefore, "a political reconceptualisation of the inner city as a spatial coincidence of more fundamental social, economic and environmental problems" (Ward, 1994, cited in Couch et al., 2011:3). These authors also show that there are different approaches to urban regeneration depending on who is undertaking it. Therefore, it is informed by the urban policy of any government of the day. The origins of urban regeneration can be traced back to the 1960s and early 70s when the decline of Britain's inner urban areas was first brought to light. It had resulted from an array of factors, including the aftermath of the Second World War and economic recession of the time—all of which led to the realisation that some form of economic and spatial restructuring and policy intervention was necessary. This intervention had to go beyond just the clearance of slums and physical redevelopment but had to address broader social and economic issues. One of the earliest appearances of the word "regeneration" was in the policy document developed for the Merseyside

County Council in 1975. The document outlined a new strategy for the renewal of urban areas. It would give birth to a new field of policy called "urban regeneration" (Couch et al., 2011:3). It would, among others, promote investment in the county to improve ageing housing, encourage economic expansion and reverse the decline in urban areas.

Therefore, an all-encompassing definition of urban regeneration emphasises the sustainable enhancement of people's economic and social life in their localities and neighbourhoods within improved built environments/urban areas. Zheng et al. (2014:272) write that urban regeneration is a "comprehensive integration of vision and action aimed at resolving the multi-faceted problems of deprived urban areas to improve their economic, physical, social, and environmental conditions." As a strategy for urban transformation, urban regeneration is characterised by upgrading target areas in housing, infrastructure, services, mobility systems, and public and private buildings. These target areas are typically dilapidated inner-city zones and deserted industrial or residential premises. Urban regeneration is a multi-dimensional joint effort between city planners, urban designers, economists, architects, engineers, civil servants, entrepreneurs, and even politics. These stakeholders collaborate at local, sub-national, national and occasionally even international levels. Sometimes urban regeneration takes the form of flagship projects as part of a lasting legacy of major sports events (such as the Olympic games or football world cups), artists' works and performances (Investor Trading Academy, 2016: video). This would involve building, among others, stadia and arenas around which an overall urban regeneration project is centred.

12.2.2 Inclusionary Housing

Housing is a basic need (Nahrin, 2018:20), and this applies to everyone regardless of social or economic status. This implies that housing provision should be affordable for and inclusive of everyone. Lerman (2006:383) writes that the term Inclusionary Housing (IH), also known as Inclusionary Zoning (IZ), is an American concept that refers to the attempt to counter former practices by towns and cities in the United States that were designed to discriminate against and exclude low-income residents from "higher income" neighbourhoods and settlements. In other words, it is an approach to housing that seeks to combat the segregation of different income groups and racial groups through exclusionary zoning. IH is also a practice that attempts to remedy the affordable housing problem, and in fact, Lerman asserts that it is the "optimal approach to affordable housing."

"In its simplest form, an IH program might require developers to sell or rent 10 to 30 percent of new residential units to lower-income residents. IH policies are sometimes referred to as IZ because this type of requirement might be implemented through an area's zoning code; however, many programs impose similar requirements outside the zoning code" Jacobus (2015: 7). Jacobus further writes that developers are normally encouraged to implement this approach through the use of incentives.

Furthermore, IH policies are flexible in that they are mandatory in some areas and voluntary in others, depending on their unique conditions. Developers are also given an array of options from which to choose but with the same aim of achieving inclusionary and affordable housing. The Australian Housing and Urban Research Institute (AHURI) (2017: online) defines IZ simply as "a land-use planning intervention by the government that either mandates or creates incentives so that a proportion of a residential development includes a number of affordable housing dwellings". Similar to Jacobus, the AHURI identifies two types of IZ, namely: (a) the mandatory model, under which housing developments *must* include a number of affordable/low-income homes as a condition for planning approval; and (b) the voluntary incentive model which encourages affordable housing through cost reduction for developers. IH is usually implemented at the county/municipal level.

12.2.3 Gentrification

The concept "gentrification" was coined in 1964 in London, by Ruth Glass, a British sociologist in her book "London: Aspects of Change" (Lees, 2019:6; Viney, 2015: online). It was originally a low scale neighbourhood-based process which saw the arrival of middle/upper-class persons into a historically disinvested inner-city neighbourhood, whose residents are low-income/working-class groups, the result of which was the "rehabilitation of properties, an increase in property values, and the displacement of the working classes" (Lees, 2019:6). Van Gent and Hochstenbach (2019:2) describe it as "a process of capital reinvestment in previously poor neighbourhoods, reflected among other things in house price appreciation, rising rents, and increasing owner-occupancy rates." Therefore, gentrification involves a process of neighbourhood change, characterised by (a) change in the demographic and racial make-up of the area (mixture of low-, middle- and high-income residents) (b) upward economic change, whereby there is investment in real estate and arrival of high-income residents, developers, and investors (c) may include changes in land use and character of neighbourhood and (d) is often associated with adverse social and economic effects for lower-income groups. Lately, however, Lees (2019:5–6) argues, gentrification has evolved into a more global process beyond the neighbourhood; comprising largescale urban redevelopment driven and/or induced by the state. Gentrification is not necessarily limited to the redevelopment of historic neighbourhoods but may include newly built modern neighbourhoods. This means that gentrification has taken a broader form on the world stage, and governments have taken a major role when previously it was undertaken primarily by private developers. This reveals similar and crosscutting themes in urban regeneration and gentrification, which are discussed later in the chapter.

Gentrification carries with it the famous debate over whether it is good or bad for communities, or if both, which one outweighs the other. Some see it as benefiting communities in a number of ways, while others perceive it as a disruptive practice that destroys communities and neighbourhoods. Indeed, Butler (2007:162) indicates that for some, gentrification is "a process of colonising the city, for others a manifestation of belonging." On the one hand, it is seen to be a process of urban development, while on the other, it is labelled as "an active and careless process of socio-economic exploitation." This shows a rift between proponents and opponents of gentrification and the two-faced nature of the practice itself. Lees (2019:5) writes that gentrification typically results in the "displacement and disenfranchisement of low-income groups in favour of wealthier in-movers." In fact, the increase in property values and prices normally instigates complaints that gentrification is actually exclusionary and favours wealthy people over the low-income class; this has often led to protests and even vandalism of property by aggrieved residents. This was seen in Berlin in 2015 when protesters claimed that the increased prices marginalised residents and "represented a form of social cleansing" (Viney, 2015: online).

12.2.4 Policy and Legislation Review in the South African Context

The South African Constitution grants every citizen of the Republic "the right to have access to adequate housing" (South Africa. Constitution of the Republic of South Africa, 1996:11); but does not specifically state what form this housing should take, nor how to achieve it. It only assigns government the responsibility to ensure the realisation of this right through "legislative and other measures." Subsequent legislation has therefore shaped and clarified the direction to be taken in this regard. The Spatial Planning and Land Use Management Act (SPLUMA) goes on to mandate developmental norms and standards that "promote **social inclusion**, spatial equity, **desirable settlement patterns**, rural revitalisation, **urban regeneration**, and sustainable development" (South Africa. Spatial Planning and Land Use Management Act, 2013:20). The Housing Act and Social Housing Act take it a step further by both charging government with the responsibility to prioritise minority, poor, and otherwise disadvantaged groups in housing developments and programs (South Africa. Housing Act, 1997:5–6; South Africa. Social Housing Act, 2008:8).

From this legislative stance taken by South Africa, we observe that all pieces are in place for the implementation of urban regeneration, IH and gentrification in the Republic. Firstly, it appears to be primarily the role of government to shape the urban landscape, human settlements and housing provision. This is in line with modern practice in urban regeneration, IH and gentrification, as observed in previous sections that governments have taken a leading role as the scale of their implementation has expanded. Secondly, the legislation, in its wording, specifically mentions urban regeneration and inclusion and prioritisation of disadvantaged groups in housing, human settlements, and overall urban development. With the addition of the necessary resources, South Africa is well on its way to successfully implementing these beneficial urban practices. Therefore, it seems fitting that the City of Johannesburg (COJ) in particular has taken a major step towards IH, recently passing its

Inclusionary Housing Policy that would incentivise private developers to allocate 30 percent of the units in their housing developments to low-income people. The policy has been hailed as a step forward in the spatial planning practice, despite being met with mixed reviews (Mabasa & Mthetwa, 2019: online). However, with low-income units dedicated to families earning at least R7000 a month, the policy has been criticised for failing to account for people with the greatest need for affordable housing, the authors further show.

12.2.5 Linking Urban Regeneration, Gentrification and Inclusive Housing

Having observed the three practices individually, there is a correlation that urban regeneration is about the wider city as a whole; gentrification goes a notch lower and focuses on human settlements and neighbourhoods, while IH focuses on actual housing for every social and economic class of people. Gentrification and IH are linked in the sense that the former could form part of an IH programme whereby residents of different income levels not only move into a given area but have a certain proportion of housing units allocated for each class. This also intertwines with urban regeneration because the redevelopment of historically disinvested neighbourhoods; and the development of new mixed housing projects both result in improvement of urban areas, whether by repairing old damage or instituting new development. Together, these three practices inform a comprehensive, inclusive, sustainable and resilient direction towards urban development, economic development, and social well-being.

Questions have often arisen over whether gentrification is actually urban regeneration in a sense, or whether it is a form of "social cleansing" as Viney (2015: online) alluded to. What is similar about the two and where is the line drawn? Both are similar in that they transform and uplift urban areas, improving them economically, physically and socially. The difference is that; while urban regeneration aims to improve the economic, physical, social, and environmental situation of dilapidated urban areas at a wider city level, gentrification aims to make the same improvements in historically disinvested regions that were never "well off," and at a more local, neighbourhood level. Gentrification also emphasises housing a lot more than other aspects, while urban regeneration is more about improving the entire make-up of the city.

Having noted earlier that gentrification uproots disadvantaged people who in fact are in most need of housing, it seems fitting that IH programs form part of, and effectively work hand-in-hand with any gentrification project. It must be a mandatory requirement that any gentrification project is coupled with an IH policy that guarantees a portion of housing units for low-income groups. This will not only solve the problem of dilapidating inner urban areas but will also improve the local economy,

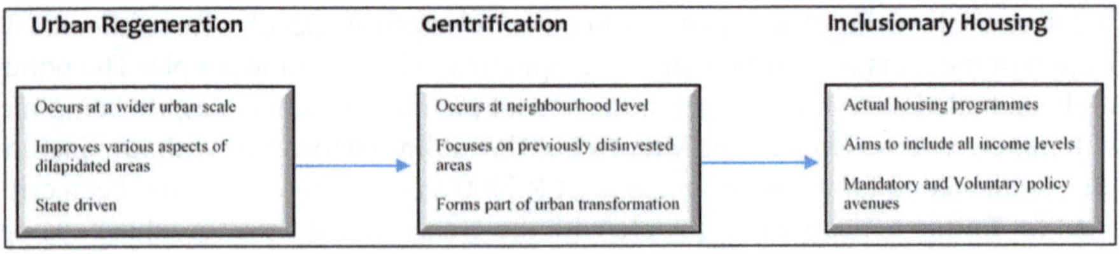

Fig. 12.1 The interlink between urban regeneration, gentrification and inclusionary housing. *Source* Authors' own construction (2024)

while securing and bettering the lives of the poor and maintaining social cohesion all at once. Figure 12.1 summarises this interlinkage.

12.3 Area of Study: Bloemfontein, South Africa

The study area is a largely undeveloped 500-hectare site with the following notable features: it is bordered by (i) low-income residential areas across the road to the east; and higher-income suburbs further west and northwest; (ii) the Hamilton industrial area to the northeast, (iii) Vista Park 2 and 3 to the north—an area planned for future residential use; and (iv) a landfill site and cemetery immediately west. Within the site itself is a section of the Hamilton Industrial area on the north-eastern edge, the University of the Free State (UFS) South Campus and the Technical and Vocational Education and Training (TVET) college along the Oliver Tambo (M30) main road, the UFS experimental farm on the southwestern corner, and a water reservoir atop a hilly area on the north-western edge.

Figure 12.2 shows that the study area (demarcated by a red dashed boundary) is located about 9 kms south of the Bloemfontein central business district (CBD), within the larger the Mangaung Metropolitan Municipality (MMM), in the Free State (FS) province. Figure 12.3 highlights the major notable aspects of the site.

12.4 Research Methodology

The method used for this research project was primarily twofold. It comprised desktop research and the application of analytic and creative skills to learn as much as possible about the study area and its surrounding urban fabric to produce a relevant proposal that responds to the area's unique needs. The desktop research included gathering all secondary data (policy papers, plans, frameworks, and legislation) pertaining to the MMM and Bloemfontein proper; and analysing the municipality's vision for its development. The application of analytic and creative skills included extensive study of statistical and geographic information systems (GIS) data on Bloemfontein and

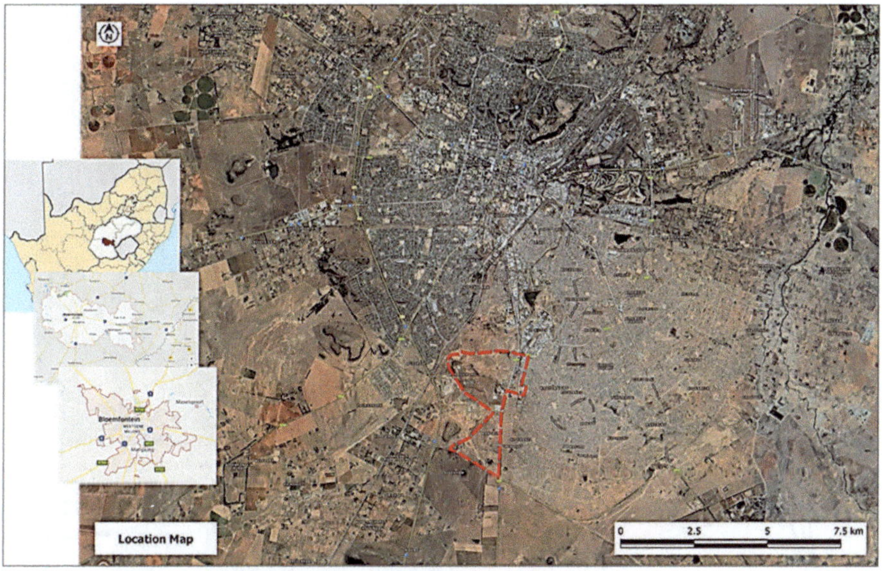

Fig. 12.2 Location of the study area. *Source* Authors' construction from Google imagery (2024)

the wider MMM. This study included manipulating various datasets and presenting such data in tabular, graphical, and map form to analyse it and gain a comprehensive understanding of the area and thus respond accordingly in proposing ideal developments. Thus, in the end, the use of this twofold method (which included analysis of both qualitative and quantitative data) together with a comprehensive literature review and review of other works and papers, frameworks, and regulations on best practices in spatial planning and human settlements in South Africa, the authors were able to arrive at a good proposal for the development of the site.

12.5 Spatial Analysis

This section chronicles an extensive spatial analysis to learn more about the site and its wider locality in order to come up with a plan that responds to the unique conditions ranging from ownership of the land on site, national and municipal policy and legal documents, the socio-economics of Bloemfontein and the larger MMM, the wider locality of the study site, and the mapping of statistical data (which gives insight into the key characteristics of the population) as well as environmental data.

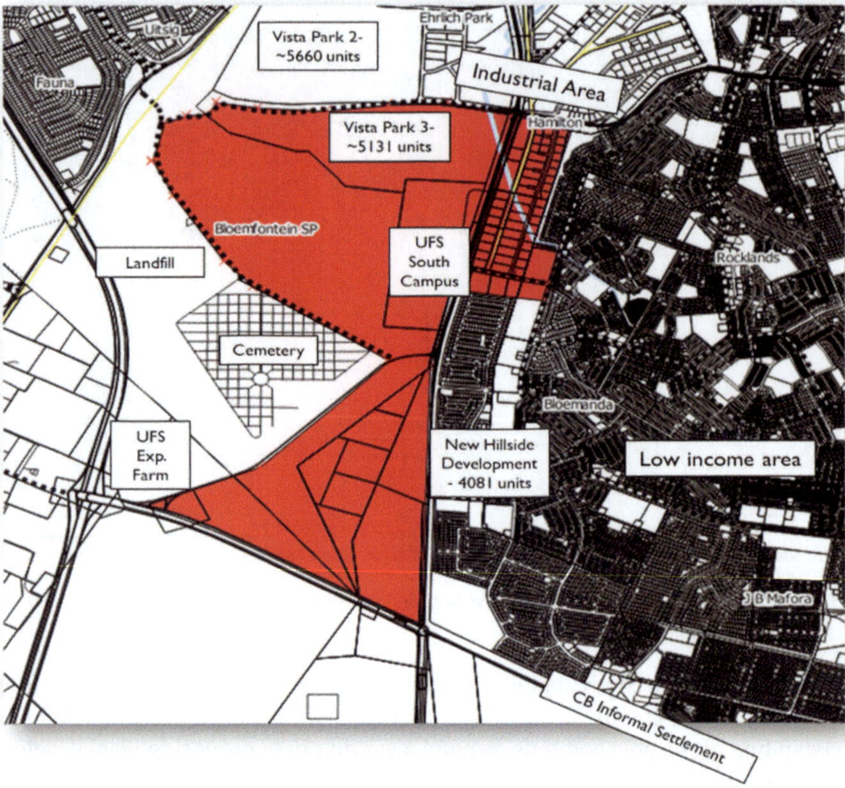

Fig. 12.3 Notable features on and around the site. *Source* Denoon-Stevens (2019)

12.5.1 Land Ownership

The MMM owns the majority of the land on the site; any developments on this land will therefore likely be headed by the municipality. The university owns the areas that house the UFS South Campus and experimental farm and any further developments on those areas will be determined by the UFS in liaison with the municipality. The rest of the site is privately owned by others (Fig. 12.4).

12.5.2 Legislative and Policy Context at the National Level

It is important to look at South Africa's approach to spatial planning, development and land use at the national level because the national stance is what informs what happens at the municipal level.

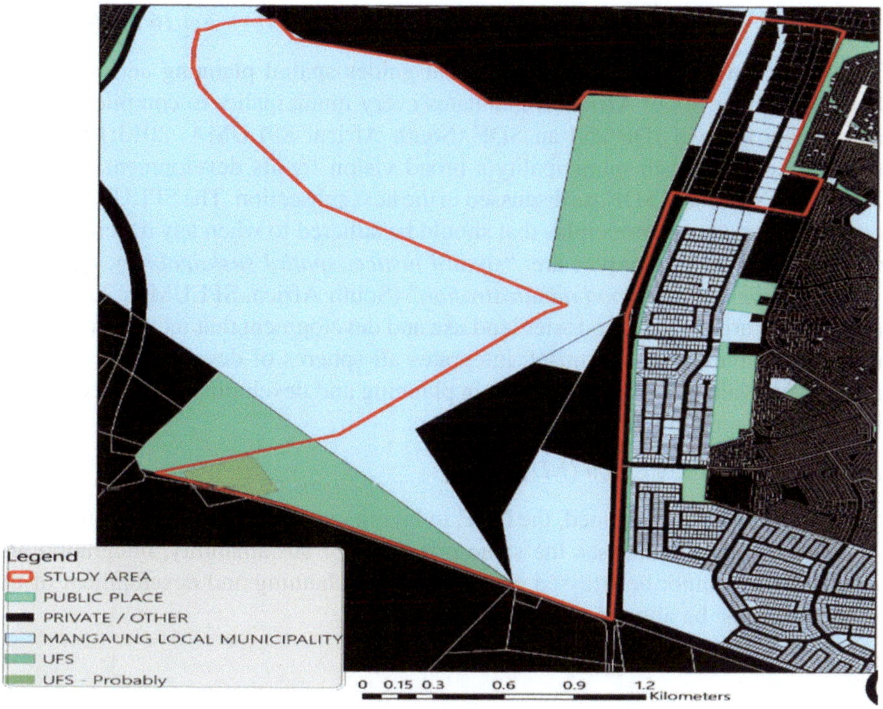

Fig. 12.4 Ownership of land on the site. *Source* Denoon-Stevens (2019)

Integrated Sustainable Human Settlements (Breaking New Ground [BNG])

This was a new strategy formulated by the South African Department of Human Settlements (DHS) in the early 2000s, aimed at "promoting the achievement of a non-racial, integrated society through the development of sustainable human settlements and quality housing." Housing and services provision has formerly failed to keep up with the emergence of new households due to high unemployment rates. The DHS realised that subsidised housing for the poor did not help. Therefore, the department planned to increase housing delivery as a poverty alleviation measure. BNG intends to transition from mere housing provision to more sustainable human settlements. It also aims to bolster the integration of informal settlements into the wider urban sphere in an attempt to curtail spatial, social and economic exclusion. It seeks to: promote densification and integration; enhance spatial planning; develop social and economic infrastructure (DHS, 2004). This line of discussion is aimed at supporting this author's proposition that the development of the subject area must adopt a spatial planning style that emphasises the principles of sustainability, densification, integration, and inclusion. BNG housing must be part of various housing typologies on the site.

Spatial Planning and Land Use Management Act (SPLUMA), Act 16 of 2013

This is the single piece of legislation that guides spatial planning and the use of land throughout South Africa. It mandates every municipality to compile, approve and review its own IDP and an SDF (South Africa. SPLUMA, 2013:16). These documents detail each municipality's broad vision for its development. As such, Mangaung's IDP and SDF are discussed in the next subsection. The SPLUMA further outlines development principles that should be adhered to when any development is undertaken. These principles are: "*spatial justice, spatial sustainability, efficiency, spatial resilience, and good administration*" (South Africa. SPLUMA, 2013:18)—all of which are designed to foster land use and development that includes everyone, minimises environmental impact, integrates all spheres of development, and optimally uses available resources. Thus, in planning and developing the subject area, it is binding to remain within these parameters.

National Development Plan (NDP)

Similar to the aforementioned, the NDP, in laying out South Africa's national vision of development, emphasises the same principles of sustainability, integration and inclusivity. It cannot be stressed enough that the planning and development of the study area must be aligned with these principles.

12.5.3 Policy Documents Specific to Mangaung

The following are specific policy documents that inform the developmental route that the MMM intends to take, they in turn, guide some of the arguments for the development of the study area. These strategic documents (IDP, SDF and BEPP) complement each other and should be read and studied together (MMM, 2018:48) to establish a better comprehension of Mangaung's vision and make sound arguments for development. Table 12.1 shows the interrelatedness of the strategic documents.

Mangaung's Integrated Development Plan (IDP)

Among others, the IDP shows that Vista Park Phase 3 is planned for mixed development; that all engineering and planning designs are in place, and redirection of water and sewer lines has commenced. This portion of the site is planned to yield a variety of house types to accommodate all income levels, as well as supporting uses including schools, business, health, recreational and public areas. This proposition is viable, but arguments can be made against the proposed filling station, seeing as there are already two of them along the Oliver Tambo Road. The number of parks (which in the IDP stands at 14) can also be reduced, and instead, include pockets of agricultural land use in order to have a local source of food.

Table 12.1 IDP, BEPP and MSDF alignment

IDP strategic objectives	BEPP elements	MSDF (urban network)
Poverty eradication, rural and economic development and job creation	– Informal settlement prioritisation and upgrading – Economic nodes	– Nodal development – Marginalisation area development
Financial sustainability e.g. revenue enhancement, clean audit	– Long terms financing strategy – Institutional arrangements	– Capital investment framework
Spatial development and the built the built environment	– Catalytic land development programme	– Strategic land parcels, CBD regeneration
Basic service delivery	– Inclusive and equitable basic services	– Integration zones – Underserviced areas
Integrated human settlement	– Integrated mixed development mega projects	– Implementation of CRUs, social housing, FLISP, BNG
Integrated public transport, environmental management and climate change	– Transit oriented development – Alignment of human settlements and public transport	– Implementation of Mangaung IPTN (phase 1) – Linkage with human settlement programme
Social and community services	– Inclusive city	– Integrated community development

Source MMM (2018:48)

Mangaung's Spatial Development Framework (SDF)

As aforementioned, the SDF is mandated by SPLUMA, and its purpose is to guide the future use of land. It is a diagrammatic depiction of the spatial form a municipality aims to achieve over a long period of time. The MMM through its SDF seeks to do away with "the low density, sprawling, fragmented and largely mono-functional settlement forms that resulted under apartheid; and instead establish settlement patterns which create benefits accessible to all the people of Mangaung." The SDF shows that there is a decline of the Bloemfontein CBD, with people moving outwards to the suburbs and outskirts. This invariably means we can expect the study site to be occupied by residential land use in the future. It is therefore advisable to pre-empt this counter-urbanisation and start planning a large part of the site for residential use. Significant residential development of the site is therefore proposed, with supporting commercial and social land uses.

Mangaung Built Environment Performance Plan (BEPP)

This is a strategic plan aimed at enhancing the long-term performance of Mangaung metro's-built environment (MMM, 2018:47). The MMM's approach to development is one that looks to "*integrate Residential, Recreational, Retail, Industrial and Commercial (RRRIC) elements in any given development.*" A number of land parcels have been singled out in strategic areas for this very purpose of mixed-use development aimed at establishing integrated human settlements. These are Vista Park 2, Vista Park 3 and Hillside View. "MMM's strategic approach into the future is the Implementation of mixed developments. This strategy departs completely from prioritising the building of houses in isolation; to building more inclusive communities with access to various other amenities." Therefore, based on this premise, we advocate for a mixed-use development of the subject area, with emphasis on residential use, and the other RRRIC elements playing supporting roles, for a well-integrated development.

Preliminary Proposals

(a) The hilly area on the western end of the site should not be developed. The first reason is that the area houses a water reservoir that would feed water to the site once it is developed. There would be no sense in demolishing this reservoir for anything else. The second reason is that a steep terrain is significantly difficult and expensive to develop.
(b) Seeing as the UFS South Campus is located on the site, part of the development of the site should go towards student housing.
(c) Some measures should be taken to buffer the newly proposed residential area from the Hamilton industrial area on the northeast. This can be in the form of a green buffer, where the barrier would be an open space like a park.
(d) An argument can be made for the area around the centre of the site to be used for a variety of offices, labs and other buildings to complement the research endeavours of the UFS South Campus. Lastly, it would be a possible idea for the land to be used for agriculture, seeing as the Bloemfontein landscape is dominated by a built environment. This would not only provide food security for residents in the locality but also contribute towards enhancing Mangaung's primary sector of the economy which, in subsequent sections of this chapter, is demonstrated to be lagging.

12.5.4 Socio-Economic Analysis of Bloemfontein and the Wider Mangaung Area

This section looks at the socio-economic situation in the broader locality of the study site, by analysing 3 variables, namely—the main problems facing the municipality, growth in Gross Value Added (GVA) by sector, and the dwelling types within which people live, respectively. These analyses are conducted in comparison with larger geographic areas; that is, the Free State province (FS) and South Africa (RSA) as a whole.

Main Problems Facing the MMM

Analysis of these problems was important in identifying what the municipality is struggling with the most, and therefore inform the ideal development avenues to be prioritised on the site.

Figure 12.5 reveals that, based on public opinion from community surveys, inadequate sanitation services are the biggest problem facing the MMM. This is 11%

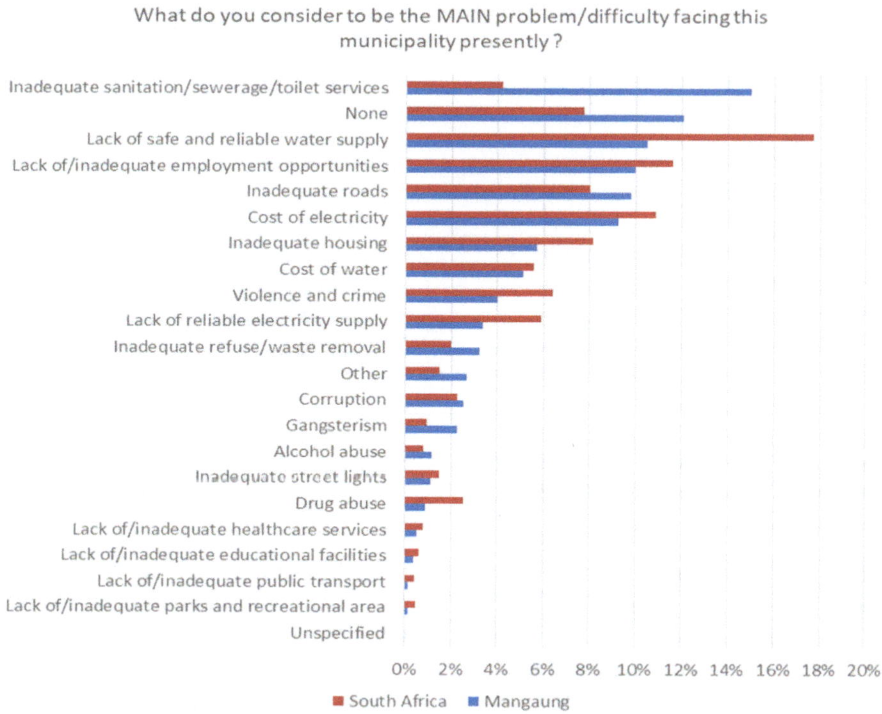

Fig. 12.5 Main problems faced (MMM & SA). *Source* Authors construction from Stats SA data (2024)

more of a problem in the MMM than in the broader RSA. Other major problems in the MMM are unsafe and unreliable water supply, poor employment opportunities, inadequate roads, high cost of electricity and inadequate housing. However, Mangaung is doing better than the RSA in most of these, with the exception of roads and sanitation.

All this means that, in the process of making proposals for developing the site, as well as the actual development itself, efforts should be made at establishing a better network of sanitation services; improving the water supply system; improving accessibility through better roads and road networks; prioritising housing and planning for more dwelling units, especially for people of lower-income. With regard to improving employment, it may be worthwhile to invest in and allocate some land to agriculture and manufacturing. This would reduce unemployment while simultaneously boosting the weak primary sector of Mangaung.

Per Annum Growth in GVA by Sector: 1993–2015

GVA data inform us which sectors are thriving and contributing more to the economy and which are lagging behind. This might inform how we ought to prioritise land use

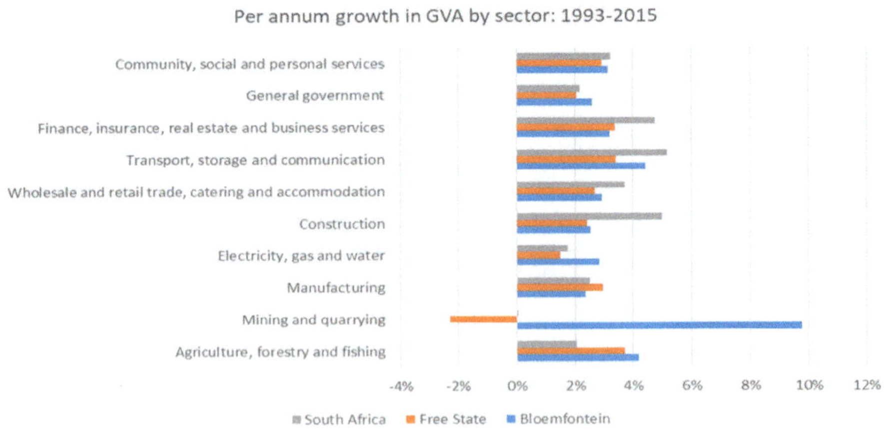

Fig. 12.6 GVA growth by sector (South Africa, Free State, and Bloemfontein). *Source* Authors construction from Stats SA data (2024)

when developing the study area to boost ailing sectors of the economy. Figure 12.6 shows GVA growth by sector (South Africa, Free State, and Bloemfontein).

From Fig. 12.6, over the 22-year period, Bloemfontein has seen the most growth in GVA in the mining sector, and its growth rate far surpasses that of the RSA and FS, which has in fact seen a decline. However, this information may be misleading if misinterpreted. This growth should not be assumed to mean the mining sector dominates the economy; it only means that the sector has seen the most growth. The tertiary sector was found to be the most dominant in Bloemfontein's economy, as illustrated by Fig. 12.7.

Based on all the GVA analysis above, it would seem viable to propose that in as far as improving Bloemfontein's economy when developing our site, some of the land be dedicated to formal agricultural practices. This would be aimed at improving the primary sector while also improving food security for the people in existing surrounding settlements and the new one that would be developed.

The data from Fig. 12.8 show that formal houses are the overwhelmingly dominant type of dwelling in both the MMM and the RSA, with the former bettering the latter by 8%. The other house types lag behind quite significantly. In light of this data, this author's proposal for residential development of the study area is centred around three aspects: variation of house types, densification and IH. The bulk of the dwelling units developed on the new site should be apartment-style units and semi-detached housing, so we will densify the residential area and cover less land, thus reducing the environmental footprint. Also, these residential units make IH easier because they provide for people of different income levels within one apartment complex or settlement. This would improve the variety of house types in the study area as well as the wider Bloemfontein area, while providing residence for everyone, and covering less area for the same number of people than would be the case in single-family detached housing.

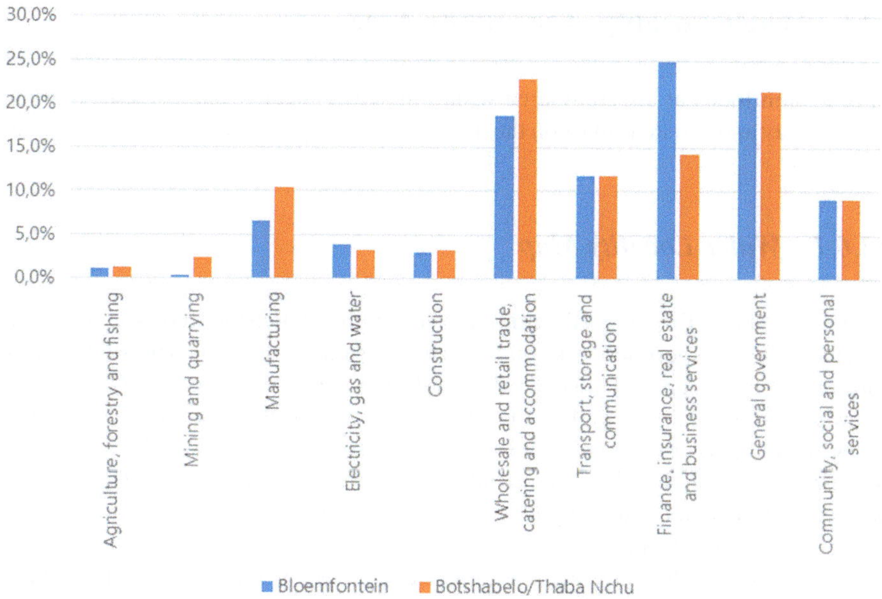

Fig. 12.7 Percentage contribution to GVA by sector in 2015. *Source* Authors construction from Stats SA data (2024)

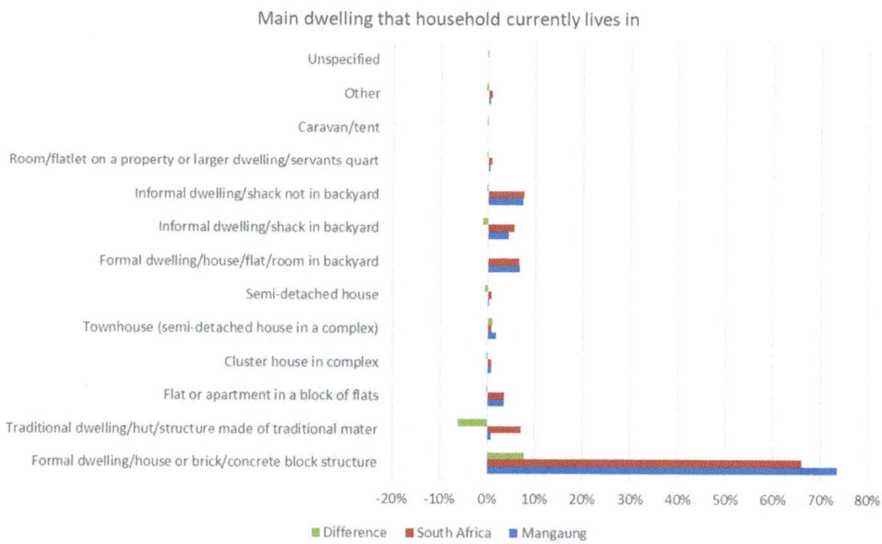

Fig. 12.8 Comparison of house types. *Source* Authors construction from Stats SA data (2024)

12.6 Wider Locality of the Study Area

This section looks at the study area within its wider locality and presents how the locality may and should impact on the development proposals and final precinct plan.

12.6.1 Wider Locality Map

Figure 12.9 shows the wider Bloemfontein area within which the study site is located. It depicts the transport networks and the river system in the area. The study area is surrounded by an arterial road on the north, the N1 freeway to the west, the N6 freeway to the South, as well as the M30 main road on the east edge. This makes access in and out very smooth. The study area also has access to a railway line that begins/ends in the Hamilton industrial area on the northeast corner. The site also has access to a water source by way of a river that goes through Hamilton into the residential area on the east.

We observe that the study area is very well connected by road and rail, which makes transport of people and goods to and from very easy. This supports the proposal made earlier that agriculture should be prioritised in order to improve the primary sector of

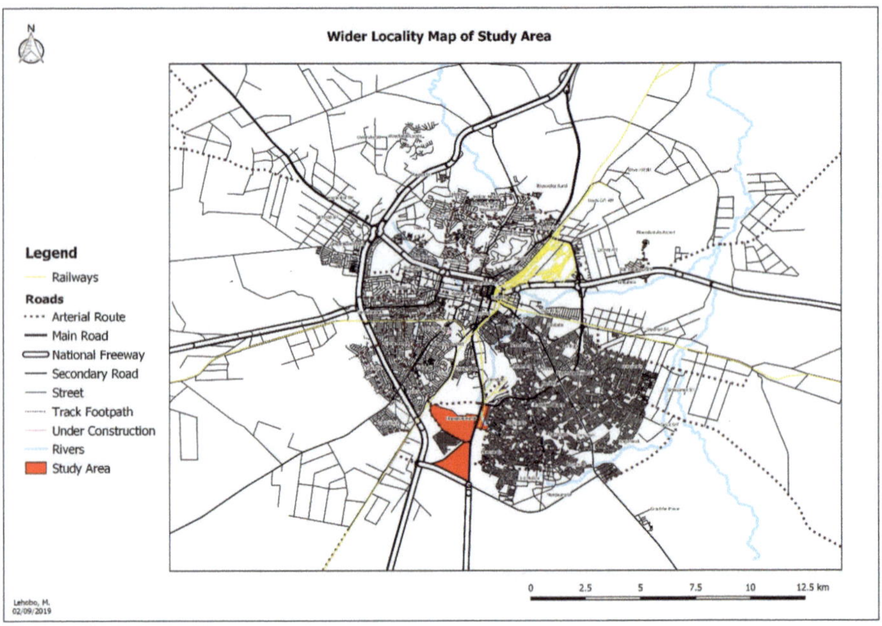

Fig. 12.9 Wider locality map of the study area. *Source* Developed by Authors using readily provided GIS data (2024)

the economy, employment, and food security. This would be assisted by easy transit of the agricultural produce to the city as well as other towns and cities for sale. The river system should be capitalised on to improve water supply, which was seen to be a problem in the previous section.

12.6.2 More Emphasis on Surrounding Land Uses

In the vicinity of the site are the Hamilton industrial area, a low-income residential area on the east, a middle-to-upper-income residential area on the northwest, and commercial activity in the CBD to the north. This study site is well-positioned to implement the IH proposition that forms the core of this report. It is located between low and middle-upper-income residential areas, and can, therefore, serve as a bridge between these two and accommodate new residents from both, if, in the course of a new residential development, a compulsory IH requirement is mandated. This would make the site attractive to and accommodative of everyone. This phenomenon of the site's ideal location is illustrated in Fig. 12.10.

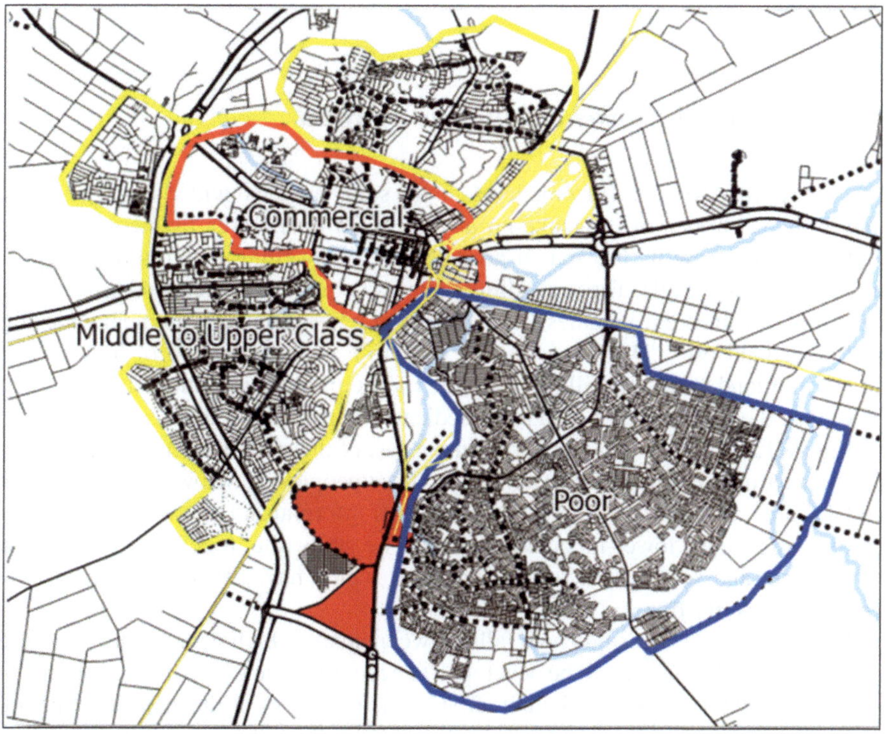

Fig. 12.10 Surrounding land uses. *Source* Denoon-Stevens (2019)

This section displayed maps of the study area and briefly discussed its immediate vicinity and wider locality. It showed that the site is well connected in terms of transport networks, is well located in terms of implementing policies of inclusion and has an additional prospective water source by virtue of a river. The proposition was, therefore, that: (a) primary sector land uses such as agriculture be implemented in order to capitalise on the transport system to sell products, create employment, improve food security, and thus boost the economy, (b) implement inclusionary housing and (c) improve the water supply system.

12.7 Mapping Census Data

This section analyses Stats SA 2011 data for the Bloemfontein area. The data are presented by electoral wards in the form of heat maps, the analysis of which will inform the planning and development of the study site and have a bearing on the final precinct plan. It looks at the percentage of households living below the poverty line and tenure status, assessing the percentage of people who actually own a house. Based on that information, recommendations are made for how the site ought to be developed.

12.7.1 Annual Household Income by Ward

It was important to assess income levels in the study area and wider locality because they provide an idea of the make-up of the Bloemfontein area by income classes, and therefore show us the spatial arrangement or dwelling patterns of people of different income levels. These dwelling patterns help inform where and how to cater for people of different incomes when developing new human settlements/housing, and other land uses. That is to say, we seek to ascertain the type of housing and supporting land uses that will best provide for the particular population of the area, as well as the proportion of land to be dedicated to residential use.

Figure 12.11 shows that the wards on the east of the study area have over 60% of families earning below R38200 annually. These people fall below the poverty line, and this area, being a township with a mixture of informal and upgraded settlements, is known to house lower-income residents. The study site, together with all areas north, west and south of it, fall under wards which have 29–46% of households below the poverty line.

From Fig. 12.11, the study area falls between areas of high poverty and those that are better off. The site should be developed to accommodate lower-income groups by implementing IH; this should include BNG housing, seeing as the majority of people in surrounding areas are the lower-income class. Furthermore, as already alluded to, it may be advisable to plan for land uses in agriculture and manufacturing to provide employment for lower-income workers. Seeing as poverty is typically associated with

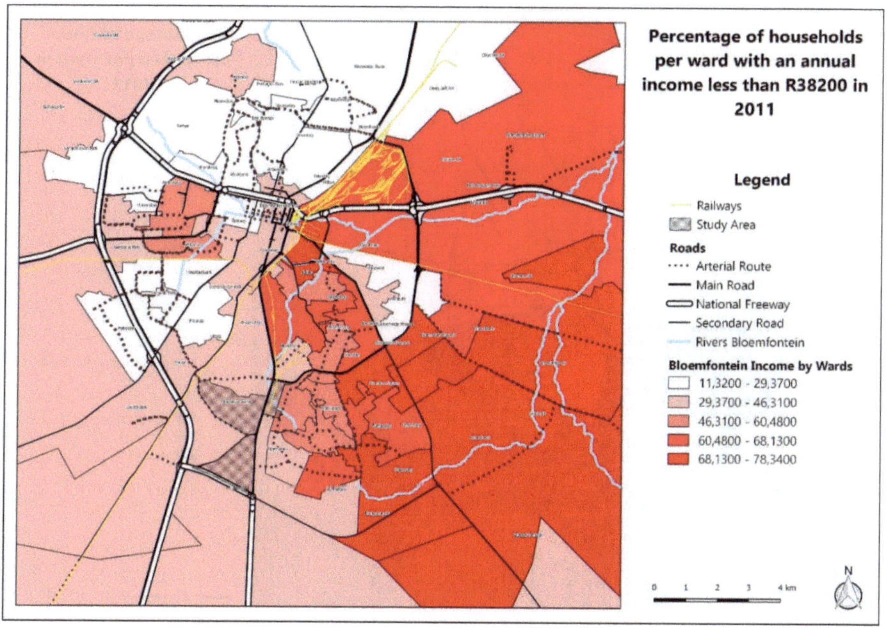

Fig. 12.11 Bloemfontein annual household income by ward, 2011. *Source* Developed by Authors using Stats SA Data (2204)

lower education, this makes sense, rather than dominating the site with tertiary sector activities which normally has jobs that require highly skilled workers. Heavy tertiary sector developments would not demographically make sense for this particular area.

12.7.2 Tenure Status; House Ownership by Ward

Since any new development will invariably seek to house people in a proper settlement, studying the trends in house ownership/tenure status in the areas surrounding our study site makes sense. Figure 12.12 shows the percentage of households who own their own houses in each ward. Ownership here includes both houses that are owned and fully paid for, as well as those which are owned but not yet fully paid. It is observed that all areas immediately surrounding the study site have over 60% house ownership (with the rest either renting or occupying informally). The study area itself falls within a ward with 59–63% house ownership.

Figure 12.12 suggests that land and housing south of the CBD may be cheaper than the north, which may explain the high house ownership rates. This is also aided by informal settlement upgrades in the township which continue to formally house people and grant them secure tenure. The recommendation is that the majority of the

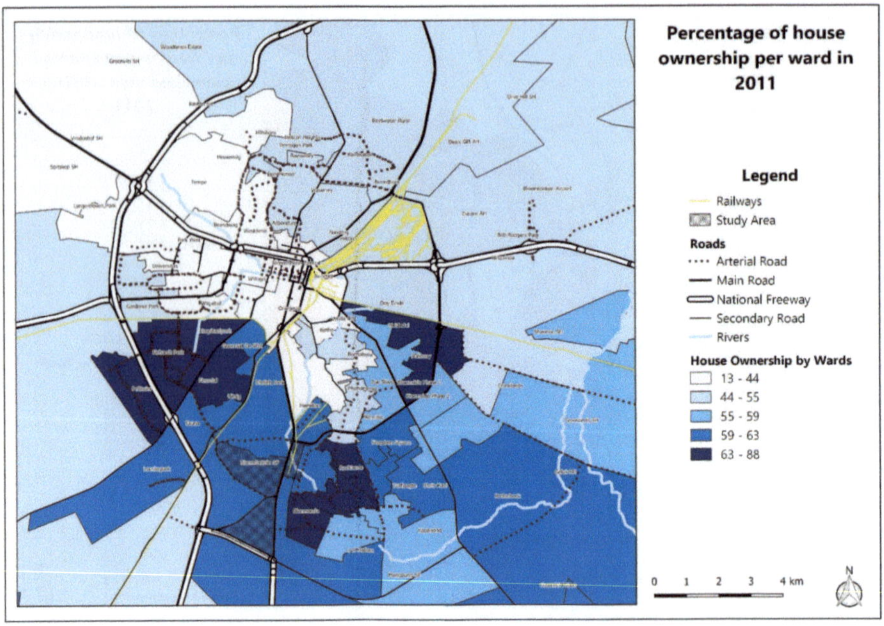

Fig. 12.12 Bloemfontein house ownership by ward, 2011. *Source* Developed by Authors using Stats SA Data (2024)

site be planned for residential use to capitalise on these conditions and implement IH to give everyone an opportunity at homeownership.

12.7.3 Mapping Environmental Features on the Study Site

It is vital to heed environmental features that may hinder or aid any new developments on a given site. Without environmental analysis, we run the risk of severe impacts on human and non-human life and the environment itself. This section displays and analyses the main environmental features on the study site to determine the areas suitable for development and those not. The environmental features are hydrological bodies, a landfill site, and the topography/slope of the site. The second task is to single out, with reason, all undevelopable areas on the site.

Figure 12.13 is a map depicting the study site's main environmental features and surrounding areas. These include water bodies such as dams, reservoirs and rivers; a landfill site; as well as the topography of the site. The mapping of these features reveals sensitive areas upon which development would be ill-advised. These areas would either be hazardous or expensive to develop. The analysis of these features will inform how a new development ought to take shape.

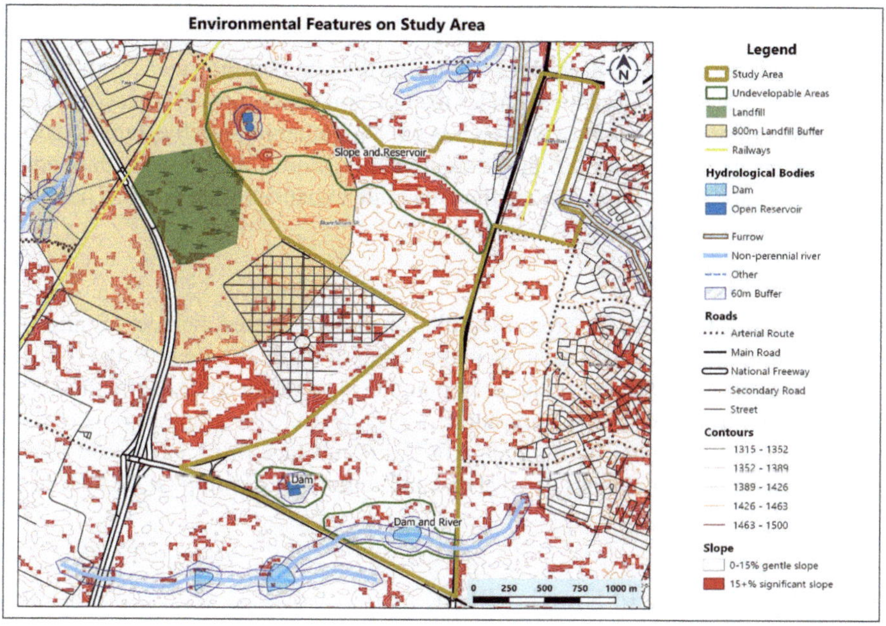

Fig. 12.13 Main environmental features on the study site. *Source* Developed by Authors using readily provided GIS data (2024)

Hydrological Bodies

There is a water reservoir atop a hill on the far north-western extreme of the site, two other dams and a river on the South. These are potential water sources for whatever new development that may materialise on the site. Therefore, destroying or developing them would be expensive and senseless. They should rather form part of an efficient water supply system for the site, and in fact, the reservoir on the hill is already well stationed to feed water downslope. A 60-m buffer is enforced on each water body to indicate that any developments should be at least 60 m away.

Landfill Site

A landfill site is located on the west, just outside the study site. Landfills, being waste disposal sites, are an environmental and health hazard (Laner et al., 2009:3674), perhaps even if well maintained. It would, therefore, be ill-advised to have developments on, or near them. More importantly, there is an 800-m buffer imposed on the landfill site, signifying that developments should be at least 800 m away. Part of this buffer encloses the western section of the study site that already includes the aforementioned reservoir, further fortifying its "undevelopable" status. The buffer also covers some 42 hectares of flat, developable terrain just south of the hilly area, which might have problematic implications for developments thereon.

Slope

Much of the area along the north boundary of the site is quite steep, with heights between 1400 and 1500 m. This includes the hill atop which a water reservoir stands, as aforementioned. Developing on slopes is not altogether impossible, but it is costly. Any construction on steep terrain is significantly more difficult, and thus costly (Bommer et al., 2010). This renders the area quite expensive to develop. The rest of the site is generally flat and developable.

Undevelopable Areas

From the analysis above, the areas identified as undevelopable include: (a) all water bodies within the site as well as areas within 60 m of them: (b) areas on the site with steep slopes above 15%; and (c) areas within the site that are enclosed by the 800-m landfill buffer. It is not strictly impossible or prohibited to develop on some of these areas, but it would either be too dangerous, difficult or costly; and would require extremely compelling reasons to go ahead and remove or develop on, or near them.

12.8 Synthesis and Spatial Plan

This section is a synthesis of all aspects covered in the spatial analysis. It brings everything together and presents how the discoveries inform the final spatial plan proposed herein.

12.8.1 Recapping the Theoretical Framework and Spatial Analyses

After a research project comprising a comprehensive literature review; assessment of national and municipal policy and legal documents; and spatial analysis of the subject area and its wider context—this section presents the culmination of all work undertaken in the effort to produce a suitable development plan of the aforementioned study area. It is a recap of all preparatory work done, which resulted in the discovery of various planning issues, opportunities and subsequent proposals, all of which were arrived at after a collective assessment of the literature review, policy documents and spatial analysis.

The literature review centred on urban regeneration, gentrification, and IH. Although an effort was made to link the three, particular interest is placed on the latter for purposes of this research project. The study area for the said project was found to lie in between low-income residential areas (popularly known as "townships") and middle–higher income suburban areas. This would inform the housing/human settlements approach to be employed in developing this site. The next step was to review municipal documents to get a sense of the nature/state of the wider area

Table 12.2 Summary of spatial analysis

Socio-economic context	Wider locality	Demographics	Environment	Existing uses on site
Inadequate infrastructure (water, sanitation, roads, and electricity) • Unemployment is prevalent • The economy is dominated by the tertiary/service sector • Primary and secondary sectors are weaker	• The study area is connected to road and rail networks • The close proximity of the study area to CBD • The study area lies between low and middle-upper class residential areas	• The lower-income areas live below the poverty line • Middle-upper class live above the poverty line • Correlation between Settlement/house types and income was observed	• Water bodies on the northwest, and South of the site • Steep terrain on northern areas of the site • Landfill and cemetery outside the western boundary of the site	• Hamilton industrial area • UFS South Campus • TVET College • UFS Experimental Farm • A few unidentified buildings

Source Authors' Construct (2024)

within which the study area is located and the intended developmental direction that the MMM wishes to take. It was established that the MMM seeks to establish integrated human settlements, mixed-use developments, inclusive and equitable access to services, and economic development that among others prioritises job creation and poverty eradication. It was also determined that the Bloemfontein CBD is declining, and development is moving outwards. Measures would need to be taken to respond to this. The third step was conducting spatial analyses of the site itself, as well as its immediate and wider vicinity. This was done by analysing statistical and environmental data using various techniques such as comparative charts, and heat maps. Table 12.2 is the summary thereof.

From all this information, a synthesis map was then produced to establish developable and undevelopable areas; as well as break down the study site into "character areas" which would inform the shape of the precincts of the final spatial plan.

12.8.2 Synthesis

This section presents a synthesis of the aforementioned analyses and identifies key constraints and opportunities for the development of the site. First, a structures map is presented, and then an in-depth synthesis identifies individual "character areas" on the study site and their capacity for development. These analyses are herein used to present a synthesis of all key aspects (constraints and opportunities) of the study area and its immediate vicinity.

From Fig. 12.14, we note that developments that already exist on the site include parts of the Hamilton industrial area, the UFS South Campus, TVET College, the UFS

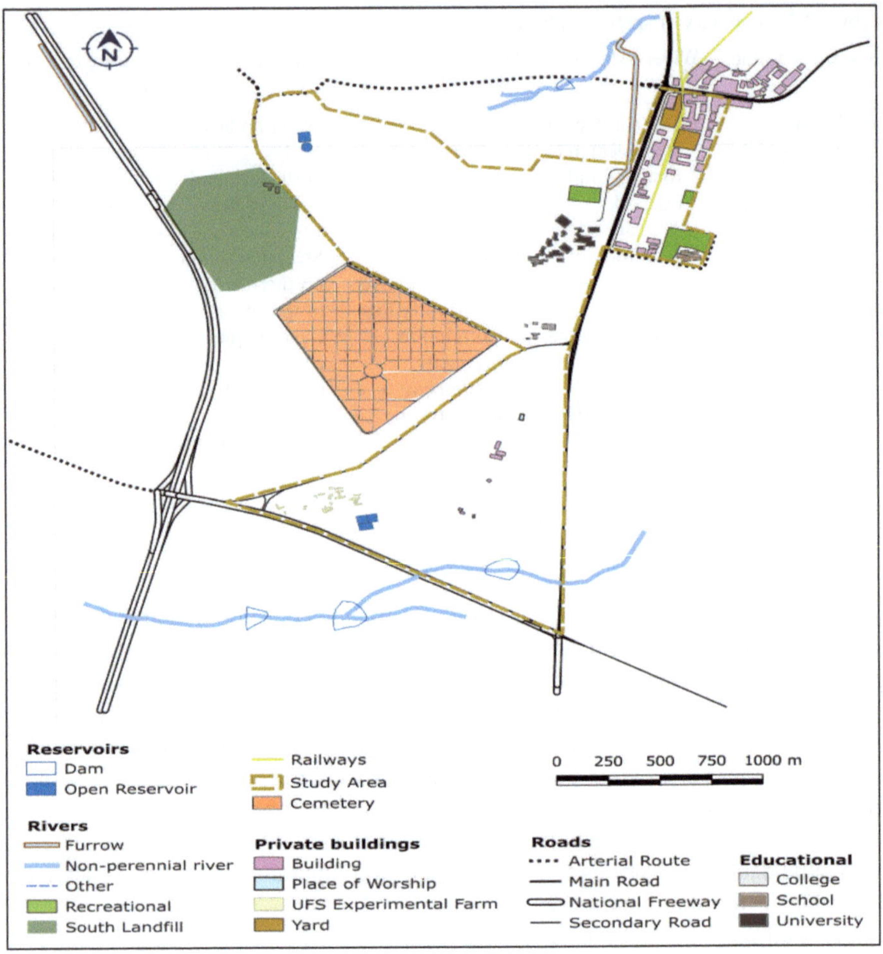

Fig. 12.14 Structures map. *Source* Authors' Construct (2024)

experimental farm, a water reservoir, a place of worship and a few other unidentified buildings. A cemetery lies just outside the study area to the west.

12.8.3 Description of Character Areas and Synthesis Map

In the process of preparing a synthesis map based on the features on and around the site, the same was broken down into individual character areas. This subsection describes each of these character areas, their sizes and potentials and/or constraints for development.

Areas 1 and 3 which total 149 ha (30% of the study area) are undevelopable; or would require more expenses or removal of other land uses in order for development to be possible. Areas 2, and 4 through 10 make up a total developable land area of 355 ha (70% of the study area). This, however, is minus the land area already occupied by all existing buildings on site.

From Table 12.3 and Fig. 12.15 above, we observe that Area 1 is rendered undevelopable as a result of steep terrain, the landfill buffer zone, and the reservoir atop the hilly area. This means that viable land uses in this area include public open spaces (POS), agriculture, recreation, and other land uses that involve little or no construction. Area 2 is the Hamilton industrial area and is largely occupied. Possible developments are further industry, wholesaling or office complexes to serve as administrative centres for the industries. Area 3 is undevelopable because of the dam and river thereon. Area 4 can be used for residential purposes to complement Vista Park. Area 5 is the UFS experimental farm and will most likely continue being used by the university for experimental purposes.

Area 6 is entirely vacant and can be developed for residential or commercial purposes. If used for the former, then measures would be taken to erect a barrier between the settlement and the nearby cemetery. Area 7, being already zoned for educational land uses, can be used to house students (which is perhaps not advisable so close to the industrial area), extend south campus recreational facilities, or be used for office and business complexes. Area 8 can also be used to extend the university or college facilities, student housing or commercial services to cater for students and nearby settlements. Area 9 is the largest undeveloped area and can hold an array of land uses. Area 10, being near crossroads, can be used for services oriented towards

Table 12.3 Breakdown of the study area

Character area	Description	Area (ha)	Percentage of total area (%)
1	Slope, reservoir and landfill buffer	126	25.0
2	Industrial	59	11.7
3	Dam and River	23	4.6
4	Vacant land by Vista Park 3	10	2.0
5	Experimental farm	32	6.3
6	Vacant land by cemetery	46	9.1
7	Educational 1	23	4.6
8	Educational 2	33	6.5
9	Vacant land adjacent to hillside view	141	28.0
10	Vacant land south of dam	11	2.2
		504	100.0

Source Authors (2024)

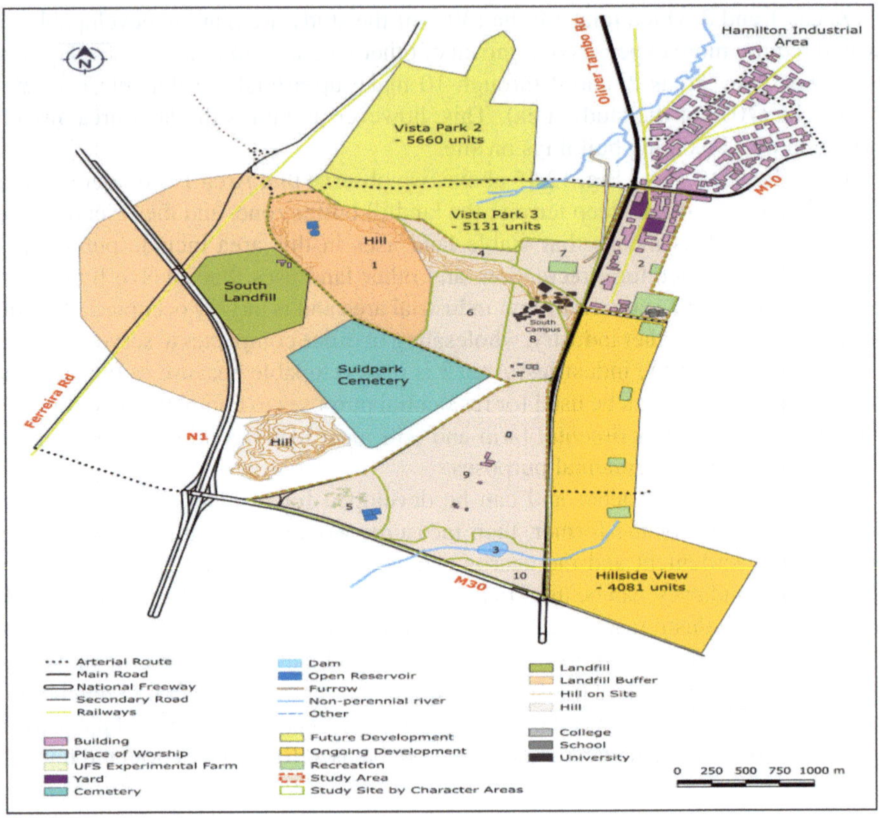

Fig. 12.15 Synthesis map. *Source* Authors' Construct (2024)

motorists, such as a petrol station, shopping complex, and some residence further inland.

12.8.4 Proposals and Rationale

As it lies between settlements of different income groups, the study area seems to be the perfect place to implement IH to bridge this gap and promote the integration of different income classes in one neighbourhood. This trendy approach to housing will help revamp the urban fabric of this area while meeting the needs of people at all levels of society. The declining CBD means preparation should be made by prioritising residential land uses in the outer skirts in order to accommodate increasing populations. A solar farm was considered for the undevelopable land that falls within the landfill buffer zone, but it was not the best use of the land. Instead, solar panels can be mounted atop each house, and the land used for agriculture instead, as a local food source and primary sector activity. Office complexes are proposed in order to consolidate the industrial and business sectors. Commercial land uses are essential for a neighbourhood and should be planned for, and accessible from all parts of the neighbourhood. All undevelopable areas should form part of public open spaces and should be used for aesthetic and recreational purposes, which are necessary for a well-functioning neighbourhood. Supporting land uses such as schools, a hospital, and student housing should be incorporated into the proposed development of the site. Existing land uses should be left as they are, and efforts should be made to integrate them into the new proposed development.

12.9 Proposed Spatial Plan

This section presents the proposed spatial plan, a broad land budget, and a justifica tion of how the proposed plan responds to the issues identified in the early stages. Table 12.4 is the land budget for the proposed development of the site.

Table 12.4 Broad land budget

Land use	Area (ha)
Residential	160
Commercial	5
Public open spaces	115
Student housing	20
South campus	20
TVET college	6
UFS experimental farm	30
Industry	37
Primary schools × 2	7
Crèche	1
Hospital	7
Office complexes × 2	20
Agriculture	70
Access roads and streets	2
Total	**500**

Source Authors' Construct (2024)

Fig. 12.16 Proposed spatial plan. *Source* Authors (2024)

12.9.1 Proposed Plan

Figure 12.16 depicts the proposed development of the site together with the motivations after all analyses and considerations were made. The next subsection provides a detailed rationale for this plan.

The public open spaces on the plan are both active and passive. The active POSs are the proposed extension to the South Campus sports grounds and the parks. The large park is placed along the watercourse to create an open space system because the area is otherwise not suitable for aggressive development. The water bodies and park are complementary for aesthetic, recreational and sustainability purposes. The smaller pocket park is located to provide recreational use for residents located further inland and in the northern segment of the study area, as well as for the anticipated student population. The passive POS is made of the entire area of steep terrain along the northern boundary of the site, and the proposal is for this area to be a botanical garden, which would also bring in some revenue.

The vacant area between the South Campus and TVET College is allocated for student housing. This is ideally located for students of both institutions to be in close proximity to their respective campuses. The majority of the site is allocated for residential use, with an IH approach being proposed, where housing of all income levels is catered for. The higher income residential use is placed near the steeper area because it is more of a challenge to develop. The flatter terrain is reserved for middle to lower-income residents so that no extra cost is incurred during development. Group housing is also proposed at the south end of the site.

Commercial land uses are strategically placed at intersections for convenience of motorists, passengers and pedestrians alike; and are evenly distributed for ease of access from anywhere within and around the study area and adjacent neighbourhoods. Office complexes are proposed near the industrial area for administrative purposes and along the Oliver Tambo Road to provide rented office space for willing businesses. In addition to the existing primary school that serves the low-income areas further east, a new primary school and crèche are proposed in the middle of the study area and would serve both the low and middle-income residents of the new neighbourhood. A hospital is also proposed on the site near the main road to serve the new neighbourhood and areas to the east, seeing as the next nearest hospital is Rosepark Hospital 8 kms away. This is an essential service for the wellbeing of the residents of the proposed neighbourhood.

Lastly, the portion of the study area that falls within the landfill buffer zone cannot be developed as long as the landfill is active. The best way to make productive use of this land while simultaneously avoiding human settlement and aggressive development is by using it for agriculture. This would do three things – the first is to provide a local organic food source, and the second is to make some contribution to the weak primary sector of Bloemfontein (as was revealed during the analysis phase); while employing the low-class labour force, and thus addressing the unemployment problem as well.

12.10 Conclusion

In conclusion, the housing crisis facing low-income communities in Bloemfontein and other cities in South Africa and around the world is a complex and pressing issue. The proposed inclusionary housing and mixed-use development strategy offers a promising approach to address this challenge by promoting sustainable, inclusive, and equitable urban development. The chapter provided a detailed analysis of the housing crisis in Bloemfontein, highlighting the impact it has on low-income residents. It has also explored the theoretical and empirical foundations of inclusionary housing and mixed-use development, examining their potential benefits and challenges based on case studies from other cities. The chapter then applied this knowledge to the selected precinct of Bloemfontein, proposing a set of guidelines for the design and implementation of inclusionary housing and mixed-use development that considers its unique needs and conditions. The proposed strategy has the potential to address the housing crisis in Bloemfontein by increasing the supply of affordable housing, promoting economic and social diversity, and fostering sustainable urban development. However, its success will depend on effective public–private partnerships, community participation, and sustainable financing mechanisms.

The chapter contributes to the ongoing debate on how best to address the housing needs of the urban poor in South Africa and beyond, by proposing a strategy informed by theoretical and empirical evidence and taking into account Bloemfontein's specific contextual factors. By promoting inclusive and sustainable urban development, the proposed strategy seeks to create more equitable and liveable cities that offer opportunities for all their residents to thrive. As the proposed strategy is implemented and evaluated, it is important to monitor its progress and make adjustments as needed to ensure that it achieves its goals of providing affordable and inclusive housing within the context of sustainable urban development. By working together, government, private sector, civil society, and residents can create more liveable and thriving cities that offer opportunities for all their residents to flourish. Future research could focus on the impact of the proposed strategy on the residents of the selected precinct and on the wider implications for urban development in South Africa.

References

Australian Housing and Urban Research Institute (AHURI). (2017). Understanding Inclusionary Zoning: Utilising land use planning systems to deliver affordable housing. Retrieved June 20, 2019 from https://www.ahuri.edu.au/policy/ahuri-briefs/Understanding-inclusionary-zoning

Bibri, S. E., Krogstie, J., & Kärrholm, M. (2020). Compact city planning and development: Emerging practices and strategies for achieving the goals of sustainability. *Developments in the Built Environment, 4*, Article 100021.

Bommer, C., Phillips, M., & Arenson, L. U. (2010). Practical recommendations for planning, constructing and maintaining infrastructure in mountain permafrost. *Permafrost and Periglacial Processes, 21*(1), 97–104.

Butler, T. (2007). For gentrification? *Environment and Planning, 39*, 162–181.

Calavita, N., & Mallach, A. (2010). An international perspective on inclusionary housing. In N. Calavita & A. Mallach (Eds.), *Inclusionary housing in international perspective: Affordable housing, social inclusion, and land value recapture* (pp.1–14).

Couch, C., Sykes, O., & Borstinghaus, W. (2011). Thirty years of urban regeneration in Britain, Germany and France: The importance of context and path dependency. *Progress in Planning, 75*, 1–52.

Denoon-Stevens, S. (2019). *URMD6808: Analysis.* Unpublished Presentation. University of the Free State.

Department of Human Settlements. (2004). *Breaking new ground: A comprehensive plan for the development of sustainable human settlements.*

Gouverneur, D. (2015). *Planning and design for future informal settlements.* Routledge.

Jacobus, R. (2015). *Inclusionary housing, creating and maintaining equitable communities.*

Klug, N., Rubin, M., & Todes, A. (2013). Inclusionary housing policy: A tool for re-shaping South Africa's spatial legacy? *Journal of Housing and the Built Environment, 28*, 667–678.

Laner, D., Fellner, J., & Brunner, P. H. (2009). Flooding of municipal solid waste landfills – An environmental hazard? *Science of the Total Environment, 407*(12), 3674–3680.

Lees, L. (2019). Planetary gentrification and urban (re)development. *Urban Development Issues, 61*, 5–13.

Lerman, B. R. (2006). Mandatory inclusionary zoning—The answer to the affordable housing problem. *Boston College Environmental Affairs Law Review, 33*, 383–416.

Lopes, A. C., Rodrigues, E. B., & Vera-Cruz, R. (2019). Tourism gentrification. *IOP Conference Series: Materials Science and Engineering, 471*(9), 092025. IOP Publishing.

Mabasa, N., & Mthetwa, A. (2019). The City of Johannesburg takes a cautious step towards inclusionary housing. Retrieved June 23, 2019 from https://www.dailymaverick.co.za/article/2019-03-15-the-city-of-johannesburg-takes-a-cautious-step-towards-inclusionary-housing/

Mangaung Metropolitan Municipality (MMM). (2018). *Built environment performance plan 2018/19–2020/21*. MMM.

Nahrin, K. (2018). Analysis of inclusionary housing as an urban planning instrument of the North in the South: The context of Dhaka. *Urban Development Issues, 58*, 19–27.

Nzimande, L. (2020). *The challenges and benefits of a regulated versus a negotiated approach to inclusionary housing in Johannesburg*. Doctoral Dissertation.

Onatu, G., & Baloyi, V. K. (2020). The use of mixed-income housing development to address poverty and inequality, based on Delphi empirical approach: A case of South Africa. *Journal of Advanced Research in Social Sciences, 3*(3), 62–77.

South Africa. (1996). *The constitution of the Republic of South Africa, Act 108 of 1996*. Government Printer.

South Africa. (1997). *Housing Act, Act 107 of 1997*. Government Printer.

South Africa. (2008). *Social Housing Act, Act 16 of 2008*. Government Printer.

South Africa. (2013). *Spatial planning and land use management Act, Act 16 of 2013*. Government Printer.

Thaden, E., & Wang, R. (2017). *Inclusionary housing in the United States: Prevalence, impact, and practices*. Lincoln Institute of Land Policy.

United Nations. (2014). *The right to adequate housing*. Retrieved May 29, 2023 from https://www.ohchr.org/sites/default/files/Documents/Publications/FS21_rev_1_Housing_en.pdf

Van Gent, W., & Hochstenbach, C. (2019). The impact of gentrification on social and ethnic segregation. *Handbook on Urban Segregation*.

Viney, S. (2015). Is gentrification urban renewal or a form of social cleansing? Retrieved June 25, 2019 from https://www.abc.net.au/news/2015-10-11/gentrification-urban-development-or-exploitation/6818440

Wardner, P. (2014, January). Explaining mixed-use developments: A critical realist's perspective. In *20th Annual Pacific-Rim Real Estate Society Conference*, Christchurch, New Zealand (pp. 10–22).

What is Urban Regeneration? (2016). Video. YouTube: Investor Trading Academy.

Zheng, H. W., Shen, G. Q., & Wang, H. (2014). A review of recent studies on sustainable urban renewal. *Habitat International, 41*, 272–279.

Monaphathi Timothy Lehobo holds a master's degree in Urban and Regional Planning, an honours degree in Spatial Planning and a bachelor's degree in Urban and Regional Planning. He is affiliated with the South African Council for Planners as a Candidate Town and Regional Planner. He is currently a Lecturer in the Department of Urban and Regional Planning at the university of the Free State. He has had both practical and academic experience in Town Planning; and has a few peer-reviewed publications ranging from housing to urban health. His interests are in housing, informal settlements, geographic information systems, and township development.

Stuart Paul Denoon-Stevens is an academic and town planner who is dedicated to understanding how to create better places to live and work. His approach is a unique blend of research, policy development and practical application. He is a Planning Policy and Research Officer at Nottingham City Council, United Kingdom, and a former Senior Lecturer at Nottingham Trent University. He also previously lectured in the Department of Urban and Regional Planning at the University of the Free State, South Africa. He is a registered professional town planner, with research interests including municipal land management and regulation, planning practice, mining towns, and a variety of others.

Johannes Bhanye is an Urban Planner and Postdoctoral Research Fellow at the African Climate and Development Initiative (ACDI), University of Cape Town He holds a Ph.D. in Social Sciences (Migration and Land Settlement), an M.Sc. in Social Ecology, and a B.Sc. in Urban and Regional Planning both from the University of Zimbabwe. He is also a former Research Fellow at the University of the Free State the Andrew W. Mellon Foundation and the Centre for Applied Social Sciences (CASS) at the University of Zimbabwe. Dr. Bhanye is very passionate about Southern and Postcolonial Urbanism with a particular focus on Spatial Justice, Urban Informality, Vulnerabilities in African Cities, Urban Health, Urban Governance, Complexity and Systems Thinking, Climate Risk and Urban Resilience, and Migration and Statelessness. He has practically conducted research projects in Zimbabwe, South Africa, and Lesotho using Innovative Participatory and Ethnographic Methodologies in understanding the lived experiences of people surviving on the margins in African cities.

Chapter 13
(De)Constructing Tenure Prospects in Emerging Housing Schemes—A Case Study of Harare

Lovemore Chipungu and Hope Hangwelani Magidimisha-Chipungu

Abstract The fast-track land reform programme in Zimbabwe (in 2000) provided a platform for popular movement of people to access land in and around cities legally and illegally. This mass movement, analysed from an urban perspective, signified a determined occurrence by the general populace to shrug off decades of denial to the city and its associated benefits. This chapter analyses the acquisition of land and the eventual emergence of housing on such land. These houses, while accepted by beneficiaries, were scorned and criticised by others who saw the informalisation of urban space under the guise of social justice. At the centre of these dynamics were tenure issues which, despite the government's determination to protect beneficiaries in the newly acquired space, still left a cloud of uncertainty. This was not surprising given that successive Zimbabwean governments have failed to embrace popular modes of housing production in the past. Informal housing schemes have always been treated with apathy and indifference—hence whenever they mushroomed, they were ruthlessly destroyed. What could have changed the scenario in 2000? In response to this question, this paper analysed emerging housing schemes in Harare and noted that the tenure status of beneficiaries was varied with "chance" and "determination" being the most determinant among most of them. In a bid to unravel the structural determinants of these developments, the study's main focus was in the period 2000 to 2009. The author employed both qualitative and quantitative research methods using mapping, interviews and observations in order to collect data. The selected housing schemes of Nehanda Housing Scheme provided cross-cutting perspectives that enabled the extrapolation of key findings at both city and national levels. The findings emanating from this study show that the FTLRP provided a platform for innovativeness among key stakeholders and was fuelled by the unpronounced national urban vision and the prowess of the political leadership—thus guarantying low-income households' the right to the city despite prevailing challenges of infrastructure.

L. Chipungu (✉) · H. H. Magidimisha-Chipungu
School of the Built Environment and Development Studies, Howard College, University of KwaZulu-Natal, Durban, Republic of South Africa
e-mail: Chipungu@ukzn.ac.za

H. H. Magidimisha-Chipungu
e-mail: MAGIDIMISHAH@ukzn.ac.za

L. Chipungu et al. (eds.), *Contemporary Models for Pro-Poor Housing in Urban Sub-Saharan Africa*, https://doi.org/10.1007/978-3-032-05489-0_13

Keywords Land reform · Housing · Cooperative · Tenure: land · Occupation

13.1 Introduction

The Fast Track Land Reform Programme (FTLRP) was a stage for land tenure deconstruction in as much as it was a stage for land tenure reconstruction. From its inception, it provided a platform for contestations of existing tenure systems as the government in power violently confiscated large-scale commercial farms that were in the hands of the few commercial farmers for redistribution to poor peasants and the landless. In the process, this saw the beginning of the deconstruction of existing tenure systems which were mainly in the hands of commercial farmers, while sowing seeds for the reconstruction of tenure systems for the poor peasantry and landless people in urban areas. This chapter pursues this de/reconstruction discourse of tenure systems that emerged at the peak of the FTLRP and repositions its arguments from the perspective of the urban poor who were in search of housing opportunities. Hence the stabilization of the urban poor who invaded land for housing provides a provocative platform where emerging tenure systems usher in both challenges and opportunities to contextualise their relevance in the wake of huge housing backlogs, which conventional housing production systems could not clear. In its search for the truth, this book chapter unfolds through a number of captivating sections which start by analysing the concept of de/reconstruction of tenure reviewing existing literature and illustrating contextual land contestations through the window of Nehanda Housing Scheme (NHS) in Harare South. The key theme being interrogated throughout these sections is the quest for the contextual relevance of emerging tenure systems and how they safeguard properties of beneficiaries in contested housing schemes.

13.2 A Case for De/Reconstruction of Tenure Systems—Background

There are a number of issues that underpin the FTLRP and these range from its unorthodox manner through which land for redistribution was acquired to its sheer failure to put in place a sound financial model to fund the redistribution programme (Chipungu and Masiiwa, 2004). However, in the interest of this book chapter, two similar critical issues that are singled out revolve around the tenure system. On the one hand, is the deconstruction of land tenure systems that was prompted by the manner in which land was acquired and its occupants were forcefully thrown out of their properties. In the ensuing process, the legality of tenure holders could not support them as they either became "instant squatters" on their land or were rendered homeless. However, as they held on to their legal tenders, (title deeds), and through protracted legal battles, it took years for them to realise the protective nature of their security of tenure. But even then, they could not return to their land, instead, they had to settle for partial compensation which recognized their capital developments only on land whose soil was "deemed stolen" and therefore, could not be part of the financial compensation settlement. Indeed, on the part of the over 1300 land owners, especially white commercial farmers; this gross political manipulation of legal processes represented deconstruction and total collapse of the once formidable land tenure system. As Kadzere (The Herald 2024) observed—even by 2024, the US$3.5 billion compensation fund they were supposed to receive under the Global Compensation Deed Deal was still outstanding.

On the other hand, the FTLRP saw the government's protracted effort to protect massive land acquisitions through invasions and its people through rushed court appeals and the repeal of some existing legal instruments. Irrespective of the political tussle between the party in power, the Zimbabwe African National Union—Patriotic Front (ZANU PF) and its major opposition; The Movement for Democratic Change (MDC), the government, in a bid to protect what it gained, argued that the people who took land were demonstrating against landlessness and therefore were rightfully so to be where they were. This was a firm decision taken in such a way that no state security could be released to protect those whose security of tenure was being violated or vice versa. Besides, from an organizational point of view, this was a dicing situation given that farm invasions were spearheaded by War Veterans (whose patron was the then President Robert Mugabe) who led ZANU PF supporters during the land invasions—therefore any action against the status quo would mean undermining the powers that be. The judiciary even enacted new laws that were meant to protect the new occupiers of land through the Rural Occupiers' law of 2008. Although this piece of legislation protected the rural community, even those who occupied peri-urban land felt covered and therefore immune from any decision to unsettle them.

Therefore, housing schemes which emerged during the FTLRP should be analysed from diverse perspectives but with a focus on those critical issues under spotlight. The same dynamics that were at play at the peak of the FTLRP should be revisited in another tone if the dilemma that befell the urban poor during Operation Murambatsvina (OM) is to be understood and avoided. This book chapter focuses on the period from 2000 to 2009—just before the onset of the Government of National Unity. However, while the Zimbabwean context provides such compelling evidence of competing factors during the time in question, it is equally important to provide an academic conceptualization of tenure systems from a purely housing perspective in order to provide a framework for analysis. Section 13.2 delves into such conceptual issues in order to offer insight into the fundamentals of tenure dynamics.

13.3 Conceptualising Tenure Systems

Housing tenure serves as a foundational concept in understanding how individuals and communities' access, secure, and manage their housing. The typologies of housing tenure—ranging from ownership, rental, and cooperative models to informal tenure systems—each reflect distinct socio-economic dynamics, legal frameworks, and cultural contexts. Critically examining these typologies reveals the tensions and possibilities inherent in housing systems, particularly in contexts marked by inequality and resource scarcity.

Ownership has traditionally been regarded as the most secure and aspirational form of housing tenure, rooted in its association with wealth accumulation and stability (Haffner et al., 2011). However, this perspective often obscures the inequities reinforced by property markets, especially in developing countries where land ownership is concentrated among elites. Rental housing, while flexible and accessible to mobile or economically vulnerable populations, is frequently characterised by precarious conditions and exploitative practices (Gilbert, 2016). Informal tenure, which encompasses squatting and unauthorised settlements, challenges conventional notions of legality and property rights but often serves as a critical mechanism for shelter provision in urbanizing regions (Roy, 2005).

Theorists like John Turner and Hernando de Soto have provided pivotal frameworks for understanding tenure security. Turner (1976) emphasised the functionality of informal settlements, arguing that self-help housing allows marginalised populations to progressively improve their living conditions despite limited formal recognition. While Turner's perspective celebrates grassroots agency, it has been critiqued for romanticising informality without adequately addressing the systemic drivers of housing exclusion (Watson, 2009). Conversely, de Soto (2000) argued that formalising property rights in informal settlements could unlock economic potential, as secure tenure enables access to credit and investment. However, Bromley (2009) entails that this perspective has faced significant criticism for its neoliberal underpinnings, which overlook the complexities of urban land markets and the risks of commodifying land.

Housing cooperatives introduce a distinctive tenure model that challenges the individualistic and market-driven logic of ownership and rental systems. Cooperatives promote collective ownership, democratic governance and affordability, offering a pathway to secure tenure for low- and moderate-income households (Lang & Mullins, 2015). Unlike private ownership, which prioritises profit maximisation, cooperatives emphasise community control and equitable access to resources. This model aligns closely with social and economic rights frameworks, particularly in contexts where the state fails to provide adequate housing. Nevertheless, Ganapati (2010), states that cooperatives face operational challenges, including the need for robust legal recognition, financial sustainability and effective member participation.

Critically, the conceptualisation of housing tenure must transcend simplistic dichotomies between formality and informality, individualism and collectivism. The dynamic interplay between these dimensions reflects broader socio-political structures and the contested nature of urban spaces. In this light, housing tenure is not merely a technical or economic matter but a deeply political one, with implications for social equity, environmental sustainability and human rights.

13.3.1 Tenure Systems in Developing Countries

Developing countries are contending with unprecedented challenges in the realm of urbanisation, which has led to the explosive growth of informal settlements and the widening gap between the wealthy and the impoverished (Huchzermeyer, 2008). In cities across Sub-Saharan Africa, South Asia and Latin America, the rapid influx of rural populations to urban areas has overwhelmed existing housing infrastructures, leading to a proliferation of informal and often illegal forms of tenure. These settlements, often constructed without access to basic services like water, sanitation and electricity, emerge in the absence of robust urban planning frameworks (Davis, 2006). According to Pimlott (2017), housing tenure in this context, is not merely a question of legal ownership but of survival, as individuals in informal settlements often negotiate their right to live in urban spaces through informal agreements that lack legal recognition, exposing them to eviction or forced displacement.

However, the role of housing tenure in perpetuating or mitigating urban inequalities is multifaceted. On the one hand, secure tenure can be a powerful tool for reducing poverty and inequality by allowing residents to invest in their homes, build wealth and access public services (Durand-Lasserve, 2012). On the other hand, the absence of formal tenure exacerbates the plight of the urban poor, making it difficult for them to secure credit or access state support (De Soto, 2000). Informal tenure systems often operate in parallel to formal markets, creating a dual system of housing provision where the affluent benefit from well-regulated ownership markets while the poor remain marginalised in unregulated, informal settlements. According to Nijenhuis (2019), this bifurcation exacerbates social stratification, with tenure insecurity contributing to social exclusion.

The tension between formal and informal tenure systems presents a complex dynamic. Formal markets tend to exclude lower-income groups who are unable to meet the financial or legal requirements for homeownership, resulting in reliance on informal tenure arrangements (Schreiner et al., 2014). The lack of integration between these systems fuels' inequality, as informal tenure holders lack the legal protections afforded to formal landowners (Parker, 2011). While many urban poor seek to formalise their tenure, bureaucratic barriers, lack of political will, and entrenched corruption often prevent this process from being realised, leaving a majority of the urban population without the benefits of secure tenure.

13.3.2 Policy and Governance

The policy frameworks designed to address housing tenure in developing countries are often inadequate, fragmented and poorly enforced, which exacerbates tenure insecurity. In Sub-Saharan Africa, for example, land tenure systems are often based on customary law that conflicts with national legal frameworks (Yeboah and Shaw, 2013). These dual legal systems create a complex, unpredictable environment where land rights are unclear, especially for women and marginalized groups. In Latin America, while the implementation of progressive land reform programs in the twentieth century made significant strides in addressing tenure insecurity, many countries have failed to maintain momentum, and informal settlements continue to proliferate. Most policies aimed at regularising informal tenure, though promising, often falter in their execution, as they are stalled by inefficiencies and corruption within land administration systems.

Moreover, these inadequate policies often fail to align with global housing rights frameworks like Habitat III and the Sustainable Development Goal (SDG) 11 on sustainable cities and communities. While these frameworks call for the provision of adequate and affordable housing for all, in many developing countries, they remain aspirational rather than operational (UN-Habitat, 2016). The gap between policy formulation and implementation is huge (Baker, 2012). Despite high-level commitments to inclusive urban development, political and economic realities often skew priorities toward market-driven solutions that favour private developers and the wealthy, leaving vulnerable populations in the periphery.

Furthermore, housing policies in many developing countries are not sufficiently attuned to the informal and evolving nature of urbanisation. The tendency to frame housing as a static problem, focusing on the formalisation of land titles or providing subsidies for homeownership, fails to account for the dynamic needs of rapidly growing informal communities. As such, tenure policies often perpetuate the same issues they aim to solve, creating a cycle of exclusion (Huchzermeyer, 2011). Policies must be recalibrated to engage with informality, allowing for flexible tenure solutions that can be adapted to the lived realities of urban poor populations.

13.3.3 Socio-Economic Implications

The socio-economic implications of insecure housing tenure in developing countries are profound. Lack of secure tenure limits access to basic services, as residents in informal settlements often have no legal claim to the land they occupy. This not only hampers their ability to demand services but also exposes them to eviction and displacement with little recourse for compensation or relocation (Alemu, 2015). Without tenure security, households are reluctant to invest in improvements to their homes, leading to substandard living conditions and poor health outcomes (Mitlin, 2004). Furthermore, tenure insecurity contributes to the fragmentation of urban spaces, making it difficult for cities to plan for infrastructure development or disaster preparedness. Harris (2003), states the inability to predict where people will settle or remain stymies the effectiveness of public investments in services, which further marginalizes vulnerable groups.

13.4 Research Methodology

Nehanda Housing Scheme (NHS) is among the first housing cooperatives which emerged through land invasions at the height of the FTLRP in 2000. Located 19 km outside the Harare city centre close to Dzivarasekwa (see Fig. 13.1), this housing scheme came into existence through land invasions and formal allocation of land to beneficiaries.

The development of NHS is a tale of conflict and contestation among various stakeholders. For that reason, the research methodology that was used to solicit information in the area evolved in response to the dynamics of on the ground. It must be remembered that this housing scheme emerged under the leadership of war veterans who spearheaded the FTLRP supported by landless people who were looking for housing land in urban areas. The situation was aggravated by the political and violent nature of events that unfolded during the FTLRP. For these reasons, multiple research techniques and approaches were used in a bid to gather comprehensive information about the housing scheme.

This study employed a mixed approach to gather data. Both qualitative and quantitative methods were used with a variety of research tools. The use of the qualitative approach was essential for it allowed creativity in an environment that was filled with uncertainty and fear. Techniques such as observations coupled with unstructured interviews of random beneficiaries guaranteed access to reasonable quantity of information. For instance, most people were concerned about how information was going to be used despite being assured about the purpose of the research. More so, some beneficiaries belonging to cooperatives were under stern warning not divulge

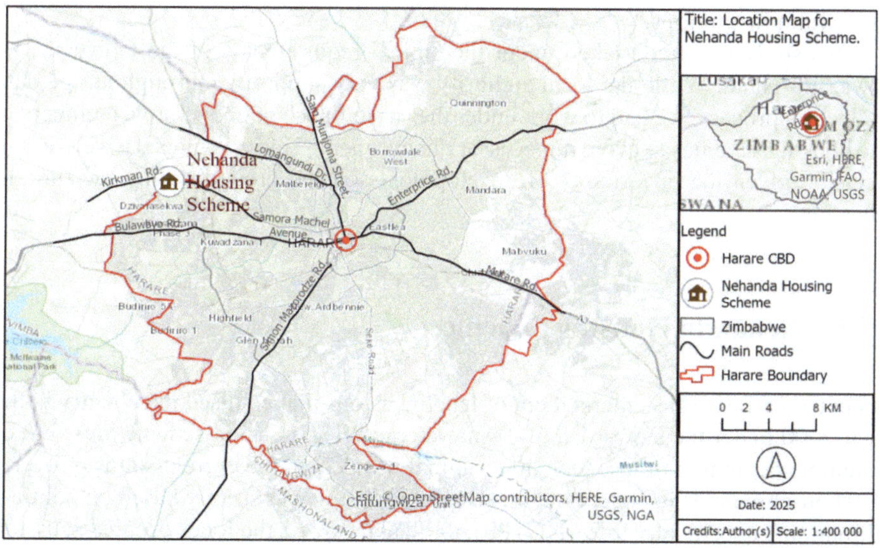

Fig. 13.1 Locality Map of Harare North Housing Scheme. *Source* Authors (2025)

information unnecessarily to strangers. Hence creativity proved to be a key tool to gather critical data.

However, detailed quantitative data were also obtained through a household survey. This survey was limited to 100 households (which is above the 10% of the target population) who were selected randomly from 750 households who were already onsite at the time of fieldwork. These were interviewed using the same structured household questionnaire in order to yield a uniform data set which allowed for easy presentation and analysis. Where necessary, gaps detected during household interviews were noted and addressed through open-ended questions provided within the household questionnaire.

The official position of this housing scheme was equally important. This was obtained by collecting qualified opinions from different key informants representing the Town Planning Department in Harare City Council, the Harare Provincial Government, the Department of Physical Planning and some officials from the housing cooperatives. Information from key informants provided insight into the prevailing contestations among various stakeholders and the justified legitimacy of the housing scheme.

A combination of these different approaches supported by a variety of data collection techniques were very instrumental in yielding a fairly representative data set for this area. Both qualitative and quantitative data proved to be flexible under different data analysis techniques. Through excel, quantitative data yielded graphs in various formats, that were extracted to explain patterns of behavior and collective opinions though from individual households. These were strategically placed in sections that were thematically presented and discussed. The thematic approach was easily supported by insights provided by key informants and other qualitative tools which further provided an enriched narrative about dynamics at work among beneficiaries and within institutions.

This study was undertaken under the ethical requirements of the University of KwaZulu-Natal. While the local authority was very supportive throughout the data collection process, we had to work under the surveillance of cooperative members in the housing schemes to avoid unnecessary harassment. This was indeed a welcoming decision given the uncertainty of developments on the ground during the time in question.

13.5 Nehanda Housing Scheme

As noted earlier, NHS emerged out of land invasions that eclipsed the country in the year 2000 under the stewardship of war veterans. The housing scheme grew out of a humble beginning of less than 300 original invaders to the current status of over 1 000 households. At its inception, this housing scheme was strictly based on political affiliation—specifically to ZANU (PF) political party. As the focus group discussion noted, potential beneficiaries were supposed to be either active party members (as evidenced by possession of a party card) or those willing to join the party. These conditions were not a hindrance to many people who saw this as an opportunity to acquire land and build their own houses.

The land acquisition process also determined the settling and stabilisation of beneficiaries. During the initial days of land occupation, beneficiaries were allocated plots as they joined the housing scheme using rudimentary means and were advised not to build permanent structures. Some beneficiaries opted to continue staying in their rented accommodation adjacent to the housing scheme. A security team made up of youths from "land invaders" guarded the site from rival land seekers. In a move to formalise their housing scheme, a housing cooperative was formed under the leadership of war veterans. Beneficiaries were required to attend cooperative meetings and make financial contributions. The transformation of the housing scheme into a registered housing cooperative strengthened the beneficiaries' vision. A voluntary professional team made up of builders, carpenters, and other artisans was selected and mandated to steer the cooperative towards a proper housing scheme following government regulations. Investment in development was supported by monthly contributions of Zimbabwean US$40 from households into the housing scheme. Using these resources, the cooperative was able to engage professionals such as town planners, architects, land surveyors and engineers who prepared a residential layout and building plans. A residential layout with plot sizes ranging from $300m^2$ to $800m^2$ was approved in 2006 with provision for a mixed development comprising land for residential, commercial, and public developments. However, the building of four-roomed family core-houses began earlier in 2002 when the residential layout was still under consideration.

While the housing scheme was driven by political forces with many beneficiaries being supporters of the ruling party, it also attracted people from diverse backgrounds. The landless people who were part of the forces driving the FTLRP were actually bona fide urban dwellers who could not afford land for housing despite their marital, educational and employment status. Most of them were renting in the nearby residential areas of Dzivarasekwa, Mufakose and Glenview—thereby giving them an opportunity to be part of "land invasion protesters" (as the late President Robert Mugabe remarked) with the intention of getting an opportunity for a housing plot. This was an opportune time as most of them indicated that they were tired of being "slaves" to landlords whose rental charges and other demands were becoming unbearable. It should be remembered that at the time in question, Zimbabwe was also experiencing economic challenges. Hence while 68% of the respondents interviewed were formally employed, only 40% earned above US$300 which in this case, represented the apex income of the low-income group. This was expected given that 70% of the respondents' highest education was just an Ordinary Level (O'Level) qualification as shown in Fig. 13.2.

The situation was aggravated by the fact that 18% of the respondents were informally employed—a situation which meant that their income was erratic, thereby making it difficult for them to plan in advance (see Fig. 13.3 for employment status). Hence, the FTLRP provided an opportunity for tenants to escape from *'rental slavery and claim their right to freedom'* and the city through acquisition of land for housing.

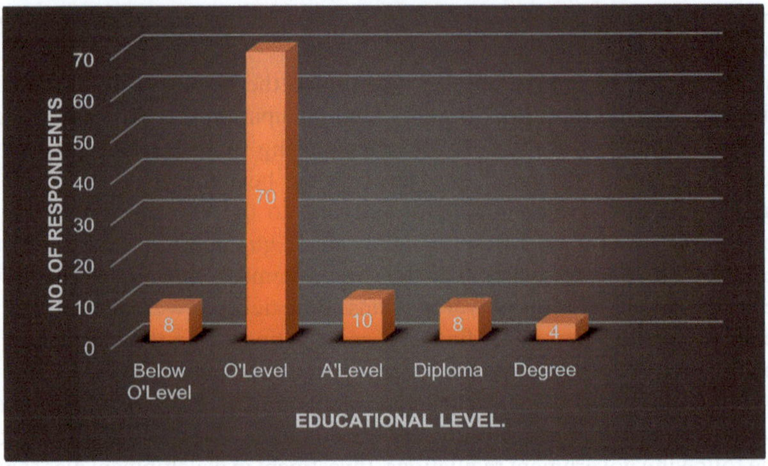

Fig. 13.2 Educational Status of Beneficiaries in NHS. *Source* Fieldwork (2011)

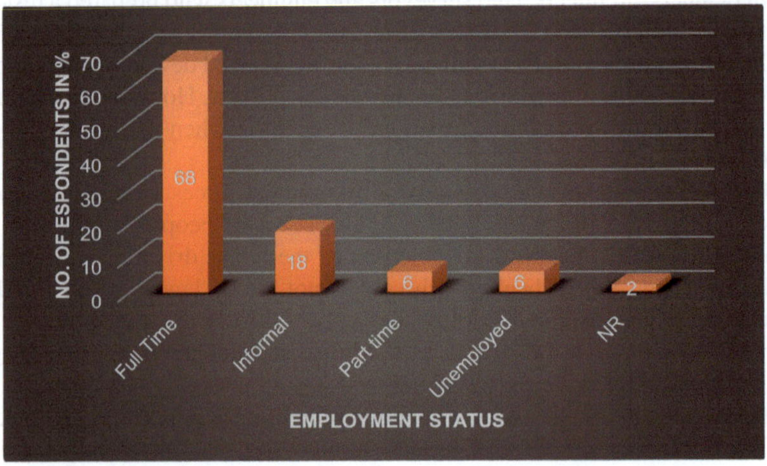

Fig. 13.3 Employment status of beneficiaries. *Source* Fieldwork (2011)

However, being part of a housing cooperative was beneficial to households since they were able to achieve remarkable milestones in their search for decent housing. Beneficiaries of NHS submitted building plans to the local authority (Harare City Council) for approval before embarking on the construction process. These building plans were based on a 4-roomed core house plan which was proposed and prepared by the cooperative's technical team and submitted to the LA for approval. This became the template upon which all beneficiaries constructed their houses. While the core house was built by the cooperative, any further extensions were done by individual beneficiaries. This enabled the construction of proper structures using appropriate building materials as prescribed by the local authority. As shown in Fig. 13.4, houses were built of standard bricks under the guidance of municipal building inspectors who came for stage inspections.

While many beneficiaries were happy that they finally got decent housing, security of tenure was not clear among many of them. This partly stems from the controversial nature of the FTLRP where land was invaded and under contestations from previous land owners. This explains why 12% of the respondents simply said that the housing scheme was controversial and they were not sure about security of tenure (as 6% of them stated). The lack of response from 38% of the beneficiaries further sends signals to the nature of the housing scheme as far as security of tenure is concerned (see Fig. 13.5 for further responses).

Despite the controversy surrounding the FTLRP and emerging housing schemes, some beneficiaries (about 38%) were unfazed since they had hope and faith in the development despite not having title deeds. More so, they were convinced that having receipts (32%), which showed their contribution to the housing cooperative was equally enough to safeguard their acquisitions. Their membership to both the housing cooperative and its affiliation to the ruling party (ZANU-PF) provided sufficient security of tenure as witnessed by responses such as "safe and hope". More so, the non-response of some of the beneficiaries has a silent meaning, especially in an environment fraught with violence and retribution, more so where people did not want to openly reveal their political affiliation. It remained to one's imagination to conclude that they were either ruling party cadres or mere opportunists who were not bothered by the situation.

What is emerging so far is that most beneficiaries were not clear about issues of security of tenure given the controversial nature of the FTLRP. While others pointed to deeds of sale, title deeds, and certificates of occupation, the reality of the matter is that at the time when this research was conducted, some of these documents were not

Fig. 13.4 Houses in NHS. *Source* Fieldwork (2011)

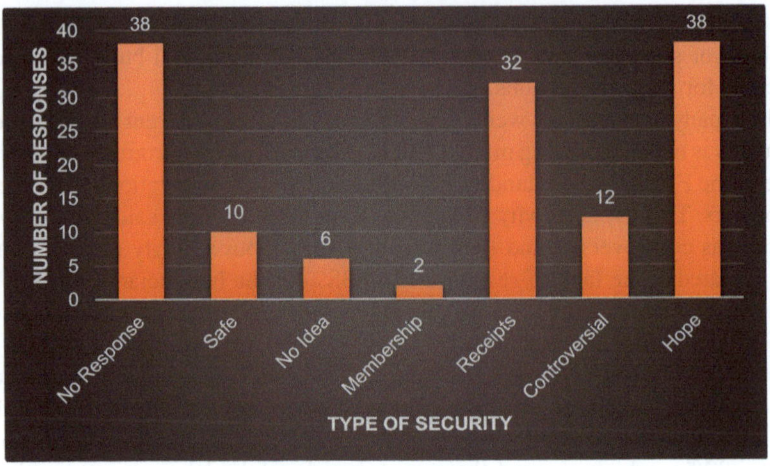

Fig. 13.5 Ownership guarantee. *Source* Fieldwork (2011)

yet to be made available to beneficiaries. Therefore, under these circumstances, what guarantee of security did beneficiaries of these housing schemes have? Sect. 13.5 below analyses emerging tenure systems in the context of dynamics at work during the time in question.

13.6 Emerging Issues: De/constructing Tenure Issues

Since the inception of NHS, there has been an interplay of ignorance, fear, and uncertainty among some beneficiaries of this housing scheme. Indeed, these were genuine concerns emanating largely from how the housing scheme emerged at a time when the land question was being contested. However, there is need to elaborate on how these underlying tones in order to establish the extent to which beneficiaries are guaranteed security of tenure. Irrespective of factors that were at play at the inception of the development, it is critical to understand at this juncture that security of tenure is not static—but it is dynamic and responsive to changes in socio-economic and political landscapes. Pettitore (2021) in his World Bank Commentary clearly points out this aspect by noting that:

> Land tenure exists along a spectrum from strong too weak with numerous factors moving the needle between the two.

Indeed, in the context of housing schemes which emerged at the height of the FTLRP in Zimbabwe, there has been a gradual move from a point of tenure deconstruction to the current stage of tenure reconstruction. These housing schemes have since evolved in terms of tenure to provide testimony on how beneficiaries are guaranteed security of security. In this regard, there are a number of factors that should be considered, and these range from mere individual hope to existing legal evidence provided by both beneficiaries and local authorities.

13.6.1 Land Occupation

The development of housing schemes in question began with the occupation of land. Land occupation is an act of physical dispossession that guarantees immediate access to a piece of land. As process, it defies and redefines existing authority in favour of occupiers. This is a significant step because occupation is an expression of interest which manifests itself physically whereby new occupants settle on a particular piece of land that does not belong to them. Hence occupation of space, does not only lead to the demarcation of boundaries by occupants, but also provides the basis for identity in the initial and subsequent stages of the development process. The Black's Law Dictionary (1910) categorically states that the word "occupation" when applied to real property is ordinarily equivalent to possession. However, in so doing, it can be a potential source of conflict in its initial stage before evolving into a magnetic point as more people are attracted to the occupied piece of land. While this is completely detached from possession, (due to lack of legal evidence of ownership), it creates a potential platform for future legal contestations as original owners fight against dispossession.

Occupation is the most powerful tool at the disposal of landless and homeless people who engage in informal housing. In the Zimbabwean context, the contested nature of the FTLRP provided a window of opportunity to the homeless people who used it to occupy land for housing development. What made this strategy more effective was the ability of households to showcase their interest by not only occupying, but also building houses and planning future developments of the whole neighbourhood. To safeguard against any form of intrusion or intimidation on the site, youths were employed whose task was to ensure that land the designated area was occupied in an orderly manner under the guidance of the cooperative. During the formative stages where land was under a lot of contestations, the youth protected the interests of cooperatives against strangers who were not affiliated to the cooperative and to the ruling party. Thus occupation, quickly evolved to become a guarantor of people's interests in housing development because their presence on site became a physical barrier to any form of physical displacement.

Depending on existing development control mechanisms, land occupation can lead to eviction or stabilisation of occupiers as seen through rezoning and reconstruction. This is a narrative which is associated with contested housing developments such as informal housing. More so, it is a narrative which fits in contemporary housing schemes that emerged during the FTLRP in Zimbabwe—and NHS is no exception. In substantiating the value of occupation, Mahoney et al. (2007:11) argue that:

> …. the significant point is that security of tenure ……can be and frequently is enjoyed without any concrete evidence of title other than occupation. In those countries where individual property rights are recognised and the rule of law prevails, the courts will uphold occupation against anybody—including the state or the government—other than a person who can prove a better right. In fact, provided that nobody else can produce evidence establishing some right, the courts will not require any proof from the occupier, for there is much truth in the old saying that 'possession is nine points of the law.

Indeed, based on this observation by Mahoney (2007), NHS is among many housing schemes whose status has been challenged largely because of how the land acquisition process was handled. Under the cover of ZANU PF and the housing cooperative, NHS developed to what it is today despite contestations from the local authority at the time of occupation. But more so, it is important to note that the FTLRP was the first land reform programme whose poor implementation process allowed for mass land invasions countrywide. So, it came as a shock to most people and institutions that had known formality as the normal development process in Zimbabwe. In the face of prevailing contestations, it became the first stage in the deconstruction of existing tenure yet forming the cornerstone in the formation of a new chapter in tenure systems. The recently announced Farm Title Deed Programme meant to replace the 99-year leases introduced by the Zimbabwean government on farms occupied during the FTLRP (Mike Juru, January. 17, 2025) points to some contestations and dynamics associated with security of tenure.

13.6.2 Hope as a Guarantor of Security of Tenure

Can hope really be a guarantor of security of tenure? It is interesting to note that beneficiaries' reference to the biblical notion of hope and faith emerges as one of the factors, which kept them committed to their housing scheme. This must be understood from a historical context knowing that most Zimbabweans are spiritual people with strong faith in their traditional and Christian values. The first Christian missionaries found the indigenous people entrenched in their traditional beliefs. While many missionary centres sprouted around the country, spreading Christian values, some people remained rooted to their traditional values. It is not by coincidence that the 1st and 2nd *Chimurenga* (war of liberation fought between 1896-1897 and 1967-1979, respectively) were fought under the spiritual guidance of Nehanda (a powerful spiritual medium of the Shona people). The heroic leadership of Nehanda Charwe Nyakasikana (also known as Mbuya Nehanda—who was the spirit medium) against British occupation during the First Chimurenga is well documented (Dube, 2018; Beach, 1998). Despite being executed by the British South African Company in 1898, her legacy and the spirit of Nehanda became associated with spiritual guidance and resistance as seen during the Second Chimurenga when the spirit of Nehanda was invoked (Alexander, 2006; Helliker et al., 2021). Similarly, during the FTLRP, (which is also called the 3rd Chimurenga) the spirit of Mbuya Nehanda was rekindled by ex-combatants of the Zimbabwean War of Liberation who led the FTLRP. It is therefore not by coincidence that this housing scheme is called Nehanda Housing Scheme since it was under the leadership of war veterans who endured the brutal war of liberation as they clung to spiritual hope and faith. This strong belief was candidly explained by one beneficiary when asked about security of tenure:

Hatizivi, asi ticharamba tichinamata kuna Mwari naana Mbuya Nehanda kuti vatitungamirire parwendo runo sekutitungamirira kwavakaita muhondo (we don't know, but we will continue praying to God and Mbuya Nehanda to guide us as they did during the war of liberation).

Indeed, this cannot be ignored because it is a continuation of the country's liberation struggle with land being at the core of resistance against colonial rule. Its capitalisation into housing is symbolic of the core value of the *Third Chimurenga* (FTLRP) whose call was for economic emancipation through land acquisition. In this regard, faith and hope provided a platform for resistance and negotiations towards recognition and formalisation of emerging housing schemes. While they did not provide tangible evidence or guarantee for ownership as universally known in real estate, it is important to note that the existence of housing cooperatives is a collective gesture of ownership and resistance especially in an environment of political instability as witnessed at the peak of the FTLRP. This is substantiated by Pimlott (2017) who argues that housing tenure is not a mere question of legal ownership but also a question of survival. In this context of faith and hope, tenure security exhibits itself through resilience and resistance to any intervention that would distort the existing status quo of the housing scheme—thus further confirming the theme of tenure deconstruction as a tool to solidify the acquisition of land and subsequent capital development.

13.6.3 Political Affiliation as a Tenure Guarantor

It has been documented by many authorities that the FTLRP was largely driven by political dynamics emanating from the entry of the Movement for Democratic Change on the political scene (see Chipungu and Magidimisha, 2021; Alexander, 2006; Helliker et al., 2021). The FTLRP, while being a government initiative, was led by War Veterans who, at a later stage, did not only occupy influential decision-making portfolios on land but were also at the helm of most housing cooperatives that emerged during this time in question. It must be remembered that War Veterans were members of ZANU-PF and their organisation was affiliated to ZANU-PF as seen by its patron (the President of Zimbabwe) who was at its helm. In this regard, the ruling party, ZANU-PF continued on its trajectory of being the liberator of the country from both political and economic perspectives.

Therefore, association with the ruling party was critical during this period. The names of many housing cooperatives (such as Ushewokunze, Sally Mugabe Heights, Chimurenga, Kadungure, Border Gezi, etc.), which emerged during this period showed this connectivity with the ruling party. In fact, as observed in the case of NHS, one was guaranteed access to land if only he/she was a member of the ruling party or committed to join the ruling party. Indeed, there was no question to doubt the significance of this association given the historical success of the party despite its challenges in post-2000. Chipungu and Magidimisha (2021), quoting Castells' (1983:294) equally observe that:

> Without affiliation to a political party, the demands and desires generated by social movements do not prevail in the production of social space and institutional change.

This substantiates the powers of War Veterans as leaders who were at the helm of the FTLRP. Hence it is this affiliation, which gave beneficiaries leverage to occupy land and build houses knowing very well that they were under the protection of the ruling party. The presence of the national flag on many offices of housing cooperatives served as a reminder about this affiliation. It is therefore not surprising to note that 10% of the beneficiaries openly said that they were not worried about the future of the housing scheme or their security of tenure since they were strongly linked to the ruling party.

While there were those who cited their developments as controversial or chose not to respond, still they had a sense of assurance knowing very well that they were connected to the ruling party. Hence affiliation to the ruling party (ZANU-PF) became the smoke-screen through which many cooperatives sprouted as people searched for avenues to access land for housing construction.

13.6.4 A Cooperative as a Guarantor

As already noted, once land was occupied, leaders of NHS transformed the occupations into a housing cooperative in line with the Cooperative Societies Act Chapter 24:05 revised in 1996 (Government of Zimbabwe). The formation and subsequent registration of the NHS as a cooperative provided another critical milestone in the process of establishing security of tenure for all beneficiaries. Cooperatives in this regard are instrumental in providing collective ownership which forms the basis for security of tenure for all beneficiaries since land and all capital developments are registered under the cooperative as a legal entity.

However, being a member of a housing cooperative is also a responsibility where beneficiaries should abide by rules of the institution. To ensure this guarantee, (in terms of security of tenure), members of the cooperative should commit themselves to payment of monthly subscriptions and adhere to other obligations as per the cooperative requirement. Some beneficiaries lost their affiliation and subsequently their properties due to their failure to comply with cooperative rules. This explains why some of the respondents in NHS pointed out with confidence that they are keeping records and receipts of their financial contributions to the cooperative since these guarantee their security of tenure.

Therefore, security of tenure is not underscored by one legal document such as title deeds—instead, these legal documents exist on the plane which ranges from deeds of sale, certificates of occupation, and lease agreements. Under the umbrella of the cooperative, all these documents provide an avenue not only to own a piece of land, but also to enlist support from the cooperative to partake in some of the benefits which accrue to all members. This notion is supported by the government which is more than willing to sell land to cooperatives at intrinsic value for housing development. The Harare City Council was equally very clear on the authenticity and financial capacity of cooperatives since this allowed them to connect to bulky infrastructure with minimal financial and technical support.

13.6.5 The Township Establishment Process

One interesting observation about NHS is that it is located in an area which was designated for residential development since it falls under Gillingham Local Development Plan No. 47 (GLDP-47) of 2005 (Harare City Council). It is important to note that the initial occupation of land began in 2000 at a time when the local plan was still under development. This, to some extent, shows that the land acquisition process for NHS was not an ad hoc process, but a calculated move on vacant land covered by the Harare Combination Master Plan of 1992 (Harare City Council) which designated the areas for housing developments despite the absence of the local plan. This superseded the Salisbury Rural Planning Scheme (Western Section) of 1981 (Harare City Council) which zoned the area as a green belt with the intention of arresting urban sprawl and preserving economic farming units. Therefore, from a purely technical perspective, the identification and subsequent development of the area prepared a platform for future recognition and regularisation of the housing scheme. It is therefore not surprising that the NHS prepared a residential layout which was approved and laid the ground for housing development.

The residential layout and other land use schemes are critical inputs into the creation of security of tenure. Apart from ensuring that land is used efficiently and in accordance with municipal policies and legislations, these also play a crucial role in managing township development. These controls are essential for maintaining a balance between residential, commercial, and public land use while supporting efforts to formalize the township and secure land tenure for its residents. It is this component which most respondents were oblivious of given that it is part of the technical town planning process. Therefore, this further confirms that security of tenure is based on administrative recognition, legal recognition, and township establishment. Commenting on these tenure mechanisms, the Urban Land-Mark (2010a, b:14) observes that:

> these modalities lay the basis for plot identification, recording of occupants, land use management, and provision of services.

Therefore, as noted above, the level of ignorance among residents in relation to this technical town planning process is understandable given that 70% of the beneficiaries had only O'Level as their highest academic qualification. This level of education coupled with the chaotic nature of the FTLRP which was fanned by political gimmicks disconnected beneficiaries from reality. The prevailing environment created at the time was not for questioning but for grabbing opportunities. In the process, appropriate information about the legality and illegality of housing schemes was not communicated.

13.6.6 The Construction Process

Another important dimension of the emerging housing scheme which points to the security of tenure is associated with the housing construction process. The construction process provides insight into the administrative and technical process, which guarantees security of tenure to beneficiaries as prescribed by the Urban Councils Act (*Chapter 29:15*) (Government of Zimbabwe). The approval of building plans, followed by stage approvals during the construction process by building inspectors from the local authority authenticates both the building plan and the structure (or house) which is supposed to be built using building materials prescribed by local authorities.

In this regard, it is not only occupation which guarantees security of tenure, but also the administrative and technical recognition, which is endorsed by officials from local authorities, which authenticates it. Once more, lack of knowledge on the part of beneficiaries speaks to ignorance and lack of advocacy on the part of the housing scheme leadership. The authenticity of this observation is shown by the survival of this housing scheme when Operation Murambatsvina left a trail of destruction when illegal developments were demolished by the government in 2005 (see Chipungu & Adebayo 2013).

Hence, housing production under the guidance of local authorities is a technical process that is governed by existing building regulations (such as building lines, building materials, and infrastructure requirements). Once beneficiaries comply with such regulations, their properties are recognised and approved as safe for human habitation. It is on the basis of such recognition and approval (among other factors) that certificates of occupation are issued. The importance of compliance (among other factors was highlighted in the legal battle between Nehanda Housing Cooperative and the Harare City Council (Case No. 748 Of 2022) in the High Court of Zimbabwe where the local authority testified that:

> .... the rampant non-compliance by most land developers in completing roads, water, and sewer connections, they as council, have devised a new tightening of screws strategy of demanding at the forefront the presentation of certificates of compliance by any land developer before the issuance of a Rates Clearance Certificate.

Therefore, it must be remembered that in new developments, certificates of occupation are key documentary evidence that contributes towards connectivity to municipal infrastructure, payment of rates, and eventually provision of title deeds. This, in essence, is a major guarantor of security of tenure. In summary, it must be emphasised that what makes security of tenure effective is a portfolio of evidence as provided by the Urban Land Mark (2010a, b:15) which attested that:

> Tenure security is increased when there is evidence to support a claim. Evidence can take many forms from oral, community backed assertions to documentary evidence or shack numbering.

13.6.7 Concluding Remarks

Emerging tenure systems in Harare's emerging housing schemes show a recalibration of tenure systems in line with the reality on the ground yet underpinned by a veil of uncertainty rightly informed by political dynamics, governance challenges and ignorance. A combination of all these factors creates both optimism in the future punctuated by fear on the part of beneficiaries yet anchored by resilience and resistance. These housing schemes (especially NHS) stand as a testimony of both factors (resilience and resistance) in an environment clouded with political and governance uncertainties largely emanating from the divide between the central and local governments. As it is, one would argue that security of tenure in housing schemes in Harare is a dynamic process gradually trending on a trajectory from deconstruction to reconstruction. Mass movement at the height of the FTLRP, saw the deconstruction of tenure systems when land invasions led to land occupation without any legal means which by default, created a safe haven for the government whose intention was to stabilise invaders on occupied land. This legally formed the first initiative to in protect land occupiers at the expense of land owners who had legal documents of ownership. However, in some instances, this stabilisation was temporary as the tenure deconstruction process continued under Operation Murambatsvina. It was only after Operation Murambatsvina that tenure reconstruction emerged under diverse modalities as shown in this analysis of NHS. Indeed, tenure systems are not static, but dynamic since they respond to various socio-economic and political factors at play at any stage in the process of the country's development.

References

Alemu, G. (2015). The role of secure tenure in alleviating urban poverty in Ethiopia. *Journal of Housing and the Built Environment, 30*(4), 553–572.

Alexander, J. (2006). *The unsettled land: State making and the politics of land in Zimbabwe 1893–2003*. James Currey.

Baker, J. (2012). Housing policy in the global south: Urban informality and the market in Latin America. *Habitat International, 36*(1), 1–9.

Beach, D. N. (1998). An innocent woman, unjustly accused? Charwe, Medium of the Nehanda Mhondoro Spirit, and the 1896–97 Central Shona Rising in Zimbabwe. *History in Africa, 25*, 27–54. https://doi.org/10.2307/3172179.ISSN0361-5413.JSTOR3172179.S2CID162261666

Bromley, R. (2009). Critiques of Hernando de Soto's ideas on land rights and formalization. *Journal of Urban Policy, 15*(2), 123–136.

Castells, M. (1983). *The city and the grassroots. A cross-cultural theory of urban social movements.* London, The UK.
Chipungu, L., & Adebayo, A. A. (2013). Policy and Planning divide: An evaluation of housing production in the aftermath of operation Murambatsvina in Zimbabwe. *Journal of Housing and the Built Environment, 28*(2), 381–396.
Chipungu, L., & Magidimisha, H. H. (2021). *Housing in the aftermath of the fast track land reform programme in Zimbabwe.* Routledge.
Davis, J. (2006). The role of housing cooperatives in securing tenure rights. *Urban Studies, 43*(7), 1147–1165.
Dube, Z. (2018). The ancestors, violence and democracy in Zimbabwe. *Verbum et Ecclesia, 39*(1), 3. ISSN: (Online) 2074–7705. https://doi.org/10.4102/ve.v39i1.1875.
Durand-Lasserve, A. (2012). The role of secure land tenure in urban development. *Urban Studies, 49*(1), 87–103.
Ganapati, S. (2010). Enabling housing cooperatives: Policy lessons from Sweden, India and the United States. *International Journal of Urban and Regional Research, 34*(2), 365–380.
Garner, B. A. (1910). *The Black Law's Dictionary* (2nd ed.). Thomson Reuters.
Gilbert, A. (2016). The return of the slum: Urban uprisings in the Global South. *International Journal of Urban and Regional Research, 40*(3), 682–690.
Government of Zimbabwe. Cooperative Societies Act Chapter 24:05 revised in 1996. Harare, Zimbabwe.
Government of Zimbabwe. Urban Councils Act (*Chapter 29:15*) of 2005. Harare, Zimbabwe.
Haffner, M., et al. (2011). Housing studies in transitioning markets. *Housing Policy Studies, 18*(5), 1234–1251.
Harare City Council. Salisbury Rural Planning Scheme (western Section) of 1981, Harare, Zimbabwe.
Harare City Council. Harare Combination Master Plan of 1992. Harare, Zimbabwe.
Harare City Council. Gillingham Local Development Plan No. 47 (GLDP-47) of 2005, Harare, Zimbabwe.
Harris, R. (2003). A double irony: The originality and influence of John FC Turner. *Habitat International, 27*, 245–269.
Helliker, K., Bhatasara, S., & Chiweshe, M. K. (2021). *The fast track land occupations in Zimbabwe.* Springer Nature. Switzerland.
Huchzermeyer, M. (2008). *Slums of hope or slums of despair: Changing perspectives on informal settlements in developing countries.* UCT Press.
Huchzermeyer, M. (2011). *Cities with 'slums': From informal settlement eradication to a right to the city in Africa.* Juta and Company Ltd.
Juru, M. (2025). *Farm title deeds programme bold step.* Southern Eye. Zimbabwe.
Kadzere, M. (2024). *Over 1300 farmers sign new compensation plan.* The Herald, Harare, Zimbabwe.
Lang, R., & Mullins, D. (2015). Bringing real localism into practice through co-operative housing governance: The prospects for community-led housing in England.
Mahoney, R., Dale, P., McLaren, R. (2007). Land markets—Why are they required and how will they develop? United Kingdom. International Federation of Surveyors Article of the Month.
Masiiwa, M., & Chipungu, L. (2004). Land reform programme in Zimbabwe: Disparity between policy design and implementation. In M. Masiiwa (Ed.), *Post-independence land reform in Zimbabwe: Controversies and impact on the economy* (pp. 1–25). Harare: Friedrich Ebert Stiftung and Institute of Development Studies, University of Zimbabwe.
Mitlin, D. (2004). Urban informality and tenure security in Africa: Evidence from case studies. *Environment & Urbanization, 16*(2), 21–40.
Nehanda Housing Co-operative Society Limited v Harare City Council (748 of 2022) [2022] ZWHHC 748 (19 October 2022)
Nijenhuis, R. (2019). *The effect of housing prices on the relatedness and complexity of Dutch municipalities* (Master's thesis).

Parker, S. (2011). *Homelessness among immigrant women in Ireland: A qualitative exploration.* Doctoral dissertation, School of Social Work and Social Policy, Trinity College Dublin.
Pettitore, P. (2021). *Conflict and land tenure security: What is the relationship? Commentary.* United Nations.
Pimlott, M. (2017). Negotiating tenure: Informal settlements and the urban poor in Sub-Saharan Africa. *International Journal of Urban and Regional Research, 41*(1), 47–61.
Roy, A. (2005). Urban informality: Toward an epistemology of planning. *Journal of the American Planning Association, 71*(2), 147–158.
Schreiner, M., Matul, M., Pawlak, E., & Kline, S. (2014). Poverty scorecards: Lessons from a Microlender in Bosnia-Herzegovina. *Poverty & Public Policy, 6*(4), 407–428.
De Soto, H. (2000). *The mystery of capital: Why capitalism triumphs in the west and fails everywhere else.* Basic Books.
Turner, J. F. C. (1976). *Housing by people: Towards autonomy in building environments.* Pantheon Books.
Un-Habitat (2016). *World cities report: Urbanisation and development—emerging futures.* UN-Habitat.
Urban Land-Mark (2010). Incrementally securing tenure: An approach for informal settlement upgrading in South Africa
Urban Land-Mark (2010). Development of an approach for the recognition of informal settlements and tenure security in South Africa with potential for regional applicability. Durban, South Africa.
Van der Walt, L. (2022). Beyond the cooperative ideal: Power and exclusion in collective housing. *Housing Studies, 37*(3), 421–439.
Watson, V. (2009). The planned city sweeps the poor away. Urban planning and 21st century urbanisation. *Progress in Planning, 72*, 151–193.
Yeboah, E., & Shaw, D. (2013). Customary land tenure practices in Ghana: Examining the relationship with land-use planning delivery. *International Development Planning Review, 35*(1), 21–39.

Lovemore Chipungu holds a PhD in Town and Regional Planning, a Master's Degree in Rural and Urban Planning as well as a Bachelor of Science Degree (Honours) in Rural and Urban Planning. He is an active member of the Zimbabwe Institute of Regional and Urban Planners as well as a Corporate Member of the South African Planning Institute. He is currently an Associate Professor in the School of the Built Environment and Development Studies, at the University of KwaZulu-Natal. Professor Chipungu is both a practitioner and academic. In addition, he is an editorial board member of various journals and the Managing Editor of the Journal of Inclusive Cities and the Built Environment. He has published widely in peer-reviewed journals and books; and his areas of interest are housing policy, land-use planning and urban design. He is currently a co-co-principal investigator on a project on inclusive cities in Southern Africa.

Hangwelani Hope Magidimisha-Chipungu is a professor in town and regional planning at the University of KwaZulu-Natal. Her record of service spans but not limited to the South African Research Chairs Initiative (SARChI) for Inclusive Cities, National Planning Commission (RSA), Alternate Chairperson Scientific for the South African Council for Planners (SACPLAN), Committee Member for the International Society of City and Regional Planners (ISOCARP), City Planning Commission for eThekwini Municipality, advisory committee of Cooperative Governance and Traditional Affairs (COGTA-KZN), member of KwaZulu-Natal Tribunal and up to late a Board member of SACPLAN, Researcher at the Human Sciences Research Council (HSRC) etc.

Chapter 14
Gender Inequality and Access to Adequate Housing Provision in Buea, South West of Cameroon

Baba Adamu, Balgah Sounders Nguh, Nformi Beatrice Maluh, Bate Godwill Bate, and Christiana Vegah Abonge

Abstract Access to adequate housing is a fundamental human right of every citizen. Yet, this remains unrealized in the lives of over 300 million women in Sub Saharan Africa characterized by severe patriarchal cultural practice, unequal rights in legislative and policy frameworks and underlying systemic and structural gender inequalities norms due to acute poverty, violence, disabilities and migration status. Illustrating with Buea, a cosmopolitan city in Cameroon, this study examines the challenges faced by women in acquiring adequate housing and the gaps in housing policies implementation practices. The study adopted an explanatory and exploratory design. Field observations, interviews and questionnaires were used in data collection. Both descriptive (percentiles) and inferential (correlation analysis) statistical tools were used to process the data to establish the results. Findings revealed significant gender inequalities in the housing sector in terms of land tenure and housing ownership, access to services and policy formulations. Ultimately, the study calls for a rethink of housing policy and services provision through the promotion of equality, dignity, finances and worth for all while accommodating women's difference and effective participation in decision-making to ensure housing equality in Sub-Saharan cities.

Keywords Gender equality · Housing provision · Accessibility · Affordability · Challenges

B. Adamu (✉) · B. S. Nguh
Department of Geography, Faculty of Social and Management Sciences, University of Buea, Buea, Cameroon
e-mail: baba.adamu@ubuea.cm

N. B. Maluh
Department of Town Planning, National Advanced School of Public Works, Yaoundé, Cameroon

B. G. Bate · C. V. Abonge
Department of Women and Gender Studies, Faculty of Social and Management Sciences, University of Buea, Buea, Cameroon
e-mail: bate-godwill@lukmefcameroon.org

L. Chipungu et al. (eds.), *Contemporary Models for Pro-Poor Housing in Urban Sub-Saharan Africa*, https://doi.org/10.1007/978-3-032-05489-0_14

14.1 Introduction

Housing is a necessity for creating inclusive, equitable, safe, resilient, and sustainable cities (Habitat for Humanity, 2021). Adequate housing means more than just having a roof over one's head. It also means having adequate privacy, space, physical accessibility, security, security of tenure, structural stability, and durability. It includes proper lighting, heating, and ventilation, as well as adequate basic infrastructure. It also encompasses suitable environmental quality and health, and accessibility. All of these factors should be available at an affordable cost (UN-HABITAT, 2003). Home is a place where women socialize, work, and care for the children. A home provides women with a secure place to live, ensuring privacy and protection against harassment or attacks. In many developing countries, women spend more time supporting home-based activities (Habitat for Humanity, 2021). However, with more than half of the world's population already living in cities (Bilfinger et al., 2019; UN Habitat, 2019; UN Women, 2021), continued urbanization presents significant challenges for women's ability to access adequate and affordable housing (World Bank, 2020; UNICEF, 2020). How cities respond to this growth will have a profound impact on urban sustainability (World Bank, 2020). Achieving gender equality and the empowerment of women is the basis for reaching all SDGs including Agenda 2063 of the African Union (UN Women, 2021). The concept of gender refers to the social differences between men and women, which are determined by social, historical, cultural, rational, societal norms, or religion. Gender influences people's experience of and access to social services, including housing (WHO, 2023).

Equality emphasizes that all human beings are entitled to all human rights on an equal basis with one another (UN Women, 2021). This is closely related to the principle of non-discrimination. In relation to gender, equality means that men and women should enjoy all human rights equally (WHO, 2023). Equality has two distinct aspects: equality in practice and equality in legislation. Many constitutions recognize equality between men and women in law, including the Islamic Holy Quran and the Hadith. These religious texts affirm that all humans have the same origin and call on them to be kind to each other and challenge socio-economic disparities (Riffat, 1987). Land on which housing is built, the main source of tangible and intangible wealth, is described as the sole property of Allah the Almighty. Islam discourages individuals from acquiring land solely for the purpose of amassing excessive wealth, as this contradicts the belief that Allah has provided resources for the benefit of all people on Earth. Land, like rain from the sky, air, and water, is provided to all creatures alike (UNDP & Government of Baluchistan, 2022). One verse of the Holy Quran states, "O mankind!" We have created you from males and females and made you into nations and tribes so that you may know one another. Verily, the most honourable among you with Allah is the believer who is pious. "Verily, Allah is All-Knowing, All-Aware" (Al-Quran, 49:13). But in practice, it is a very different story, as women and men rarely experience such equality. There is still a long way to go towards achieving complete equal treatment of men and women in terms of access to basic facilities, services, and resources (Bilfinger et al., 2019). This is particularly apparent in housing, where more than 50% of urban women and girls in developing countries especially women live without access to clean water, improved sanitation, durable housing, or a sufficient living area (UN Women, n.d., 2021). Women often lack security of tenure and equal rights to land and property due to inheritance laws, customs, and traditions, which results in their dependence on male family members.

The United Nations Guidelines for the Implementation of the Right to Adequate Housing emphasize gender equality in housing and land. This also relates to laws concerning inheritance, access to credit, mortgages, and homeownership, as well as equal participation in decisions related to house design and construction (UN Human Rights Council, 2020). The International Covenant on Economic, Social and Cultural Rights provides for "the right of everyone to an adequate standard of living for themselves and their families, including adequate food, clothing, and housing, as well as the continuous improvement of living conditions" (art. 11.1). While the human right to adequate housing is the entitlement of every individual, regardless of gender or age, to have a safe and secure home and community in which to live with peace and dignity (UN, 2002), Germany and other Western countries have made significant advancements in achieving SDG Target 11 (which aims to improve the lives of at least 100 million slum dwellers, particularly vulnerable women, by 2030) (UN-Habitat, 2018). In contrast, developing nations face challenges, with over 70% of women living in slum conditions (Bilfinger et al., 2019). Chatham House (2023) observed priority livelihood interventions in Ukraine, noting the likelihood of a large share of female-headed families and single-earner households. They proposed intervention strategies that target employment and reskilling programs to meet individuals' socio-economic needs (Chatham House, 2023). Conversely, in Sub-Saharan Africa (SSA), where over 300 million women lack access to adequate housing, particularly in urban areas, rapid urbanization, patriarchal cultural practices, discrimination, gender inequality norms, acute poverty, violence, disabilities, and migration status have impeded progress towards providing women with adequate housing (UN-Habitat, 2018; Africa Access, n.d.).

Rakodi (2014) observed that while many countries have made improvements to their legal framework for land administration and personal law to protect women's rights to housing access, it is uncommon for all the necessary components of a progressive legal framework to be fully established. Often, conflicts arise between different sources of law and individual pieces of legislation. In addition, there are gaps between legal provisions and social norms and practices. Whether policies and laws address existing unhealthy and insecure living conditions and redress gender inequality depends on political, social, and familial relations, as well as the extent to which poor women and men have a voice in decision-making. In practice, many recent land registration, housing programs, and wider land and planning policies are not gender sensitive. Even when gender is mainstreamed, it has not always been implemented in practice. Women are often disadvantaged in terms of education, income-earning capacity, marital status, and ability to become homeowners. When they head households, they are disproportionately represented among renters (UN-HABITAT, 2012). In Ethiopia, for instance, approximately 33% of households in urban areas are headed by women. Additionally, 70% of these women live in rented homes in the capital city of Addis Ababa. Many of these women are single, divorced, or widowed (Eskezia, 2011). Consequently, numerous human rights mechanisms and stakeholders, including government agencies, have drawn attention to the effects of gender inequality and discrimination on women's enjoyment of the right to adequate housing (Eskezia, 2011).

Cameroon has recently experienced significant population growth; however, housing services for women have not increased despite significant investment and government policy interventions in housing supply (National Institute for Statistics, 2005). Access to urban housing for women was limited due to socio-economic, cultural, and political changes. Moreover, data on housing affordability and access remains unknown due to a lack of available data on materials, ownership, services, facilities, and particularly on women's inclusion and participation. Significant inequalities are nevertheless expected to rise, as is the case in similar contexts where women are involved in general (UN Habitat, 2020). Furthermore, the utilization of enhanced water and sanitation facilities is higher among households headed by males, which underscores the importance of focusing on the most vulnerable populations (WHO/UNICEF, 2017) residing in slums. There is a need to focus on the variations in access to housing services between male and female groups. Studies and research focusing on disaggregated data in Cameroon are limited, as are policy formulation and implementation efforts. The extent of these variations is not well known. Using the case study of Buea in Cameroon, this study compares the indicators of access to affordable housing provision among women, as well as their vulnerabilities and adaptations. Further, the study considers these two measures to identify where priority should be placed for interventions and improvements, as those who should be addressed first are the segment of the population with limited access to affordable housing, as they are the furthest behind. The objectives of this study are to examine gender inequalities in the provision of adequate housing services in Buea using household survey data and to identify areas where the proportion of women without access to housing remains high and vulnerable in Buea. This study on gender inequality and access to adequate housing in Buea aims to raise awareness about the challenges women face in accessing housing. It also seeks to advocate for policies and programs that can address these challenges. This is a timely opportunity to identify and target groups that have been left behind and to help achieve the targets set in the local, national, and international agendas, such as the national development strategic plan and the SDGs.

14.2 Study Area and Methods

Buea is located in the Fako Division of the South West Region of Cameroon (Fig. 14.1). It is located on the eastern foothills of Mount Cameroon at an altitude of approximately 4100 m above sea level. It is situated between latitude 4°12′ and 4°31′ north of the equator, and longitude 9°02′ and 9°09′ east of the Greenwich meridian. Buea is a historic settlement which once served as the German capital of Kamerun prior to the decolonization era. It later became the capital of West Cameroon during the Trusteeship and Federation. It also consists of steeply sloping terrain with plateaus at certain points, which have led to the emergence of various housing patterns. It is found within the tropical rainforest, where the evergreen vegetation is rapidly degraded for housing development. As the regional headquarters of the South West Region, following the government's regionalization policy in 2008, the town currently hosts numerous social facilities and services as a result of urbanization. The town's population is approximately 200,000 inhabitants, according to the Buea Council Development Plan (CDP, 2012) in 2010. It is growing at a rate of 3.6%, which is one of the highest in Cameroon, with over 7,000 people moving in each year. The target population comprised female and male household heads, as well as the municipal authorities, within Buea's urban and rural settlements who have lived in the town for more than 10 years. Due to limited statistics on the number of household heads, the study resorted to simple random sampling of household heads from the study neighbourhoods. Both primary and secondary data were obtained to achieve the results of this study. A total of 200 households, consisting of 100 female-headed and 100 male-headed households, were sampled across seven neighbourhoods in both urban and rural areas. The sampling was based on convenience and the willingness of participants to take part in the study. Primary data were generated through structured and unstructured questionnaires and interview guides administered to stakeholders and household heads, respectively.

Questionnaires were administered at the respondents' homes. Secondary data sources were obtained through a literature review of materials from internet sources, as well as a review of government documents related to the subject matter. Data were analysed using descriptive and inferential statistics with the aid of the Statistical Package for the Social Sciences (SPSS Version 23.0) and thematic analysis. Qualitative data were analysed using thematic content analysis, with some quotations presented to enhance understanding. The results were presented using figures and tables. Ethically, the responses of the respondents were kept confidential, along with the informed consent that was obtained during the field observation phase.

14.3 Results

This consists of the presentation of results and discussions on the socio-economic and demographic profile of the respondents, various dimensions of housing situations, tenure security, and strategies to ensure access to affordable housing for women, with a focus on better inclusiveness and liveability.

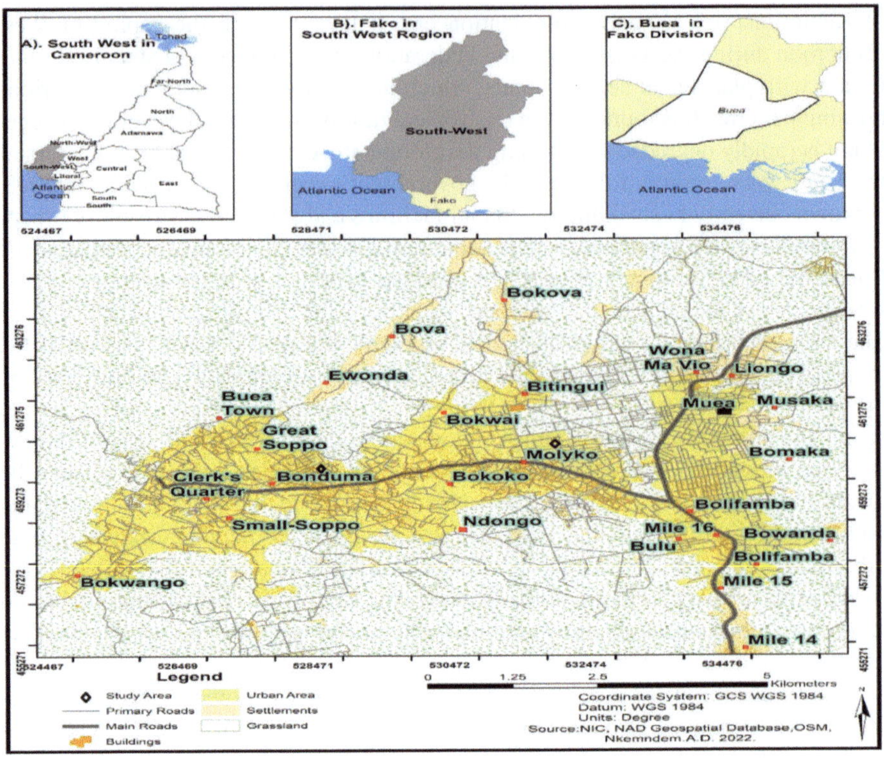

Fig. 14.1 Location of Buea in the Fako Division of the South West Region of Cameroon. *Source* Adapted from Buea Council Development Plan (CDP, 2012)

14.3.1 Socioeconomic and Demographic Profile of Respondents

The socio-economic and demographic characteristics of the respondents in the study area were multiple and varied, ranging from gender, age, marital status, among others, as presented in Table 14.1. From the analysed data, the study revealed that the majority of the sampled respondents fell within the 31–41 years age range, accounting for 42.0% (n = 84) of the participants. This was closely followed by the 41–50 years age group, which constituted 34.0% (n = 68) of the participants. The remaining respondents accounted for less than 20% (Table 14.1). In a similar approach, gender representation was equally distributed, with 50.0% (n = 100) of the participants being male and 50.0% (n = 100) being female. Marital status was relatively evenly distributed between married and single individuals, with 48.0% (n = 96) being married and 42.0% (n = 84) being single. The remaining proportion was distributed between 6.0% (n = 12) for divorced individuals and 4.0% (n = 8) for windows. Regarding the income level, the data revealed that the majority of the participants reported a monthly average income between the 50,000–100,000 FCFA ranges, which accounted for 42.0% of the participants. Additionally, 22.0% (n = 44) reported an income below 50,000 FCFA, while 18.0% (n = 90) fell within the 100,000–150,000 FCFA income range. Furthermore, 16.0% (n = 80) of the participants had an income of 150,000 FCFA or above (Table 14.1).

The educational levels of the participants varied widely. 24.0% (n = 48) had completed Advanced Level, and 22.0% (n = 44) held a degree. Non-formal education accounted for 16.0% (n = 31) of the respondents, while FSLC (First School Leaving Certificate) and O/L (Ordinary Level) were represented by 14.0% (n = 28) and 8.0% (n = 16) respectively. 16.0% (n = 32) of the respondents had completed a postgraduate degree qualification. These results suggest that the respondents in this survey are a relatively young, single, and educated group of individuals who reside in urban areas and are employed as skilled workers or farmers. The majority of respondents have a relatively low income, which suggests that they may face financial challenges in acquiring their own houses or land. Significantly, results from the field data analysis reveal that nearly half of the participants (46%) reported living in households with 4–6 individuals. A significant portion (34%) indicated that their households consisted of 1–3 individuals. In addition, 12% reported living in households with 7–9 individuals, while 8% mentioned having 10 or more individuals in their households. Based on room density, results from Table 14.2 reveal that the majority of respondents (68%) stated that 4–6 persons share a room in their houses. A smaller proportion (22%) reported having 1–3 individuals sharing a room, while only 6% mentioned 7–9 individuals sharing a room. Furthermore, 4% stated that 10 or more individuals share a room in their house. The majority of participants (68%) identified themselves as tenants, while the remaining 32% mentioned being landlords.

14.3.2 *Dimensions of Housing Situations*

Table 14.2 explores various aspects of housing in Buea, including housing status, acquisition, legal status, type of accommodation, adequacy of physical space, age of buildings, and access, among others.

Analysis of the housing situations presented in Table 14.2 showed that 68.0% of the respondents live in male-headed households, while 32.0% mentioned that their households are female-headed. Respondents relied on various individuals to acquire land or housing, with the highest dependency reported on husbands (30.0%) and fathers (34.0%). The legal status of the respondents revealed that 54.0% of them lived in unapproved houses, while 46.0% stated that their houses were unapproved. Respondents occupied various types of housing accommodations, including traditional houses (20.0%), tenements (40.0%), detached houses (20.0%), and block flats (20.0%). In terms of physical space, respondents mentioned varying levels of adequacy for different factors, such as heat (30.0%), cold (36.0%), rain (24.0%), and wind (10.0%). Respondents considered the location of their houses in terms

Table 14.1 Socio-Economic and demographic profile of respondents

Characteristics	Category	Frequency (%)
Age	21–30 years	32 (16.0%)
	31–40 years	84 (42.0%)
	41–50 years	68 (34.0%)
	>50 years	10 (5.0%)
Gender	Male	100 (50.0%)
	Female	100 (50.0%)
Marital status	Single	84 (42.0%)
	Married	96 (48.0%)
	Divorced	12 (6.0%)
	Widow/Widower	8 (4.0%)
Monthly income (FCFA)	<50000	44 (22.0%)
	50000–100000	84 (42.0%)
	100000–150000	36 (18.0%)
	150000+	36 (18.0%)
Education	Non formal education	32 (16.0%)
	FSLC	28 (14.0%)
	O/L	16 (8.0%)
	A/L	48 (24.0%)
	Degree	44 (22.0%)
	Post Degree	32 (16.0%)
Family size	4–6 Individuals	92 (46.0%)
	1–3 Individuals	68 (34.0%)
	7–9 Individuals	24 (12.0%)
	10 or more	16 (8.0%)
Room density	4–6 persons	136 (68.0%)
	1–3 persons	44 (22.0%)
	7–9 persons	12 (6.0%)
	10 or more	8 (4.0%)
Household STATUS	Tenants	136 (68.0%)
	Landlords	64 (32.0%)
Cost of rent amount (FCFA)	26,000 - 50,000 units	88 (44.0%)
	More than 25,000 units	56 (28.0%)
	100,000 units or more	32 (16.0%)
	51,000 - 100,000 units	24 (12.0%)

Table 14.2 Dimension of housing situations

Variable	Category	Frequency (%)
Household status	Male-headed	136 (68.0%)
	Female-headed	64 (32.0%)
Land acquisition dependent	Husband	60 (30.0%)
	Father	68 (34.0%)
	Male friends	36 (18.0%)
Legal status of the house	Approved	108 (54.0%)
	Unapproved	92 (46.0%)
Type of housing accommodation	Traditional	40 (20.0%)
	Tenement	80 (40.0%)
	Detached	40 (20.0%)
	Block flat	40 (20.0%)
Adequate physical space	Heat	60 (30.0%)
	Cold	72 (36.0%)
	Rain	48 (24.0%)
	Wind	20 (10.0%)
Location of the house	Employment opportunities	88 (44.0%)
	Health	68 (34.0%)
	School	44 (22.0%)
Age of the house	Less than 25 years old	104 (52.0%)
	25–50 years old	52 (26.0%)
	51+ years old	44 (22.0%)
Cost of rent amount (FCFA)	26,000 - 50,000 units	88 (44.0%)
	More than 25,000 units	56 (28.0%)
	100,000 units or more	32 (16.0%)
	51,000–100,000 units	24 (12.0%)

of access to employment opportunities (44.0%), healthcare facilities (34.0%), and schools (22.0%). While the ages of the house varied widely as well, from 52.0% less than 25 years old, 26.0% between 25–50 years old, and 22.0% 51 years or older. As for the rents, the monthly payment depicts a range of costs, with over half (52%) paying between 26,000–100,000 FCFA. A sizable proportion (28%) pay rents less than 25,000FCFA. The findings revealed that 44% reported paying between 26,000 FCFA and 50,000 FCFA per month for a one-bedroom apartment within the city centre around the University neighbourhood, Molyko while 28% paid more than 50,000 FCFA per month. Furthermore, 16% mentioned paying 100,000 FCFA or more per month for apartments in peripheral neighbourhoods like Muea, while 12% reported paying between 51,000 FCFA and 100,000 FCFA for a two-bedroom apartment in the same area. However, a two-bedroom apartment in the city centre can cost around 1 million CFA francs (approximately $1,500) per year. When asked if the cost of rent compromises their enjoyment, a significant proportion of the respondents believe that their rents have a negative impact on their lifestyle or enjoyment. The rent burden question revealed that 76 individuals (38%) feel that their rent compromises their enjoyment of their home while the majority (62%) stated that the cost of rent does not compromise their enjoyment because they prioritize high values.

The cost of housing in Buea is likely to continue rising in the future due to several factors. These include the city's growing population, limited housing supply, the city's topography, the lack of available land, and the high cost of materials and labour. In summary, the survey found that the majority of participants live in households with 1–3 individuals and share a room with 1–3 other people. They typically stay in their residence for an average duration of 1–5 years and are mostly non-locals. Their monthly house rents vary depending on the location (from the city centre to the peripheral zones) and the size of the property. However, most participants pay between 26,000 and 50,000 FCFA. A significant proportion of participants find that the cost of rent is too high for them to afford, which compromises their enjoyment, particularly for low-income households and households headed by women.

14.3.3 Gender Distribution and Satisfaction with Housing Facilities/Services

Despite the persistent efforts to ensure access for all to adequate, safe, and affordable housing and basic services by 2030, and to upgrade slums, the proportion of women in Buea remains limited, as shown in Table 14.3. Based on the analysis of the results of ensuring access to affordable basic services, the chi-square test detected a significant association between the satisfaction of water supply/pollution and the distribution of respondents by sex ($p = 0.022$). Specifically, the majority of females (60.7%) reported being very dissatisfied with the water supply condition in their houses, while the largest group of males (69.7%) expressed feeling very satisfied. This suggested that women face relatively fewer accessibility or quality issues with domestic water services. This finding corroborated the results presented by Fisher (2008) on women and access to water and hygiene and sanitation programme where the author noted persistent challenges faced by women in access to water supply and sanitation services. Similar trends emerged for electricity supply, with a highly significant association ($p = 0.000$) between gender distribution and the level of satisfaction. Females over 70% were very dissatisfied with their electricity supply level. Indicating that comparatively, women suffer from the impact of the inadequate and unreliable power supply in their households more than the men. Regarding waste management, a significant association was again found ($p = 0.005$), as females constituted the majority of those very dissatisfied (64.4%). This points to women shouldering a greater sanitation burden, collecting trash farther from homes or facing improper disposal sites along the way. Toilet facilities dissatisfaction displayed sex differences as well ($p = 0.006$), with females accounting for over 66% of those who were very unhappy. This underscored the unmet needs for safely managed sanitation, which is essential for women's health, safety, and dignity. Food preparation energy sources appeared to be unequally distributed by gender as well ($p = 0.047$), with lighting showing a similar pattern ($p = 0.004$). Dependence on polluting fuels and lack of adequate lighting exacerbate women's domestic burdens.

14.3.4 Level of Satisfaction with Tenure Security

Various aspects of tenure security related to gender dimensions of access to housing provision, including land ownership, permits, access to credit, and security, were assessed in the study. Table 14.4 revealed significant differences, as more women (40%) than men (20%) reported dissatisfaction ($\chi^2 = 16.136$, $p = 0.002$). Perhaps most startling was the gender disparity among satisfied homeowners—only 29% were female, while 71% were male. These significant correlations ($p < 0.05$) between ownership satisfaction and gender suggested that structural challenges disproportionately impede women's access to and control over land for housing. Moving on to the documentation, the chi-square value ($\chi^2 = 7.031$, $p = 0.134$) did not reach statistical significance for holders of neutral land certificates. However, trends persisted of elevated female dissatisfaction, while over 60% of the very satisfied group had documents solely under male names.

Access to housing loans proved to be the most striking example, with a highly significant association between gender and discontent ($\chi^2 = 15.818$, $p = 0.003$). Here, 67% of those who were merely dissatisfied and a concerning 39.5% of those who were very dissatisfied were women. Participation insights too revealed meaningful associations, as twice as many women reported complete voicelessness as men, despite male dominance among customs' upper echelons ($\chi^2 = 8.950$, $p = 0.062$). Regarding customary tenure systems, a significant association emerged between satisfaction and gender ($\chi^2 = 13.253$, $p = 0.010$). Over twice as many women as men expressed dissatisfaction, upholding claims of patriarchal systems marginalizing female land rights. Looking to housing finance, chi-square tests found no statistically meaningful correlations for neutral or satisfied groups regarding limited access to funding. However, it is worth considering the near-even gender split among the very dissatisfied ($\chi^2 = 4.652$, $p = 0.325$), especially considering the well-documented credit barriers that women face. When examining lack of disability accommodation, all categories yielded insignificant relationships with gender ($\chi^2 = 2.732$, $p = 0.604$). However, male representation prevailed among both the very dissatisfied and satisfied cohorts hinting at possible ableism compounded by sex-based discrimination. On minority marginalization, no meaningful associations surfaced according to chi-square ($\chi^2 = 6.967$, $p = 0.138$). Still, over half of those very satisfied with prevailing approaches were men compared to under half content women, a discrepancy requiring deeper intersectional study.

Table 14.3 Association between gender distribution and satisfaction with housing facilities

Variable	Category	Male f (%)	Female f (%)	Pearson χ^2 (p-value)
Water supply/pollution	Very dissatisfied	35 (39.3%)	54 (60.7%)	9.672 (0.022)
	Dissatisfied	38 (53.5%)	33 (46.5%)	
	Satisfied	4 (57.1%)	3 (42.9%)	
	Very satisfied	23 (69.7%)	10 (30.3%)	
Electricity supply	Very dissatisfied	17 (29.3%)	41 (70.7%)	19.095 (0.000)
	Dissatisfied	30 (49.2%)	31 (50.8%)	
	Satisfied	12 (54.5%)	10 (45.5%)	
	Very satisfied	41 (69.5%)	18 (30.5%)	
Waste management	Very dissatisfied	31 (35.6%)	56 (64.4%)	12.912 (0.005)
	Dissatisfied	45 (63.4%)	26 (36.6%)	
	Satisfied	7 (53.8%)	6 (46.2%)	
	Very satisfied	16 (57.1%)	12 (42.9%)	
Toilet facilities	Very dissatisfied	22 (33.8%)	43 (66.2%)	12.539 (0.006)
	Dissatisfied	32 (54.2%)	27 (45.8%)	
	Satisfied	18 (72.0%)	7 (28.0%)	
	Very satisfied	28 (54.9%)	23 (45.1%)	
Road accessibility	Very dissatisfied	36 (42.9%)	48 (57.1%)	5.393 (0.145)
	Dissatisfied	30 (61.2%)	19 (38.8%)	
	Satisfied	21 (56.8%)	16 (43.2%)	
	Very satisfied	13 (43.3%)	17 (56.7%)	
Security	Very dissatisfied	38 (45.2%)	46 (54.8%)	1.408 (0.704)
	Dissatisfied	42 (53.2%)	37 (46.8%)	
	Satisfied	8 (57.1%)	6 (42.9%)	
	Very satisfied	12 (52.2%)	11 (47.8%)	
Energy for cooking	Very dissatisfied	32 (42.1%)	44 (57.9%)	7.950 (0.047)
	Dissatisfied	20 (43.5%)	26 (56.5%)	
	Satisfied	18 (54.5%)	15 (45.5%)	
	Very satisfied	30 (66.7%)	15 (33.3%)	
Lighting	Very dissatisfied	16 (30.8%)	36 (69.2%)	13.092 (0.004)
	Dissatisfied	40 (56.3%)	31 (43.7%)	
	Satisfied	23 (67.6%)	11 (32.4%)	
	Very satisfied	21 (48.8%)	22 (51.2%)	

Table 14.4 Level of satisfaction with tenure security

Variable	Category	Male f (%)	Female f (%)	Pearson χ^2 (p-value)
Own of land for building	Neutral	25 (55.6%)	20 (44.4%)	16.436 (0.002)
	Dissatisfied	20 (33.3%)	40 (66.7%)	
	Very dissatisfied	26 (57.8%)	19 (42.2%)	
	Satisfied	6 (33.3%)	12 (66.7%)	
	Very satisfied	23 (71.9%)	9 (28.1%)	
Own a land certificate/zoning plan	Neutral	12 (42.9%)	16 (57.1%)	7.031 (0.134)
	Dissatisfied	23 (38.3%)	37 (61.7%)	
	Very dissatisfied	25 (55.6%)	20 (44.4%)	
	Satisfied	22 (57.9%)	16 (42.1%)	
	Very satisfied	18 (62.1%)	11 (37.9%)	
Have land certificate registered in your name	Neutral	17 (53.1%)	15 (46.9%)	9.073 (0.059)
	Dissatisfied	17 (33.3%)	34 (66.7%)	
	Very dissatisfied	28 (53.8%)	24 (46.2%)	
	Satisfied	16 (51.6%)	15 (48.4%)	
	Very satisfied	22 (64.7%)	12 (35.3%)	
Possesses a building permit	Neutral	16 (55.2%)	13 (44.8%)	3.211 (0.523)
	Dissatisfied	40 (44.0%)	51 (56.0%)	
	Very dissatisfied	18 (50.0%)	18 (50.0%)	
	Satisfied	10 (62.5%)	6 (37.5%)	
	Very satisfied	16 (57.1%)	12 (42.9%)	
Building permit registered in your name	Neutral	10 (41.7%)	14 (58.3%)	3.952 (0.413)
	Dissatisfied	19 (40.4%)	28 (59.6%)	
	Very dissatisfied	28 (57.1%)	21 (42.9%)	
	Satisfied	17 (51.5%)	16 (48.5%)	
	Very satisfied	26 (55.3%)	21 (44.7%)	
Have access to housing credit	Neutral	6 (33.3%)	12 (66.7%)	15.818 (0.003)
	Dissatisfied	18 (32.7%)	37 (67.3%)	
	Very dissatisfied	46 (60.5%)	30 (39.5%)	
	Satisfied	18 (69.2%)	8 (30.8%)	
	Very satisfied	12 (48.0%)	13 (52.0%)	
Face difficulty in securing documents	Neutral	9 (40.9%)	13 (59.1%)	6.984 (0.137)
	Dissatisfied	23 (42.6%)	31 (57.4%)	
	Very dissatisfied	29 (51.8%)	27 (48.2%)	
	Satisfied	24 (50.0%)	24 (50.0%)	
	Very satisfied	15 (75.0%)	5 (25.0%)	
Participate in land decisions	Neutral	7 (50.0%)	7 (50.0%)	8.950 (0.062)
	Dissatisfied	24 (35.8%)	43 (64.2%)	

(continued)

Table 14.4 (continued)

Variable	Category	Male f (%)	Female f (%)	Pearson χ^2 (p-value)
	Very dissatisfied	36 (60.0%)	24 (40.0%)	
	Satisfied	14 (51.9%)	13 (48.1%)	
	Very satisfied	19 (59.4%)	13 (40.6%)	
Gender base violence	Neutral	11 (61.1%)	7 (38.9%)	5.669 (0.225)
	Dissatisfied	24 (39.3%)	37 (60.7%)	
	Very dissatisfied	35 (53.8%)	30 (46.2%)	
	Satisfied	15 (46.9%)	17 (53.1%)	
	Very satisfied	15 (62.5%)	9 (37.5%)	
Customary tenure system prevails	Neutral	14 (66.7%)	7 (33.3%)	13.253 (0.010)
	Dissatisfied	22 (34.9%)	41 (65.1%)	
	Very dissatisfied	29 (58.0%)	21 (42.0%)	
	Satisfied	8 (36.4%)	14 (63.6%)	
	Very satisfied	27 (61.4%)	17 (38.6%)	
Limited access to funding	Neutral	8 (61.5%)	5 (38.5%)	4.652 (0.325)
	Dissatisfied	24 (42.9%)	32 (57.1%)	
	Very dissatisfied	31 (58.5%)	22 (41.5%)	
	Satisfied	25 (43.9%)	32 (56.1%)	
	Very satisfied	12 (57.1%)	9 (42.9%)	
Lack of disability considerations	Neutral	8 (61.5%)	5 (38.5%)	2.732 (0.604)
	Dissatisfied	29 (45.3%)	35 (54.7%)	
	Very dissatisfied	36 (56.3%)	28 (43.8%)	
	Satisfied	16 (47.1%)	18 (52.9%)	
	Very satisfied	11 (44.0%)	14 (56.0%)	
Marginalization of minorities	Neutral	6 (66.7%)	3 (33.3%)	6.967 (0.138)
	Dissatisfied	22 (37.3%)	37 (62.7%)	
	Very dissatisfied	27 (57.4%)	20 (42.6%)	
	Satisfied	28 (57.1%)	21 (42.9%)	
	Very satisfied	17 (47.2%)	19 (52.8%)	

14.3.5 Strategies to Ensure Better Housing Accessibility and Inclusion for Women

Diverse strategies and their levels of satisfaction by gender were observed during the field analysis of results to promote better inclusion of women in housing affordability provision in the study area, as shown in Table 14.5. For the majority of the strategies presented in Table 14.5, there is no statistically significant difference ($p > 0.05$) between males and females in terms of their level of satisfaction. This suggests that overall, both males and females have similar views on the potential effectiveness of these strategies. A few strategies did show statistically significant differences including guarantee equal access to credit/financial aid/documents as more females were dissatisfied compared to males, suggesting females see this as a more important strategy. Empowerment of women as more females expressed higher levels of satisfaction (satisfied/very satisfied) compared to males. This findings suggested that empowering women directly is viewed more positively by females. Promote an integrated housing development program as more males expressed higher levels of satisfaction compared to females. Males may perceive integrated development programs as a more effective strategy compared to females.

14.4 Discussions

The analysis of the gender dimension in housing situations revealed that 68.0% of respondents lived in households headed by males, while 32.0% lived in households headed by females. Husbands (30.0%) and fathers (34.0%) were the most relied-upon individuals for acquiring land or housing. The study aligned with the gender disparities mentioned in the literature, specifically highlighting male dominance in decision-making processes (UNICEF, 2020). According to the 2006 UN Habitat Urban Indicators Database on the state of world cities, it was reported that 20% of all household heads in over 160 cities worldwide, primarily from developing countries, were women. These women were living in slums and squatter settlement with acute poverty and deprivations. The analysis revealed significant gender disparities in several areas related to basic amenities and household services. In terms of water supply and pollution satisfaction, there was a significant association between gender distribution and satisfaction levels ($p = 0.022$). This study corroborated the findings of Fisher (2008 on women and access to water supply and sanitation services. Females reported higher levels of dissatisfaction (60.7%) with the condition of the water supply in their houses, while males were the largest group expressing satisfaction (69.7%). This implied that women face greater challenges in accessing domestic water services and experience more quality issues, leading to higher levels of dissatisfaction among women compared to men. Collaboratively, Cassivi (2020) noted that the poorest and most marginalized groups, especially women, are still the least likely to have access to these essential services in Malawi. The Asian Regional Consultation Department (2016) and the Asian Regional Study on Women, Peace and Security (2021) all suggested that, owing to the precarious conditions in slums and informal settlements, women are always afraid of fires or floods or concerned about the well-being of their children. Violence against women often result from inadequate, unaffordable housing conditions, unemployment and poverty. Consistent with this, WHO/UNICEF (2017) remarked that the use of improved water and sanitation facilities is greater in the male headed family, confirming the necessity to target the most vulnerable populations especially in urban slums and peri urban settlements. Similar

Table 14.5 Strategies to Ensure Better Inclusion of Women in Housing Provisions

Variable	Category	Male f (%)	Female f (%)	Pearson χ^2 (p-value)
Increase social/low-cost housing including those in emergency situations	Neutral	29 (43.9%)	37 (56.1%)	3.353 (0.501)
	Dissatisfied	50 (51.5%)	47 (48.5%)	
	Very dissatisfied	17 (63%)	10 (37%)	
	Satisfied	3 (42.9%)	4 (57.1%)	
Recognized the independent right of women to security of tenure	Neutral	21 (47.7%)	23 (52.3%)	6.157 (0.188)
	Dissatisfied	53 (48.2%)	57 (51.8%)	
	Very dissatisfied	23 (63.9%)	13 (36.1%)	
	Satisfied	0 (0%)	3 (100%)	
	Very satisfied	3 (42.9%)	4 (57.1%)	
Guarantee equal access to credit/financial aid/ documents	Neutral	40 (57.1%)	30 (42.9%)	14.001 (0.007)
	Dissatisfied	29 (34.9%)	54 (65.1%)	
	Very dissatisfied	19 (67.9%)	9 (32.1%)	
	Satisfied	11 (64.7%)	6 (35.3%)	
	Very satisfied	1 (50%)	1 (50%)	
Empowerment of women	Neutral	41 (48.8%)	43 (51.2%)	11.626 (0.020)
	Dissatisfied	32 (43.8%)	41 (56.2%)	
	Very dissatisfied	4 (30.8%)	9 (69.2%)	
	Satisfied	17 (77.3%)	5 (22.7%)	
	Very satisfied	6 (75%)	2 (25%)	
Organization of women community networks	Neutral	24 (47.1%)	27 (52.9%)	6.692 (0.153)
	Dissatisfied	43 (46.2%)	50 (53.8%)	
	Very dissatisfied	5 (35.7%)	9 (64.3%)	
	Satisfied	18 (69.2%)	8 (30.8%)	

(continued)

Table 14.5 (continued)

Variable	Category	Male f (%)	Female f (%)	Pearson χ^2 (p-value)
	Very satisfied	10 (62.5%)	6 (37.5%)	
Participate in decision making concerning land and housing	Neutral	29 (50%)	29 (50%)	5.082 (0.279)
	Dissatisfied	44 (44%)	56 (56%)	
	Very dissatisfied	18 (62.1%)	11 (37.9%)	
	Satisfied	4 (66.7%)	2 (33.3%)	
	Very satisfied	5 (71.4%)	2 (28.6%)	
Promote integrated housing development program	Neutral	35 (58.3%)	25 (41.7%)	9.963 (**0.041**)
	Dissatisfied	39 (40.6%)	57 (59.4%)	
	Very dissatisfied	21 (63.6%)	12 (36.4%)	
	Satisfied	1 (20%)	4 (80%)	
	Very satisfied	4 (66.7%)	2 (33.3%)	

trends were observed for electricity supply, with a highly significant association ($p = 0.000$) between gender distribution and satisfaction levels.

Over 70% of females were very dissatisfied with their electricity supply, indicating that women are disproportionately affected by inadequate and unreliable power supply in their households. Waste management also showed a significant association ($p = 0.005$), with females comprising the majority of those very dissatisfied (64.4%). This points to women shouldering a greater sanitation burden, such as collecting trash from more distant locations or encountering improper disposal sites. Dissatisfaction with toilet facilities also showed sex differences ($p = 0.006$), with females accounting for over 66% of those who were very unhappy. This highlighted unmet needs for safely managed sanitation, which is essential for women's health, safety, and dignity. Consistent with this, inadequate housing conditions have significant ramifications for women, especially because they are more likely to spend more time at home compared to men (Eskezia, 2011). In urban areas like Ethiopia, 33% of households are headed by women, with the majority living in rented housing and facing poverty. In a study conducted in Addis Ababa, it was found that 70% of female-headed households were living in rented housing. Among these households, 12% were single, 52% were divorced or separated, and 36% were widows (Eskezia, 2011). Women play a role in the rental housing market not only as tenants but also as landlords. Women can inherit the marital home if their husband dies (Varley, 2007). Many low-income urban dwellers, including women, rent out rooms in their homes either by creating additional accommodations or by vacating rooms during economic difficulties or when their households shrink due to children leaving home. As a result, some women also own rental accommodations that they do not live in themselves.

These findings from this study highlighted the diverse roles that women play in the housing market, including as tenants and landlords. They also sheded light on the specific challenges women face in accessing suitable housing and managing their households in the study area. Pervasive gender disparities and structural challenges impede women's access to and control over land for housing. These disparities exist in various aspects of housing, including land ownership, documentation, participation, customary tenure systems, housing finance, disability accommodation, and minority marginalization, discriminations, and inequalities. Land ownership for construction shows a gender disparity, with more women (40%) than men (20%) reporting dissatisfaction. Additionally, a significantly higher percentage of satisfied homeowners are male (71%) compared to female (29%). Gender disparities were also evident in the distribution of food preparation energy sources ($p = 0.047$) and lighting ($p = 0.004$). Customary tenure systems show a significant association between satisfaction and gender, with more than twice as many women expressing dissatisfaction. This supports the claims that patriarchal systems marginalize women's land rights. Addressing these disparities requires targeted efforts to promote gender equality and ensure equal access to housing and land rights.

Just as advanced by the UN-HABITAT (2008) study in Kenya and Nicaragua, one-third of female-headed households experienced all five shelter deprivations. This suggests that in some countries, households headed by women are disproportionately affected by inadequate housing in poor urban neighbourhoods. Consistent with the results of a study conducted in Sri Lanka on domestic violence in the plantation sector, it was found that women who live in a single room with their entire family are more likely to experience domestic violence. This is primarily due to the lack of space, privacy, and close proximity to family members. According to Habitat for Humanity (2021) strong land rights for housing for women have a catalytic effect, elevating the standard of living for entire communities. Unfortunately, data analysis regarding land ownership for construction from the respondents shows women have limited access to land for construction of houses.

14.5 Conclusions and Recommendations

The findings concluded that gender plays a significant role in housing situations, with women facing specific challenges and disparities. Female-headed households in Buea are more likely to experience inadequate housing conditions, with a higher prevalence of poverty and a greater reliance on rented housing. Women not only serve as tenants but also as landlords, inheriting marital homes or renting out rooms in their own residences. It is probable to harness efforts targeting the interventions and policies to address gender inequalities in housing. Efforts should be made to improve the accessibility and quality of housing for women, especially those who are heads of households. This includes addressing issues such as inadequate water supply, unreliable electricity, poor waste management, and inadequate sanitation facilities that disproportionately affect women. Furthermore, there is a need to provide support and resources for women to become homeowners and landlords, empowering them economically and improving their housing situations. There is need to promote gender-responsive land policies by developing and implementing land policies that prioritize gender equality, ensuring women's equal access to and control over land for housing. This may include measures such as providing preferential treatment for women in land allocation and ownership. Building more affordable is one of the strategies that if implemented will ensure better housing provision for women. This is because providing affordable housing will reduce the development of slum structures and illicit waste disposal, which have a greater impact on women. Gender dimensions should be integrated into urban planning and development programs by providing opportunities for women and men to participate in decision-making and access permits to authorize the construction of low-cost homes. Further, planning for utilities and services, employment, green space (including for agriculture), and public transport should take into consideration the distinct needs of women and men. Low-income women's and men's ability to access land and housing can be enhanced by supporting home owners' associations/cooperatives, promoting collective tenure, providing support for appropriate savings mechanisms, and potentially offering loans and subsidies.

References

Bilfinger, C., Eitze, J., Kaiser, M., Lens, S., Rieck C. E., & Weck, W. (2019). *Agenda 2030: Basic law of sustainability*. Infograpic:yellowtooPasiek Horntrich GMBH.

Cassivi, E. T., Waygood, E. O. D., & Dorea, C. (2020). Trends in access to water and sanitation in Malawi: Progress and inequalities (1992–2017). *Journal of Water and Health IWA Alexandra, 18*(5), 785. http://iwaponline.com/jwh/article-pdf/18/5/785/773706/jwh0180785.pdf. https://doi.org/10.2166/wh.2020.069

Fisher, J. (2008). Women in water supply, sanitation and hygiene programmes. *Proceedings of the ICE: Municipal Engineer, 161*(4), 223–229. https://doi.org/10.1680/muen.2008.161.4.223

Habitat for Humanity International. (2021). *Housing and the sustainable development goals: The transformational impact of housing*. https://www.habitat.org/sites/default/files/documents/Housing-and-Sustainable-Development-Goals.pdf.

Habitat for Humanity International. (n.d.). Improving land rights for women and excluded families in Bolivia. https://www.habitatforhumanity.org.uk/what-we-do/advocacy-land-and-inheritance-rights/land-rightsbolivia/. https://www.habitatforhumanity.org.uk/country/bolivia/.

Rakodi, C. (2014). Gender equality and development. Expanding women's access to land and housing in urban areas. *Women's Voice and Agency Research Series* 2014 No.8. World Bank. www.worldbank.org/gender/agency.

Riffat, H. (1987). Women in the context of change and confrontation within Muslim communities. In V. R. Mollenkott (Ed.), *Women of Faith in Dialogue* (pp. 13–28). Crossroad Publishing.

U. N. Human Rights Council. (2020). *Guidelines for the Implementation to the Right of Adequate Housing*. A/HRC/43/43-E-A/HRC/43/43 -Desktop. https://undocs.org/en/A/HRC/43/43.

UN Women. (n.d.). *SDG 11: Make Cities and Human Settlements Inclusive, Safe, Resilient and Sustainable*. https://www.unwomen.org/en/news/infocus/women-and-the-sdgs/sdg-11-sustainable-cities-communities.
UNDP and the Government of Baluchistan. (2022). The Sustainable Development Goals (SDGs) and Islam: Contextualizing the Sustainable Development Goals in the Normative Framework of Islamic Tradition. https://www.undp.org/sites/g/files/zskgke326/files/2023.
United Nations Children's Fund (UNICEF, 2020). Snapshot of global and regional urban water, sanitation and hygiene inequalities, UNICEF, New York.
World Bank (2020). *Handbook for Gender-Inclusive Urban Planning Design*. https://www.worldbank.org/en/topic/urbandevelopment/publication/handbook-for-genderinclusive-urban-planning-and-design.
World Health Organization (2023). *Gender and Health*. https://www.who/health-topic/gender#tab_2.

Baba Adamu is currently Senior Lecturer of Geography at the University of Buea, Cameroon, where he also earned his PhD in Human Geography. He is an active member of African Geographic Society (AGS). His research expertise spans urbanization and development, urban planning policy, urban water management/governance and urban social justice. He is the co-author of changing trends in water sources and related pathologies in small to medium size African cities (Springer, 2018).

Balgah Sounders is Professor in Urban and Development Studies with particular interest in Remote Sensing and Its applications to land use dynamics, natural resources and resource management/development. He has published extensively in these areas in learned scientific journals and is a co-author of two textbooks and has made significant contributions in Development studies. He is a consultant in these areas and in planning issues.

Nformi Beatrice Maluh is currently a Lecturer in the Department of Town Planning at the National Advanced School of Public Works. She holds a Ph.D. in Geography and Planning in 2021 and a master's degree in Geography in 2016 all from the University of Buea, Cameroon. Her area of research interests focuses on land use planning, urban development and planning, human geography and natural resource management. She has a wealth of publications in high impact journals on issues like land use change and planning, land use conflicts, housing development, urban green spaces and assessments of housing regulations. She was a part-time lecturer in the Department of Geography at the University of Buea from 2017 to 2022 and Landmark Metropolitan University from 2022 to 2024.

Bate Godwill Bate is a dedicated Cameroonian researcher and academic, currently serving as the Country Director for the Martin Luther King Jr. Memorial Foundation (LUKMEF) in Buea, Cameroon. With a strong background in human resources and operations, he has held various leadership roles within the organization since 2022, including National Coordinator and Human Resource Coordinator. Bate is also a seasoned research consultant and data analyst, contributing to significant studies on gender, climate variability, and urban planning. He is pursuing a Ph.D. in Gender and Development at the University of Buea, where he previously earned both his master's and bachelor's degrees in geography. His academic contributions include several journal articles addressing critical issues such as citizen participation in local development and the implications of climate change on public health and food security.

Christiana Vegah Abonge holds a BSc Women's Studies and Economics (Double Major) and an MSc in Women and Gender Studies from the University of Buea and a PhD from the University of Leeds; UK. As Associate Professor, her research interest is in the area of Gender and development, Microfinance and Entrepreneurship Development, social policy and social protection. She is equally a trainer on Gender responsive Economic Policy Management and Planning with consultancy skills on Gender mainstreaming, Gender, Trade and Development from UNIDE. Professor Abonge is currently involved in research and consultancy on Gender, Conflict and Development with particular interest on informal peacebuilding and gender mainstreaming in peacebuilding processes. She is the Head of the Department of Women and Gender Studies, Faculty of Social and Management Sciences, of the University of Buea, Cameroon.

[illegible]

Chapter 15
Backyard Dwellings in the City of Cape Town: Implications for Urban Management

Moegamad Rafeeq Le Roux

Abstract Post-apartheid South Africa has seen a significant rise of urbanisation. According to official census figures, a total of 53.7% of the South African population were living in urban areas in 1996 and increased to 62.9% by 2011. One of the consequences of this rapid urbanisation is the escalating demand for housing in urban areas. In response to the impact of government initiatives and the inadequacy of the housing market to provide sufficient affordable and sustainable housing, backyard dwellings manifest itself as an innovative response to the housing backlog. On a national scale the occurrence of backyard dwellings as a proportion of total housing has risen from 4.1% in 2001 to 4.9% in 2011. This represents an increase in absolute numbers from 459528 to 712956 (Statistics South Africa (StatsSA), Census South Africa 2001 Cape Town: StatsSA 2001; Statistics South Africa (StatsSA), Census South Africa 2011 Cape Town: StatsSA 2011), and according to (Lemanski, Habitat International 33:472–484, 2009) is increasing at a faster rate than informal settlements. In the City of Cape Town, backyard dwellings has risen from 4.3% of total housing stock in 2001 to 7% in 2011, and in absolute numbers increased from 32747 to 74957. The occurrence of backyard dwellings at this scale has a definite impact on urban structure and poses a number of urban management challenges. The paper investigates the spatial trends and the socio-economic characteristics of backyard households in light of its possible impact on urban form and management within the City of Cape Town.

Keywords Urban form and function · Urban management · Informal housing · Backyard dwellings · Housing provision

M. R. Le Roux (✉)
Worcester, Western Cape, South Africa
e-mail: rleroux@bocma.co.za

L. Chipungu et al. (eds.), *Contemporary Models for Pro-Poor Housing in Urban Sub-Saharan Africa*, https://doi.org/10.1007/978-3-032-05489-0_15

15.1 Introduction

The well-known work of Maslow identified a hierarchy of human needs of which shelter is ranked among the important physical and psychological needs essential for human survival (Poston, 2009). The need for shelter, especially in post-apartheid South Africa resulted in some peculiar spatial effects. In view of the inability of both government housing policies and the free market to provide sufficient housing for the urban poor, innovative means has been sought in an attempt to provide for this basic need, one such innovation is in the form of backyard dwellings (South African Institute of Race Relations, 2009). Lemanski (2009:472) defines backyard dwellings as *"informal shacks, typically erected by their occupiers in the yards of other properties, and are a uniquely South African phenomenon,* while Shapurjee and Charlton (2013:3) describe it as *"Backyard accommodation involves the co-habitation of landlord and tenant on the same plot, albeit in different dwellings."* The essential elements of backyard dwellings thus involve informal structures on a formal property not occupied by the primary dweller and implying a landlord and tenant relationship.

Formal backyard dwellings are a well-known international phenomenon (Chapple et al., 2012; Crankshaw et al., 2000). The general trend is that backyard dwellings in the developing worlds are informal in nature, whereas those in the developed world are formal (Chapple et al., 2012; Potts, 2006). Formal backyard dwellings are also present in the South African context although at a much smaller scale than informal structures. At a national scale formal backyard dwellings represent 0.82% of total housing in South Africa compared to 4.9% representing informal backyard dwellings (Stats SA 2011). Formal backyard dwellings in South Africa are also regulated in terms of local by-laws and building permits and is thus more predictable and manageable than the larger and sporadic spatial manifestation of informal backyard dwellings. In this article, the use of the term backyard dwellings refers to informal backyard dwellings.

On a national scale, the number of households residing in backyard dwellings increased from 459528 in 2001 to 712956 in 2011 (Stats SA 2001; Stats SA 2011) and represented 4.9% of total housing in 2011. Over the same period, the proportion of households residing in informal structures in informal settlements as a proportion of total housing decreased from 11.7% to 8.6%. In the Western Cape backyard, dwellings (as a proportion of total housing) has risen from 3.9% in 2001 to 6.4% in 2011, and in absolute numbers increased from 46800 to 105282. The bulk of these backyards are located within the City of Cape Town with a total of 32747 backyard dwellings in 2001 and 74957 in 2011.

One of the main reasons for the occurrence of backyard dwellings in South Africa has been ascribed to the trend of increasing rural–urban migration as a result of the freedom of movement that exists in a post-apartheid democratic South Africa. This creates a need for low-cost housing by migrants who often end up residing in the backyards of family or friends (Crankshaw et al., 2000). Other reasons for the occurrence of backyard dwellings identified by Lemanski (2009) and Shapurjee and Charlton (2013) include the natural growth of cities promoting neighbourhood lifecycles and backyard dwellings as a form of infilling; the imbalance between the demand and supply of low-cost housing; the disadvantages of other housing stock options available to the urban poor such as informal settlements which can be located further away from economic opportunities; and rental housing which can be more expensive. They also identified proximity to work opportunities as a contributing factor as backyard dwellings are often closer to economic centres and provide access to services and facilities due to the connection of the backyard dwelling to a formal serviced house. Backyard dwellings as a housing option available to the urban poor is therefore important in terms of spatiality, understanding its role in the dynamics of housing provision, projecting trends and attempting to reconcile housing supply with demand.

The purpose of this study is to evaluate the role and impact of backyard dwellings on urban form and management within the City of Cape Town. It firstly examines the various social, economic, environmental, and infrastructure impacts of backyard dwellings. It then provides an analysis of the extent, trends, and spatial characteristics of backyard dwellings in the City of Cape Town making use of the 2001 and 2011 census data.

Thirdly, it examines the socio-economic characteristics of backyard households in the City of Cape Town and then concludes with a discussion on the implications of backyard dwellings for urban form and management within the City of Cape Town.

15.2 Impact of Backyard Dwellings on Urban Development and Management

15.2.1 Contextual Setting

It is widely recognised that sustainability in general refers to the triple bottom line which advocates a balanced ecological, social, and economic human existence, for current and future generations (Victor, 2009). The concept of a sustainable city has to be broken down to smaller building blocks and the emphasis needs to be on the concept of a system of sustainable neighbourhoods within a sustainable city. The manifestation of sustainable neighbourhoods that is sketched by Rudin and Falk (1999) is a formalised approach to long term planning of neighbourhoods that is concentrated on quality of life, functioning economies and minimal resource utilisation. The mechanism to achieve this lies also in formal planning of neighbourhoods from street design to eco-friendly homes. The question thus is whether an informal approach to housing such as backyard dwellings can also contribute to sustainable neighbourhoods. Although the localised focus as described by Swilling (2006) and Rudin and Falk (1999) remains the ideal, overall urban management processes often does not have this luxury, and the focus is thus on a concept of a sustainable city which looks at the integration of various systems (Van Dijk, 2006). This concept is often evaluated in two opposing discourses either promoting densification or urban sprawl: the advocates of each challenging and comparing the benefits or disadvantages of either of the two options. These advantages and disadvantages of densification and urban sprawl are often evaluated in terms of economic, social and environmental efficiency (Jabareen, 2006). Swilling (2006:1) for example argues that "*Housing lower-income groups in conventional housing with no attention to a more compact, less auto-mobile-dependant city exposes them to high water and energy costs.*" An approach which recently gained a lot of popularity is the use of so-called sustainability indicators to measure levels of urban sustainability (Li et al., 2009; Moussiopoulos et al., 2010). This approach suggests that the concept of a sustainable city can be measured through a localised set of indicators to reflect economic growth and efficiency, ecological and infrastructural construction, environmental protection and social welfare and progress. The following section describes the potential impact of backyard dwellings on sustainable urban development in terms of social economic, environmental, infrastructure and urban management impacts.

15.3 Socio-Economic and Environmental Impacts

The impact of backyard dwellings on the social environment can be approached from the perspective of its influence on quality of life which is generally regarded as a measure of social health and sustainability (Marans, 2012). The concept of quality of life and satisfaction of living is however to a certain extent a matter of perspective that would differ from person to person. Naude et al. (2009) argue that some of the factors affecting quality of life are population density and its link to overcrowding, as well as the quality of place. If backyard dwellings thus contribute to overcrowding it can be viewed as impacting negatively on quality of life. The quality of place is however a relative concept. For example, if the occupant of a backyard dwelling were previously homeless, they would have been deprived of one of the basic human needs (Poston, 2009) and their occupation of a backyard dwelling could thus constitute and improved quality of place. According to Crankshaw et al., (2000:852):

> Physical conditions in the backyards are clearly not good and might be expected to generate tension between landlords and their tenants. In practice, however, there is limited evidence to suggest that this is the case.

Pharaoh (2008) further advocates that backyard dwellings contribute to and foster social connections due to the sense of community created by the physical proximity of people.

From an economic perspective the benefits of backyard housing can include possible income stream to landlords as well as savings in travelling costs to the occupants of backyard dwellings due to their relatively central location (Lemanski, 2009). From an economic perspective, backyard dwellings create a second-income stream for land owners with backyard dwellings who in other instances might not have been landlords. According to Geyer & Geyer (2013:12):

> Traditionally, value is added to housing by improving the aesthetic value of the house through home improvements and extensions; but in disadvantaged communities, additional commercial value is often also accrued through backyard informal development.

From the perspective of the backyard dwellers, it can be argued that the landlord tenant relationship and the paying of rent by the backyard tenants may place them in a vulnerable position. The benefits of backyard dwelling as compared to informal settlement living however to some extent negate the financial burden of paying rent (Lemanski, 2009).

From the perspective of responsible municipal authorities backyard dwellings are often not formally planned for and they can create a burden on resources such as electricity and water provision which in turn has an associated financial cost (Naidoo, Longondjo, Rawatlal & Brueton, 2012). Crankshaw et al. (2000) contend that backyard dwellers are often poor and their contribution to the tax base that is needed to maintain, improve and develop such utilities is therefore debateable. It could thus even be reasoned that it creates an increased need for cross-subsidies due to the increased demand and its associated economic impact that backyard dwellers place on utilities (Naidoo et al., 2012).

The occurrence of backyard dwellings can have both positive and negative environmental impacts. On the positive side, Lemanski (2009) argued that backyard dwellings supplement informal settlements and hence decrease the growth and development of Greenfield sites. Along similar lines, Naidoo et al (2012) further advocated that providing services to clustered population densities increases the efficiency of provision which in turn saves energy and resource costs thus having a positive environmental effect. From a construction perspective the physical characteristics of backyard dwellings mimic construction methods with recycled products such as discarded material e.g. corrugated iron (Crankshaw et al., 2000).

Negative environmental impacts include backyard dwelling occupancy synonymous with poor sanitation are a major contributor to the proliferation of water borne diseases and water pollution (Govender et al., 2011b); and an increased number of backyard dwellings resulting in an increased number of informal dwelling fires (Pharaoh, 2008).

15.3.1 Provision of Urban Infrastructure and Urban Management

The occurrence of backyard dwellings can have various impacts on the provision of municipal infrastructure. Of particular relevance is the issue of shared infrastructure when informal backyard dwellings are located on formal serviced properties. This results in an increased burden on the sewer network that is often not planned for by city managers, thus leaving sewer networks over capacitated and resulting in blockages (Govender et al. (2011b); and consequently, health risks associated with such events. Naidoo et al (2012) highlighted the issue of the physical characteristics of backyard structures itself and argues that the informal nature of the structures even though on a formal stand and linked to a formal structure is not conducive to improving basic living conditions. They also identify the element of tension between backyard dweller and the landlord in accessing basic services noting that *"...although most backyard dwellings have access to water in the main house, this access is controlled by a landlord."* (Naidoo et al., 2012:42). They argue that this situation in essence increased the vulnerability of the backyard dwellers.

Another basic service challenge linked to backyard dwellings is the provision of electricity. Gaunt, Salida, Macfarlane, Maboda, Reddy and Borchers (2012) found that there is a prevalence of illegal connections associated with backyard dwellings to the formal structure. This presents a challenge as the connection itself is dangerous and illegal. On the positive side they argue that, although the connection is illegal, the electricity used is still accounted for and can be linked and billed to a specific formal dwelling with the potential to formalise the activity (Gaunt et al., 2012).

In the South African municipal scenario, the triad consisting of the integrated development plan (IDP), the spatial development framework (SDF), and the municipal budget form the core of the municipal planning instruments. The exclusion or insufficient consideration of backyard dwelling statistics can impact negatively on urban planning and management initiatives. In the context of the City of Cape Town, the proliferation of backyard dwellings thus advocates its inclusion in such planning activities (City of Cape Town, 2013a, b, c).

One of the aims identified in the City of Cape Town IDP (2012–2017) is *"to extend services to so called 'backyarders' on City-owned land through a unique service point containing a covered toilet and basin as well as external electricity connections"* (City of Cape Town, 2013a, b, c:8) over the period 2012 to 2017 through the so-called *"backyarder service programme"* (City of Cape Town, 2013a, b, c:20). For the 2013/2014 financial year, the City of Cape has committed R37.5 million to specifically provide backyard dwelling households with electricity (City of Cape Town, 2013b:11). The SDF of the city further suggests that *"backyard dwellings are to be supported where appropriate"* (City of Cape Town, 2012:52).

The "guidelines for human settlement planning and design," also known as the "red book" published by the Council for Scientific and Industrial Research (CSIR) in 2005 in partnership with the then Department of Housing, is an industry standard used by government departments, municipalities and private developers in planning human settlements (CSIR 2005). Some of the suggested guidelines in the "Redbook" for planned lower income areas are 60–100 L of water per person per day; the removal of sewage by flush toilets with an average daily flow of 500 L per dwelling unit, the provision of conventional electricity to the extent of 500 to 30000 kW/km^2; and the regular removal of solid waste to a nearby municipal solid waste (CSIR 2005). It also identifies key considerations when planning for public utilities such as space standards, number of users at any one time, distribution of use through the day and through the week. In view of the findings of Crankshaw et al. (2000) that RDP housing areas are prime locations for backyard dwelling occurrence, it is imperative that its impact be factored into the initial planning of such settlements.

15.3.2 Summary of Impacts

From an urban morphology and management perspective, there are both positive and negative impacts associated with informal backyard dwellings. The contents of the available literature reviewed indicate that there are no definitive and converging opinions on the impact of backyard housing. A summary of the different positions and opinions outlined in the preceding sections is summarised in Table 15.1. These individual factors need to be interpreted, and appropriate responses defined within the context of individual local circumstances and trends.

15.4 Setting the Scene

The study area is defined as the municipal area of the City of Cape Town and is characterised by widely different demographic characteristics (such as income levels, education, and access to services) and represents grounds for interesting enquiry (City of Cape Town, 2006). The primary data used in this analysis is the 2001 and 2011 census data at city and sub-place name levels, as well as records and statistics obtained from the City of Cape Town. Informal backyard dwellings as a proportion of total housing stock have increased by 2.5% in the Western Cape and by 2.7% in the City of Cape Town between 2001 and 2011 (see Fig. 15.1). The City of Cape Town also represents the area with the largest increase in informal backyard dwellings over the 10-year period within the Western Cape increasing from 32747 in 2001 to 74957 in 2011 (see Table 15.1).

The review of available literature has highlighted that informal backyard dwellings manifest itself as a supplement to informal squatter settlements and provides various relative advantages such as being linked to a formal structure for services and being closer to areas of employment and social amenities. For comparative purposes, the analysis results thus also provide the comparative figures for informal settlements as shown in Table 15.2. At a national level the proportion of households residing in informal squatter settlements decreased from 12.3% in 2001 to 8.6% in 2011 (see Fig. 15.2).

15.5 Spatial Trends and Characteristics in Cape Town

The primary objective of the spatial analysis is to determine trends in both the spread and intensity of occurrence of backyard dwellings. Table 15.3 indicates an increase in both the spread and intensity of occurrence of backyard dwellings between 2001 and 2011. The total number of sub-places (spatial units of roughly equivalent size) containing households residing in backyard dwellings increased from 672 to 974. This increase was especially prevalent in newly emerging areas (consisting of 20 or

Table 15.1 Summary of impacts of informal backyard dwellings

Impact class	Positive impacts	Negative impacts
Socio-economic	1. Foster social connections due to the sense of community created by the physical proximity of people 2. Provide a second income stream to landlords 3. Savings in travelling costs to the occupants of backyard dwellings due to their relatively central location	1. Negative impact on quality of life as a result of overcrowding 2. Potential tension between landlord and tenant 3. Paying of rent by backyard tenants may place them in a vulnerable position 4. Informal nature of the structures (even though on a formal stand and linked to a formal structure) is not conducive to improving basic living conditions
Environmental	1. Backyard dwellings supplement existing formal and informal housing processes and hence decrease the growth of Greenfield sites 2. Providing services to clustered population densities increases the efficiency of provision which in turn saves energy and resources 3. Physical characteristics of backyard dwellings mimic construction methods with recycled products such as discarded material (e.g. corrugated iron)	1. Poor sanitation as a major contributor to the proliferation of water borne diseases and water pollution 2. Increased number of backyard dwellings lead to an increased number of informal dwelling fires

(continued)

Table 15.1 (continued)

Impact class	Positive impacts	Negative impacts
Infrastructure	1. Although most electricity connections are illegal, the consumption is still accounted for and can be linked and billed to a specific formal dwelling with the potential to formalise the activity	1.Backyard dwellings are often not formally planned for and create a burden on electricity and water provision 2. Backyard dwellers are mostly poor and their ability to contribute to the tax base that is needed to maintain, improve and develop the required services and utilities is therefore limited 3. Creates an increased need for cross-subsidisation of services due to the increased demand 4. Increased burden on sewer network that is often not planned for, thus leaving sewer networks over capacitated and resulting in blockages and consequent health risks associated with such events 5. Although most backyard dwellers have access to water in the main house, the access is controlled by the landlord and can create an element of tension between backyard dweller and the landlord in accessing basic services. This may in essence result in increased vulnerability of the backyard dwellers 6. Dangers associated with prevalence of illegal electricity connections
Urban planning and management	1. Can be regarded as a form of spontaneous infill and densification development 2. Reduces the need for Greenfield development 3. Could be viewed as form of housing delivery reducing burden on public resources to provide affordable housing	1. Difficult to plan for capacity of services and social facilities 2. Many backyard households fall within the lower income categories and can increase financial burden on municipality to provide free basic services to households who are regarded as indigent

Source Extent And Distribution of Backyard Dwellings In Cape Town (2024)

less backyard units) nearly doubling in number from 386 in 2001 to 614 by 2011. These figures also provide some evidence of increasing intensity in existing locations with the both the number (from 30 to 70) and proportion of sub-places containing more than 500 backyard dwellings increasing significantly between 2001 and 2011.

The spatial distribution of backyard dwellings in the City of Cape Town is further reflected on Maps. 15.1 and 15.2. In 2001, the highest concentration was located in Langa with 2597 backyard dwellings and in Phillipi 2589. Intermediately backyard

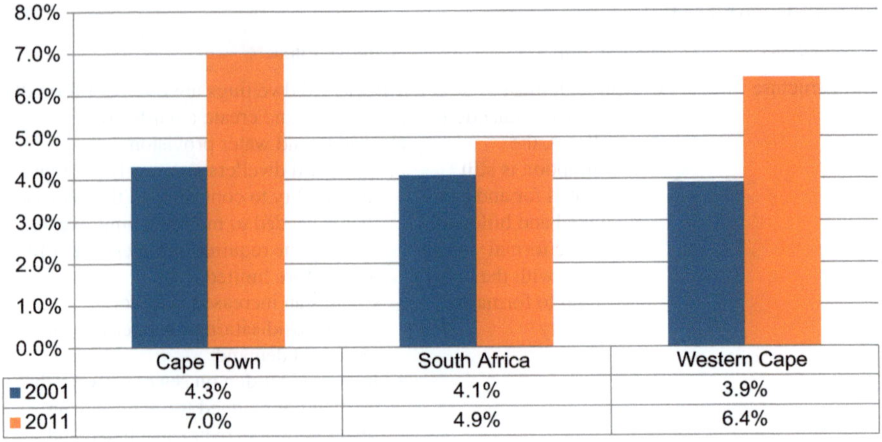

Fig. 15.1 Backyard Dwellings as a Percentage of Overall Housing in 2001 and 2011. *Source* Author's Construct using 2001 and 2011 census data (2024)

Table 15.2 Extent of backyard dwellings in absolute numbers

Dwelling Type	Cape Town		Western Cape		South Africa	
	2001	2011	2001	2011	2001	2011
Backyard dwellings	32747	74957	46800	105282	459528	712956
Informal settlements	110157	143823	142679	191668	1376707	1249777

Source Author's Construct using 2001 and 2011 census data (2024)

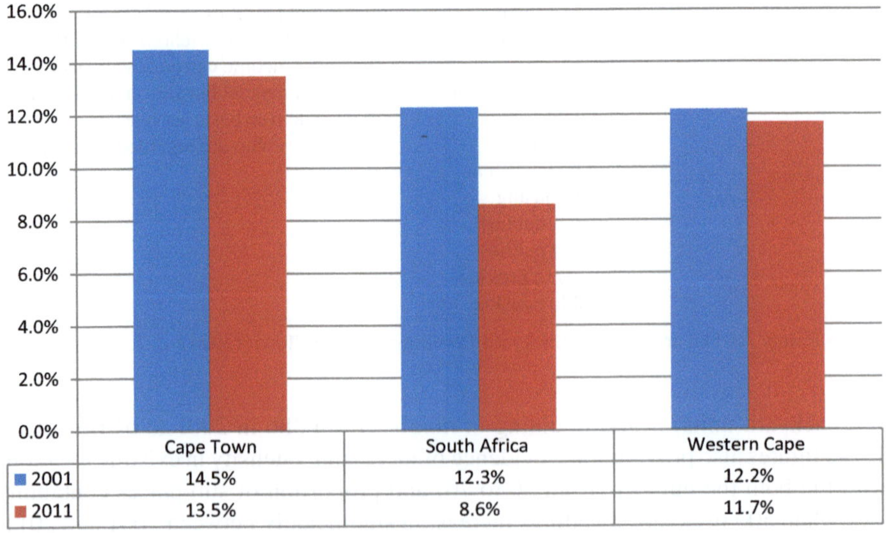

Fig. 15.2 Informal settlements as a proportion of overall housing stock 2001 and 2011. *Source* Author's Construct using 2001 and 2011 census data (2024)

Table 15.3 Number and percentage of Sub places with varying concentrations of backyard dwellings in 2001 and 2011

Concentration of backyard dwellings	Number of sub-places		% of SUB-PLACES	
	2001	2011	2001 (%)	2011 (%)
>500	30	70	4.5	7.2
100–499	96	130	14.3	13.3
50–99	60	74	8.9	7.6
20–49	100	86	14.9	8.8
<20	386	614	57.4	63.0
Total	672	974	100	100

Source Author's Construct using 2001 and 2011 census data (2024)

dwellings appear in areas such as Macassar with 370 backyard structures, Eersterivier with 268, Atlantis with 259, and to a lesser extent in areas such as Durbanville with 33 and Goodwood with 23 structures. Backyard dwellings do not appear in more affluent areas such Constantia.

In 2011, the occurrence of backyard dwellings in the City of Cape Town has increased in absolute numbers from 32747 in 2001 to 74957 in 2011 (Map 15.2). Browns Farm in Philippi is now the area with the highest amount with 6455 backyard

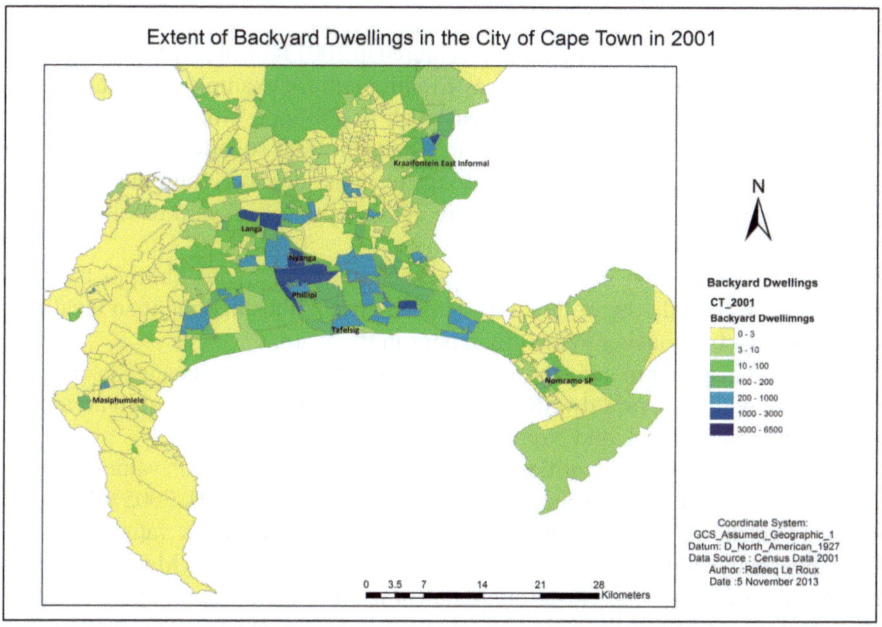

Map 15.1 Extent of backyard dwellings in the City of Cape Town in 2001. *Source* Author's Construct using 2001 and 2011 census data (2024)

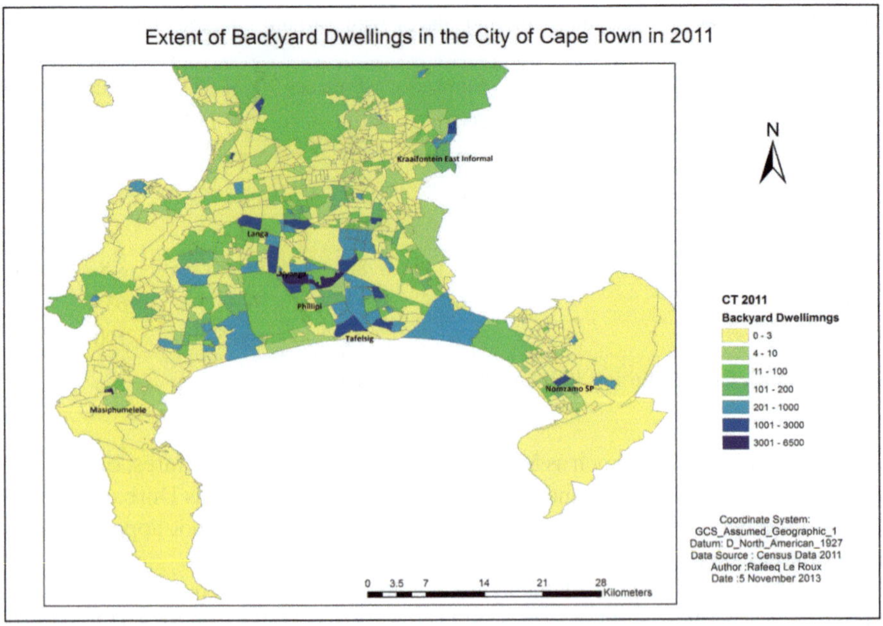

Map 15.2 Extent of backyard dwellings in the City of Cape Town in 2001. *Source* Author's Construct using 2001 and 2011 census data (2024)

dwellings and Nomzamo previously not featured in 2001 now in second position with 3666 backyard dwellings. Areas such as Kraaifontein for example are dominated by formal housing structures whereas areas such as Philippi, Nomzamo, Langa and Nyanga are a hybrid between RDP, formal and informal squatter settlements. The emerging prominence of backyard dwellings in areas such as Kraaifontein indicates that some predominantly formal areas are also adopting backyard dwellings as an alternative housing type.

Map 15.3 shows that in 2001 in areas such as Nyanga and Philippi backyard dwellings constituted 10% of the overall housing stock thus making it an invaluable source of low-cost housing to the urban poor in these areas.

In 2011, the spatial manifestation of backyard dwellings has changed somewhat (Map 15.4), with backyard dwellings have becoming more prominent in areas such as Philippi, Kraaifontein, and Nomzamo.

Getis-Ord Gi statistics identifies those clusters of spatial data with values higher in magnitude than you might expect to find by random chance (ESRI, 2005). The result is a map that shows areas of high value clustering and areas of low value clustering. A hotspot analysis (Getis-Ord Gi) of the clustering at city level (Map 15.5) reveal that the area of high value clustering manifest itself in the areas of Nyanga, Langa, Khayelitsha, and Mitchells Plain and clustering of low values in the north eastern parts of the city.

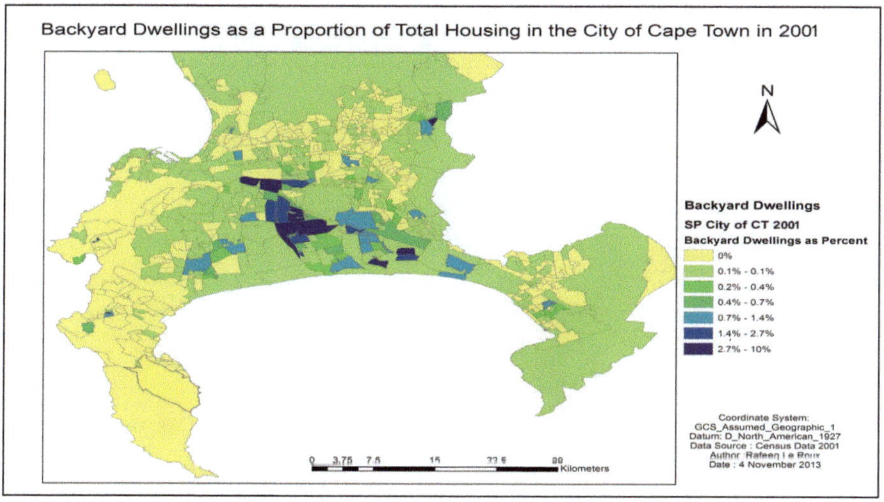

Map 15.3 Backyard dwellings as a proportion of overall housing in the City of Cape Town in 2001. *Source* Author's Construct using 2001 and 2011 census data (2024)

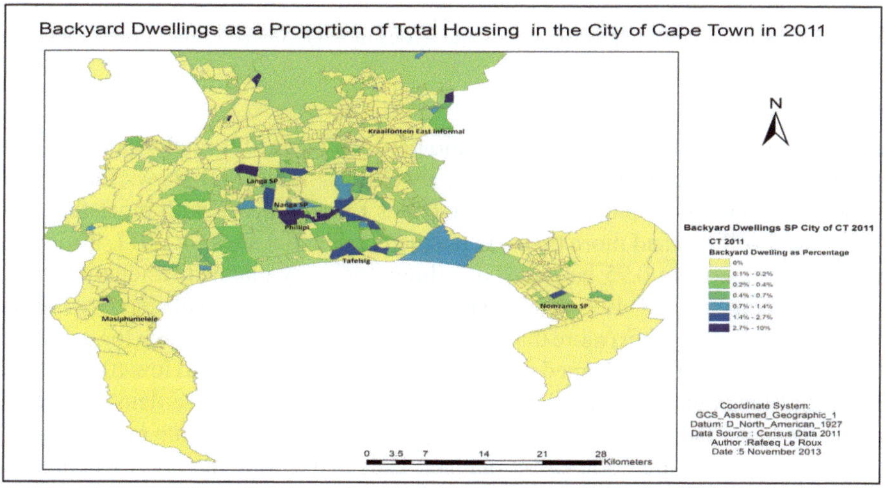

Map 15.4 Backyard dwellings as a proportion of overall housing in the City of Cape Town in 2011. *Source* Author's Construct using 2001 and 2011 census data (2024)

From the 2011 scenario, it is evident that in the City of Cape Town high- and low-value clustering does occur and that this pattern has not changed considerably from the 2001 scenario. There are thus areas that display consistently high value clustering whereas other areas display consistently low value clustering.

Against the background of the potential impact of backyard dwellings on future urban growth patterns and its implications for urban planning and management it is

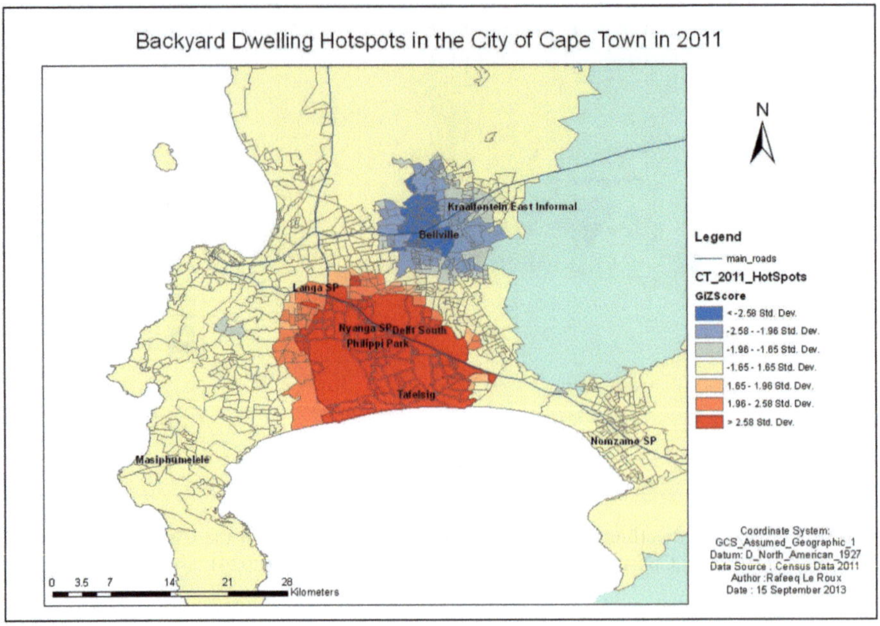

Map 15.5 Backyard dwelling hotspots in South Africa in 2011. *Source* Author's Construct using 2001 and 2011 census data (2024)

essential to gain some insight into the ratio between planned formal residential sites and the eventual occurrence of backyard dwellings. Table 15.4 shows the ratio of backyard dwellings to planned residential sites for sub places with more than 100 backyard dwellings and more that 100 planned residential sites for 2011. The mean values decrease from 0.95 backyard dwellings for every planned residential site in the lower segment (100–499) to 0.25 backyard dwellings for each planned residential site in the higher segment (more than 2500). The minimum ratio value is lowest (0.02) in the 'more than 2000' planned residential sites segment and also low (0.06) in the 1000–2499 segment. Generally, the higher the number of planned settlements per sub place, the lower the mean ratio of backyard dwellings. The maximum ratio is highest (2.48) where there are 1000–2499 planned sites and also high (2.42) where there are 100 to 499 planned residential sites. Backyard dwellings, therefore, do occur at higher ratios in sub places where there are lower and higher formally planned areas.

Map 15.6 shows the spatial occurrence of the ratio of backyard dwellings to planned residential sites for sub places that have more than 100 backyard dwellings and more than 100 planned formal settlements. Philippi shows the highest ratio with of 2.42 backyard dwellings for each planned residential site, intermediately Kraaifontein East Informal has 1.06 backyard dwellings for each planned residential site and to a lesser extent Manenberg has 0.27 and Mfuleni has 0.18 backyard dwellings for each planned residential site. The density of occurrence is therefore valuable

Table 15.4 Ratio between informal backyard dwellings and number of residential sites for all sub-places containing more than 100 residential sites and with more than 100 backyard dwellings

Number of residential sites	Maximum value	Minimum value	Mean value
More than 2500	1.07	0.02	0.25
1000–2499	2.48	0.06	0.32
500–999	1.65	0.11	0.34
100–499	2.42	0.35	0.94

Source Author's construct using 2001 and 2011 census data (2024)

statistics that should be incorporated in efficient settlement planning. If the "Redbook" (CSIR 2005) is used as a guideline, then the settlement planning should take into account these ratios when planning RDP housing for example.

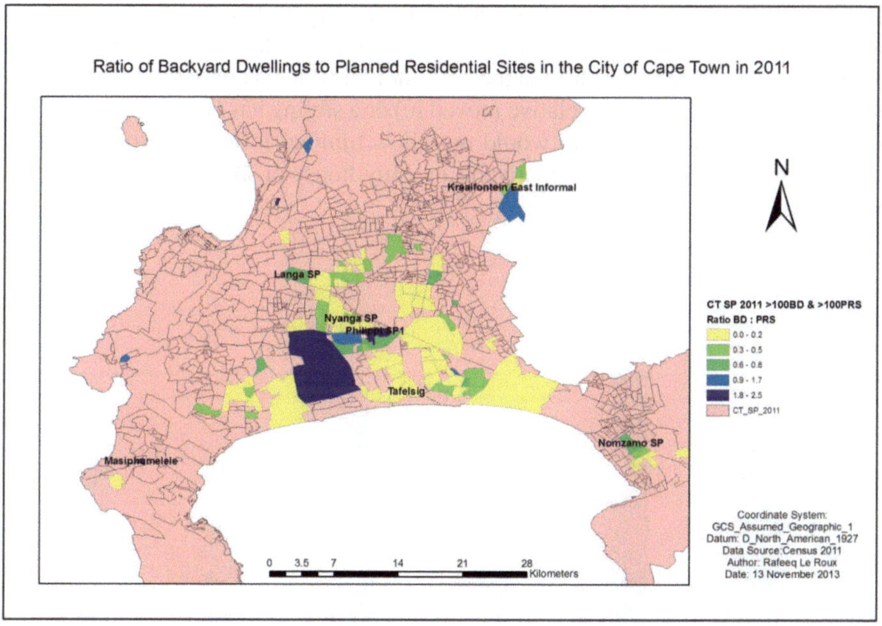

Map 15.6 Ratio of sub places with more than 100 backyard dwellings and more than 100 planned residential sites in the City of Cape Town in 2011. *Source* Author's Construct using 2001 and 2011 census data (2024)

15.6 Socio-Economic Characteristics of Backyard Households in Cape Town

15.6.1 Household Size

The general trend between 2001 and 2011 was towards smaller and more single households residing in backyards. The most notable trend over this period is the proportion of single person households that increased from 16.1% in 2001 to 21.9% in 2011. Two person households also increased somewhat from 20.2% to 21.8%, but the proportion of larger households (four people and more) declined over the same period. This is indicative of increasing numbers of single and small households residing in backyard dwellings (see Fig. 15.3).

These changes in household size can also be interpreted against the backdrop of Pacione "residential adjustment or relocation" related to human life-cycle events (Pacione, 2009: 203). It is thus plausible that increasing backyard occupancy in Cape Town is a manifestation of these lifecycle events amongst lower-income population. An example could be that a child in the household finishes secondary school, and the house is crowded (perhaps a relative moved in) or a sibling has become a parent. The low-income status of the child would not allow him/her to rent a formal dwelling such as a flat, and a logical alternative is a wendy-house or shack in backyard of the parent(s).

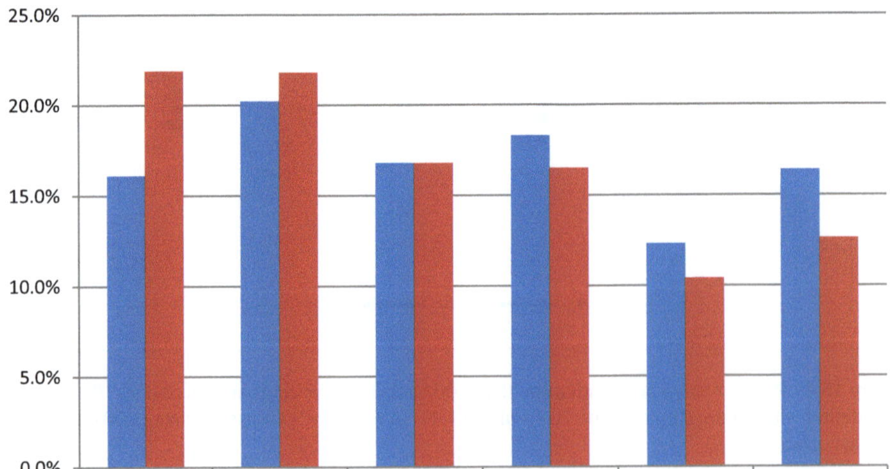

	1 Person	2 People	3 People	4 People	5 People	6 - 10 + People
2001	16.1%	20.2%	16.8%	18.3%	12.3%	16.4%
2011	21.9%	21.8%	16.8%	16.5%	10.4%	12.6%

Fig. 15.3 Average household sizes of backyard dwellers in the City of Cape Town in 2001 and 2011. *Source* Author's Construct using 2001 and 2011 census data (2024)

15.6.2 Employment Status and Household Income

The employment and income pattern of households are important factors influencing the functioning and planning of urban areas and has a bearing on how and where money is spent and can influence the retail, travelling and residential components of a city (Greene, 1980).

Figure 15.4 indicates that the unemployment rate of backyard households has increased from 18% in 2001 to 27% in 2011. The latter figure is however still appreciably lower than a comparative unemployment rate of 38% in informal settlements. Moreover, despite backyard dwellers earning a comparatively higher income than inhabitants of informal squatter settlements, they can still be regarded as a predominantly low-income segment of the housing market with 76.6% of these households earning less than R38200 (although a significant decrease from 90% in 2001). This implies that a substantial proportion of backyard dwellers can be categorised as eligible for government assistance in the form of affordable subsidy housing (based on income criteria). The City of Cape Town also provides free basic services to households who are regarded as indigent (have a household income of less than R3200.00 per month). Some of the services that indigent households qualify for are six kilolitres of water for free, a relief of R67.42 towards rates and free or rebated waste collection (City of Cape Town, 2013c).

Despite the overall low-income pattern, there has been significant increases between 2001 and 2011 (see Table 15.5). The proportion of backyard households earning more than R38200 per month increased from 10% in 2001 to 23.4% by 2011,

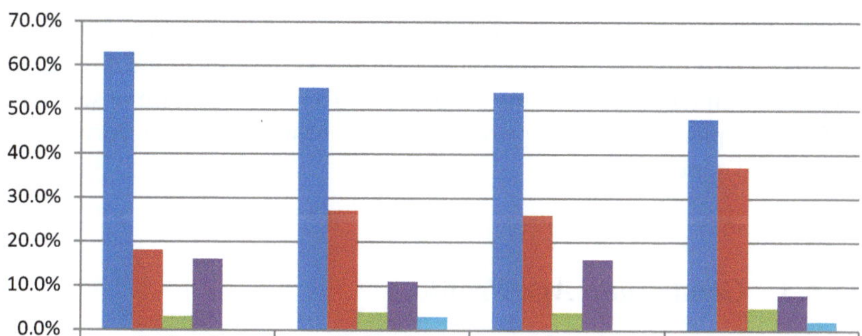

	Backyard Dwelling 2001	Backyard Dwelling 2011	Informal Settlement 2001	Informal Settlement 2011
Employed	63.0%	55.0%	54.0%	48.0%
Unemployed	18.0%	27.0%	26.0%	37.0%
Work seeker	3.0%	4.0%	4.0%	5.0%
Other	16.0%	11.0%	16.0%	8.0%
< 15 years old	0.0%	3.0%	0.0%	2.0%

Fig. 15.4 Employment statuses in percentages per housing type in the City of Cape Town for 2001 and 2011. *Source* Author's Construct using 2001 and 2011 census data (2024)

Table 15.5 Average annual household incomes in percentage by dwelling type in the City of Cape Town (Squatter and Informal Backyard) in 2001 and 2011

Income	Backyard dwelling		Informal squatter settlement	
	2001 (%)	2011 (%)	2001 (%)	2011 (%)
No income	25.0	20.4	33.0	25.7
R 1–R 9600	20.0	11.9	23.0	15.9
R 9601–R 19600	26.0	16.5	25.0	17.9
R 19601–R 38200	19.0	27.8	14.0	25.2
R 38201–R 76400	8.0	16.1	4.0	11.2
R 76401–R 153800	2.0	5.6	1.0	2.9
R 153801–R 2457601 or more	0.0	1.7	0.0	1.2
Total	100	100	100	100

Source Author's Construct using 2001 and 2011 census data (2024)

whilst households with no income decreased from 25% to 20.4% and those earning less than R9600 and from 20% to 11.9%.

The trend of increasing household income together despite increasing unemployment rates seems to be counter intuitive. This can potentially be explained by two factors. Firstly, employed backyard dwellers are now earning higher incomes and secondly those earning an income in the informal sector. These household income patterns and trends hold potentially important implications for urban planning and management purposes. It provides a clear indication of the proportion of households in backyards that could potentially be interpreted as forming part of the 'housing backlog' of the city in terms of the need for a government supported subsidised housing. The proportion of households with access to services via the main dwelling could also influence the planning for upgrading and maintenance of services, as well as financial planning for the provision of free basic services as part of the City of Cape Town's indigent policy.

15.6.3 Gender and Age Profile

Gender and age characteristics are important factors in the composition of households and its impacts on planning (Cox et al., 2004) such as resource use and movement patterns within a city, and specifically, the possible movement of people between different low-cost housing types (Clark et al., 2006).

The age profile of the head of households of backyard structures is completely dominated by the age category between 25 and 34 years of age representing 48% of male-headed households and 44% of female-headed households in 2011 (see

Table 15.6 Age and gender profile of backyard household heads in the City of Cape Town by percentage for 2001 and 2011

Age	Backyard dwelling				Informal squatter settlement			
	Male		Female		Male		Female	
	2001 (%)	2011 (%)	2001 (%)	2011 (%)	2001 (%)	2011 (%)	2001 (%)	2011 (%)
15–24	10.0	12.0	11.0	12.0	10.0	11.0	11.0	10.0
25–34	44.0	48.0	38.0	44.0	42.0	44.0	38.0	39.0
35–44	28.0	24.0	28.0	24.0	28.0	29.0	27.0	29.0
45–54	12.0	10.0	13.0	12.0	13.0	11.0	15.0	15.0
55–64	5.0	3.0	6.0	5.0	5.0	4.0	7.0	5.0
65+	2.0	1.0	4.0	2.0	1.0	1.0	2.0	2.0
Total	100	100	100	100	100	100	100	100

Source Author's Construct using 2001 and 2011 census data (2024)

Table 15.6). Moreover, this trend is becoming increasingly dominant with the proportion of households in this age category increasing by 4% and 6% for male- and female-headed households, respectively, between 2001 and 2011.

The demographic profile of households residing in informal backyards can thus be summarised as people in the young economically active age category between 25 and 34 years of age and in many instances consisting of single or two-person households. The low proportion of households in backyards older than 45 years of age further supports the concept that backyard structures are mostly utilised as a transition form of housing utilised by lower income groups in the earlier stages of life-cycle transitions.

15.7 Access to Services

Access to services and the associated use of resources is important element in assessing the impact of backyard dwellings on urban form and management. It is thus essential to understand how backyard dwellings impact services relative to other housing options and how this has changed over time (Bank, 2007).

15.7.1 Water and Sanitation

Water as a basic human need is essential in the provision of basic services as it is enshrined as a constitutional right in South Africa (Naidoo et al., 2012).

Figure 15.5 indicates that the proportion of households with access to piped water inside the yards remained relatively stable at around 60% between 2001 and 2011.

However, access to piped water inside the dwelling itself has increased from 21% in 2001 to 30% in 2011. These figures imply that by 2011 as much as 90% of backyard households had direct access to piped water, and due to the informal nature of these backyard dwellings this is most likely to take the form of an illegal connection. This trend therefore has important implications for the water consumption patterns in areas where backyard dwellings are prevalent.

The "red book" provides guidelines for different types of settlements and housing which ranges from communal to private provision of water, the type of settlement as well as the local conditions influences what type of infrastructure to use. Due to its proximity to existing infrastructure it recommends that water provision for backyard dwellings could include a combination of yard taps, communal stands or even an extension of the existing house condition with the installation of a separate meter to monitor the water use by the backyard household (CSIR 2005).

The access of backyard households to flush toilet systems connected to a sewerage system has increased from 74% in 2001 to 82% in 2011 as profiled in Table 15.7. Govender et al (2011b) however indicated that increased access doesn't necessarily

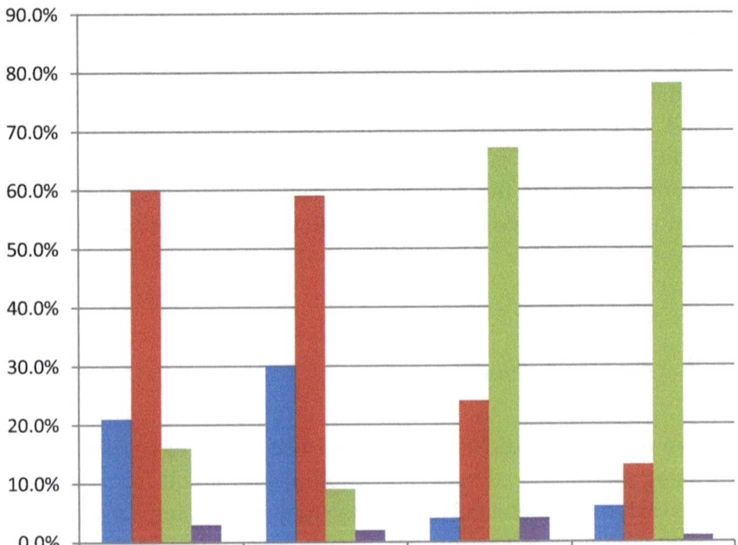

	Backyard Dwelling 2001	Backyard Dwelling 2011	Informal Settlement 2001	Informal Settlement 2011
Tap inside dwelling	21.0%	30.0%	4.0%	6.0%
Tap inside yard	60.0%	59.0%	24.0%	13.0%
Tap (communal stand)	16.0%	9.0%	67.0%	78.0%
No access to tap	3.0%	2.0%	4.0%	1.0%

Fig. 15.5 Access to water in the City of Cape Town in 2001 and 2011. *Source* Author's Construct using 2001 and 2011 census data (2024)

Table 15.7 Access to toilet facilities in percentages per dwelling type in the City of Cape Town in 2001 and 2011

Toilet facility	Backyard dwelling		Informal settlement	
	2001 (%)	2011 (%)	2001 (%)	2011 (%)
Flush toilet (connected to sewerage system)	74.0	82.0	32.0	42.0
Flush toilet (with septic tank)	3.0	3.0	2.0	5.0
Chemical toilet	1.0	1.0	1.0	7.0
Pit latrine with ventilation (VIP)	0.0	0.0	0.0	1.0
Pit latrine without ventilation	0.0	0.0	3.0	1.0
Bucket latrine	8.0	6.0	25.0	27.0
Other	0.0	3.0	0.0	4.0
None	14.0	5.0	37.0	14.0
Total	100	100	100	100

Source Author's Construct using 2001 and 2011 census data (2024)

imply lower health risk; it could be that many people have to share a single flush toilet. Table 15.7 further indicates that backyard households with no access to any form of sanitation have decreased from 14% in 2001 to 5% in 2011.

15.7.2 Electricity

The use of electricity as an energy source in backyard dwellings and informal settlements have a large impact on urban management and resource use, specifically because of illegal electricity connections (Gaunt et al., 2012). Nearly 80% of all households in informal backyard dwellings have access to electricity (see Fig. 15.6). Crankshaw et al. (2000) cited this as one of the major reasons why low-income households prefer to reside in backyard dwellings as opposed to informal settlements. The access of backyard dwellings to electricity is most likely in the form of an illegal connection and has important impacts for unaccounted electricity use, revenue derived from electricity usage and safety aspects related to illegal connections (Gaunt et al., 2012).

15.7.3 Refuse Disposal

The type of solid waste disposal depends on the available mechanisms of disposal and associated infrastructure of different housing types. As shown in Table 15.8, there are various means of refuse disposal.

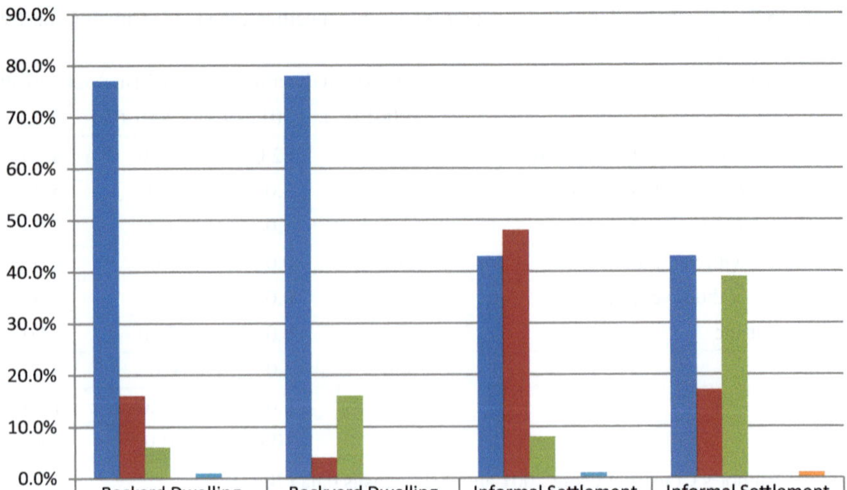

	Backard Dwelling 2001	Backyard Dwelling 2011	Informal Settlement 2001	Informal Settlement 2011
Electricity	77.0%	78.0%	43.0%	43.0%
Paraffin	16.0%	4.0%	48.0%	17.0%
Candles	6.0%	16.0%	8.0%	39.0%
Solar	0.0%	0.0%	0.0%	0.0%
Gas	1.0%	0.0%	1.0%	0.0%
None	0.0%	0.0%	0.0%	1.0%

Fig. 15.6 Access to energy sources used for lighting purposes in the City of Cape Town for 2001 and 2011. *Source* Author's Construct using 2001 and 2011 census data (2024)

Table 15.8 Backyard household refuse disposal in the City of Cape Town in 2001 and 2011

Means of refuse disposal	Backyard dwelling		Informal settlement	
	2001 (%)	2011 (%)	2001 (%)	2011 (%)
Removed by local authority at least once a week	94.0	96.0	75.0	68.0
Removed by local authority less often	2.0	0.0	4.0	3.0
Communal refuse dump	1.0	2.0	6.0	17.0
Own refuse dump	2.0	1.0	7.0	8.0
No rubbish disposal	1.0	1.0	8.0	4.0
Total	100	100	100	100

Source Author's Construct using 2001 and 2011 census data (2024)

The majority of backyard households are serviced by the City of Cape Town municipality with the removal of solid waste taking place at least once a week (94% of backyard households in 2001 increasing to 96% in 2011). This implies that informal backyard households generally receive the same level and standard of waste disposal as the main household.

15.8 Implications for Urban Form and Management

The literature has shown that the occurrence of backyard dwellings has specific socio-economic, environmental, infrastructure and urban management impacts. Some of these impacts include the role that backyard dwellings play in density and how this in turn can impact on quality of life; the economic incentive for landlords as an income source; the environmental impact in terms of promoting densification; the infrastructural challenges that backyard dwellings present and the benefits of improved access to services for backyard dwellers; and the use of backyard dwelling statistics to promote effective urban management through for example settlement planning. A summary of these implications is outlined in Table 15.9.

15.9 Recommendations and Conclusions

Based on the study, the following key recommendations are made:

- The City of Cape Town has adopted a policy of densification that is in line with national policy. Current City of Cape Town planning documents should therefore regulate and formalise the occurrence of backyard dwellings to further achieve the aim of compaction.
- The socio-economic profile of backyard households should be used in aligning available resources to the recipients thereof in urban planning documents and tools.
- The hotspot areas identified, as well as potential hotspot areas should be catered for in planning documents specifically because of their contribution to spatial impacts and resource use.
- The increase of backyard dwellings between 2001 and 2011 and the associated impacts described should be used to build a scenario for future trends and consideration should be given to backyard dwellings as a supplementary low-cost housing supply to informal squatter settlements.

In conclusion, both the literature and the findings of the census data for 2001 and 2011 have shown that backyard dwellings exhibit real implications for urban form and function within the City of Cape Town. The socio-economic trends described have shown that backyard dwellings influence urban form, function and management within the City of Cape Town. This is exemplified by the socio-economic profile of backyard dwellers and the associated environmental, economic and social impacts of this profile. The study has shown that backyard dwellings impact on service provision such as water, sanitation, electricity and refuse disposal by presenting an unplanned for demand on utilities. Based on this, the prevalence of backyard dwellings remains an urban function and management challenge within the City of Cape Town and can be seen as a serious emergent low-income housing option.

Table 15.9 Implications of research findings

Key finding	Socio-economic implication	Environmental implication	Infrastructural implication	Urban planning and management implication
1. Spatial distribution				
Limited increase in overall spread, largely occurrence has remained in areas that already had BD; some new areas such as Nomzamo are now noticeable Identification of high and low value clustering that has remained constant over the last 10 years Intensity has increased with more sub places having in excess of 500 BD	Certain areas are synonymous with this housing type and serve the urban poor as a housing option. These areas become increasingly popular. The limited increase in extent shows that people prefer to be centrally located, this has economic advantages such as time and cost associated with travelling The high value cluster areas can lead to a decreased quality of life as it could encourage crowding	Urban sprawl is limited Contributes to densification	Can provide to areas that have known BD occurrence Assist in costing the provision of infrastructure, peripheral locations might be more expensive to service, as the spread of BD hasn't increased considerably it could thus be cheaper to provide to the central locations Assists in the determination of infrastructure required to provide various services	Areas with known occurrence of BD can be incorporated into programmes such as the City of Cape Towns suggested backyarder service programme Certain areas can be prioritised High and low value clusters can be monitored, and thresholds can be established
2. Demographic profile				

(continued)

Table 15.9 (continued)

Key finding	Socio-economic implication	Environmental implication	Infrastructural implication	Urban planning and management implication
Increasing numbers of single and small households residing in backyard dwellings BD can be headed by either men or women Age profile of the head of households is completely dominated by the age category between 25 and 34 years of age Average age of BD households decreasing between 2001 and 2011 Increase in unemployment rate of BD households from 18 to 27% between 2001 and 2011 Predominantly low-income households with 76.6% earning less than R38200 per annum Proportion low-income households decreased from 90% in 2001 to 76.6% in 2011	BD is becoming a more popular housing type for smaller and younger low-income households BD increasingly utilised as a transitionary form of housing by lower income groups in the earlier stages of life-cycle transitions Employment status determines ability to pay rent	Potentially less resources consumed by smaller households Marginal increases in income can be possibly associated with increased use of resources	Potential decrease in the volume of services demanded Changing age profile can influence demand for certain type of facilities (e.g. schools) Potential provision of free basic services to BD households who are regarded as indigent (household income of less than R3200.00 per month)	Incorporate trends into planning processes to manage existing and future scenarios Changing age profile to be considered in planning of social utilities and facilities High proportion of BD households that could potentially be interpreted as forming part of the "housing backlog" of the city in terms of the need for a government supported subsidised housing Improved understanding of revenue streams and acknowledge the contribution of BD to the informal economy

(continued)

Table 15.9 (continued)

Key finding	Socio-economic implication	Environmental implication	Infrastructural implication	Urban planning and management implication
3. Access to services				
90% of BD has access to either water in the yard or in the main dwelling 82% of BD has access to flush toilet system in 2011, increasing from 74% in 2001 Nearly 80% of all BD have access to electricity and this is most likely illegal connections 96% of BD households receive waste removal service at least once a week	General access to water and sanitation increases quality of life and human dignity Potential improvement in key health aspects associated with an improved access to water and sanitation Safety risk of illegal connections Solid Waste is removed, thus contributing to quality of life	Decrease in environmental pollution Water consumed via the main dwelling can be accounted for and impact on resources better planned Positive impact on air quality as a result decreased use of fossil fuels Increased demand for electricity could contribute to cumulative impact at regional level Solid waste disposal is controlled and can be accounted for	Impact on capacity of bulk infrastructure. Increased need for cross-subsidization of services	Consider impacts of maintaining and improving services of existing BD in terms of budgeting for expansion of bulk infrastructure Consider impact of potential future BD in planning of new lower income areas Limited ability of BD households to contribute tax base to maintain and improve services

References

Bank, L. (2007). The rhythms of the yards: Urbanism, backyards and housing policy in South Africa. *Journal of Contemporary African Studies, 25*(2), 205–228.

Chapple, K., Wegmann, J., Nemirow, A., & Dentel-Post, C. (2012). *Yes, in my backyard: Mobilizing the market for secondary units*. The centre for community innovation.

City of Cape Town. (2006). The spatial distribution of Socio-Economic status, service levels and levels of living in the City of Cape Town 2001—to highlight suburbs in need. Retrieved from August 24, 2013, from www.capetown.gov.za.

City of Cape Town. (2012). Cape Town spatial development framework. Retrieved June 15, 2013, from www.capetown.gov.za.

City of Cape Town. (2013). Integrated development plan 2013/14 review. Retrieved June 15, 2013, from www.capetown.gov.za.

City of Cape Town. (2013b). City of Cape Town draft budget for the financial period 2013/14 to 2015/16. Retrieved October 15, 2013, from, www.capetown.gov.za.

City of Cape Town. (2013c). City of Cape Town credit control and debt collection 2013 / 2014 (Policy number 12416D) [Online]. Retrieved October 18, 2013, from, www.capetown.gov.za.

Clark, W., Deurloo, M., & Dieleman, F. (2006). Residential mobility and neighbourhood outcomes. *Housing Studies, 21*(3), 323–342.

Cox, K. R., Hemson, D., & Todes, A. (2004). Urbanization in South Africa and the changing character of migrant labour. *South African Geographical Journal, 86*(1), 7–16.

Council for Scientific and Industrial Research (CSIR). (2005) Guidelines for human settlement planning and design. Pretoria: CSIR.

Crankshaw, O., Gilbert, A. G., & Morris, A. (2000). Backyard Soweto. *International Journal of Urban and Regional Research, 24*, 841–857.

ESRI. (2005). Hot spot analysis (Getis-Ord Gi*) (Spatial Statistics). Retrieved October 10, 2013, from, http://edndoc.esri.com/arcobjects/9.2/net/shared/geoprocessing/spatial_statistics_tools/hot_spot_analysis_getis_ord_gi_star_spatial_statistics_.htm.

Gaunt, T., Salida, M., Macfarlane, R., Maboda, S., Reddy, Y., & Borchers, M. (2012). *Informal electrification in South Africa: Experience, Opportunities and challenges*. Westlake: Sustainable Energy Africa.

Geyer, H. S., & Geyer, H. S. Jr. (2013), Residential transformation in South Africa—Reopening the 'Dead' Capital Debate. *Urban Forum.*

Govender, T., Barnes, J. M., & Pieper, C. H. (2011a). The impact of densification by means of informal shacks in the backyards of low-cost houses on the environment and service delivery in Cape Town, South Africa. *Environmental Health Insights, 5*, 23–52.

Govender, T., Barnes, J. M., & Pieper, C. H. (2011b). Housing conditions, sanitation status and associated health risks in selected subsidized low-cost housing settlement in Cape Town, South Africa. *Habitat International, 35*, 335–342.

Greene, L. (1980). Recent trends in urban spatial structure. *Growth and Change, 11*(1), 29–40.

Huchzermeyer, M. (2001). Housing for the poor? Negotiated housing policy in South Africa. *Habitat International, 25*, 303–331.

Jabareen, Y. R. (2006). Sustainable urban forms: Their typologies, models and concepts. *Journal of Planning Education and Research, 26*, 38–52.

Landman, K., & Napier, M. (2010). Waiting for a house or building your own? Reconsidering state provision aided and unaided self-help in South Africa. *Habitat International, 34*, 299–305.

Lemanski, C. (2009). Augmented informality: South Africa's backyard dwellings as a by-product of formal housing policies. *Habitat International, 33*(4), 472–484.

Li, F., Liu, X., Hu, D., Wang, R., Yang, W., Li, D., & Zhao, D. (2009). Measurement indicators and an evaluation approach for assessing urban sustainable development: A case study for China's Jining City. *Landscape and Urban Planning, 90*, 134–142.

Marans, R. W. (2012). Quality of urban life studies: An overview and implications for environment-behaviour research. *Procedia- Social and Behavioural Sciences, 35*, 9–22.

Moussiopoulos, N., Archillas, C., Vlachokostas, C., Spyridi, D., & Nikolaou, K. (2010). Environmental, social and economic information management for the evaluation of sustainability in urban areas: A system of indicators for Thessaloniki, Greece. *Cities, 27*, 377–384.

Naude, W., Rossouw, S., & Krugell, W. (2009). The non-monetary quality of city life in South Africa. *Habitat International, 33*, 319–326.

Naidoo, N., Longondjo, C., Rawatlal, T., & Brueton, V. (2012). The provision of free basic water to backyard dwellers and/or more than one household per stand. Publication No 1987/1/11. Gezina: Water Research Commission.

Pacione, M. (2009). *Urban geography: A global perspective*. Routledge.

Pharaoh, R. (2008). Fire risk in informal settlements in Cape Town, South Africa. In M. Peilling & B. Wisner (Eds.) *Disaster risk reduction: Cases from Urban Africa* (pp. 105–125). London: Earthscan.

Poston, B. (2009). Maslow's hierarchy of needs. *The Surgical Technologist*, 347–353.

Potts, D. (2006). 'Restoring order'? Operation Murambatsvina and the urban crisis in Zimbabwe. *Journal of Southern African Studies, 32*(2), 273–291.

Rudin, D., & Falk, N. (1999). *Building the 21st century home: The sustainable urban neighbourhood*. Oxford: Butterworth: Heinemann.

Shapurjee, Y., & Charlton, S. (2013). Transforming South Africa's low-income housing projects through backyard dwellings: intersections with households and the state in Alexandra, Johannesburg. *Journal of Housing and the Built Environment.*

South African Institute of Race Relations (2009). *South African Institute of Race Relations 79th Annual Report: 1st January 2008 to 31st December 2008*. Johannesburg: South African Institute of Race Relations.

Statistics South Africa (StatsSA) (2001). Census South Africa 2001 Cape Town: StatsSA.

Statistics South Africa (StatsSA) (2011). Census South Africa 2011. Cape Town: StatsSA.

Swilling, M. (2005). *Rethinking the sustainability of the South African City*. Development Update.

Swilling, M. (2006) Sustainability and infrastructure planning in South Africa: A Cape Town case study. *Environment and Urbanization, 28*(23).

Todes, A., Kok, P., Wentzel, M., Van Zyle, J., & Cross, C. (2010). Contemporary South African urbanization dynamics. *Urban Forum, 21*, 331–348.

Van Dijk, M. P. (2006). *Managing cities in developing countries: The theory and practice of urban management*. Edward Elgar Publishing Limited.

Victor, T. (2009). The challenges of eradicating informal settlements in South Africa by 2014. In *The case of Seraleng sustainable human settlement, Rustenburg Local Municipality, North West Province*. Master's Thesis. Johannesburg: University of the Witwatersrand.

Moegamad Rafeeq Le Roux holds a master's degree in urban and regional science; honours and bachelor's degrees in geography and environmental studies with majors in urban geography and urban change & policy. He is also a professionally registered and practising natural scientist and environmental assessment practitioner. He has been employed for more than 15 years within the water sector contributing directly to land use management, developmental and various urban planning processes and decisions. He has a keen interest in urban form and function, management and development as well as urban change, having provided input into various strategy documents. He has co-published in the Water SA journal and assisted students in researched based thesis. He is a member of the Water Use Assessment & Advisory Committee and has various professional memberships such as being a corporate member of the South African Planning Institute.

Chapter 16
Housing the Urban Poor—An Epilogue

Joseph Kamuzhanje

Abstract It should be emphasised that delivering land and housing is not an end in itself, but a continuous process. This model is built on the notion of low-income housing as a process that 'migrates from informality to formality'. But even-though, the environment so created requires proper management—a responsibility which cannot be left to beneficiaries alone. Pertinent technical issues such as those of infrastructure, require skilled maintenance and management which ordinary beneficiaries cannot adequately attend to without government support. In this regard, the LA is better positioned to manage these new environments as per existing legislative and policy mandates.

Keywords Pro-poor · Artificial intelligence · Housing technology · Sustainable development · Inclusion

16.1 Introduction

This book was birthed out of a conversation that alleged that the current housing development interventions were not meant to benefit the poor. The argument therein was that there was an oddity whereby these interventions were driven by people who were not necessarily poor and who did not understand what it meant, first, to be poor, second, not to have a house to call their own and third, not to be involved in the very interventions that were meant to benefit them. As the process of developing the conceptual framework for the book evolved, it became apparent that in some countries, models had been tried to provide housing for the poor, sometimes, even with the involvement of the poor themselves.

The focus then shifted towards understanding what it was that was missing in the recipe for pro-poor housing. Going through the chapters in this book, it is very clear that indeed models exist, and they have been largely successful. However, it is also possible that the failure to address the housing challenge does stem from the fact that

J. Kamuzhanje (✉)
Cooper Zimbabwe (1992) Private Limited, Msasa, Harare, Zimbabwe
e-mail: josephkamuzhanje1@gmail.com

L. Chipungu et al. (eds.), *Contemporary Models for Pro-Poor Housing in Urban Sub-Saharan Africa*, https://doi.org/10.1007/978-3-032-05489-0_16

these models have not been inclusive in terms of the issues that need to be addressed. The models described in this book can easily end up as 'islands of success', if there is no pressure to look at the issues in a holistic and integrated manner. To illustrate this, it is not possible to have successful housing programmes where the cost of the land is prohibitive, where there is no conducive and legal and policy framework, and where interest and mortgage rates push borrowing beyond the reach of the people that are supposed to benefit. If there is agreement on this, it is therefore important to understand that the book cannot, and should not be read as separate chapters, but as one chapter which has different but closely interlinked sub-chapters.

The different authors and chapters have addressed the issue of pro-poor housing from a multi-faceted and multi-dimensional perspective. This is important because issues of pro-poor housing are not simplistic, and neither are they linear. It is important to note that there could actually be contradictions between the issues. For example, the same people that cannot afford housing are also the subject of sometimes punitive interest rates when it comes to access to finance. In Zimbabwe, for example, there has been a wide-ranging debate about the need to provide smaller stands, but there has been very little done in terms of servicing those stands and providing basic services.

Adamu et al. addresses an area of housing provision, which, although highly undesirable and unfortunate, is becoming part of the fabric of African communities. With increased civil strife, especially in countries such as the Democratic Republic of Congo, Sudan, the Horn of Africa, Chad and Nigeria, more and more people are being displaced. But there is also displacement caused by natural disasters such as flooding, droughts and climate change. Most African countries do not have well-developed systems to cope with internal displacement. Whilst they are already facing challenges providing housing and housing services in 'normal' times' the added strain and stress of civil strife and natural disasters only compounds the situation. The situations in Gaza and Ukraine, have also shown that providing adequate housing for the poor, may remain a mirage, if there is no handle on reducing the impact of internal displacements on the affected people. One of the challenges that internally displaced people face is that Governments and authorities do not have systems in place to quickly and adequately respond to internal displacements. They are also so constrained for resources that without outside financial assistance, they cannot address the challenges. According to Adamu et al. there is need for innovative ways of dealing with the problems of internal displacement. Sometimes, these do not necessarily need to stop at just providing shelter, but also incomes and livelihood options. The authors also contend that it is important to ensure that other social services, such as health and education, are provided wherever the internally displaced persons are settled. This will address issues on 'leaving no one behind' and ensuring that all forms of human rights are respected. There have also been instances where host communities do not take kindly to any assistance being given to displaced people as they think that they are actually taking resources that are meant for them.

Yankson focuses on a quite exciting approach to the issue of housing. His main argument is that all the issues to do with housing are interconnected. They are connected in such a way that dealing with one issue without addressing the others may actually result in more problems. The key challenge is that the issues are not necessarily all technical. Some of them are social, others, financial, others, environmental, and yet others, policy-driven and political. According to Yankson, the issues are not linear, and they manifest at different times and with different intensity along the spectrum. According to Yankson, informal settlements in Namibia, as would the case in other countries, are largely a result of the rapid pace of urbanisation. As the rural areas become more and more unattractive, people migrate to the urban areas in search of better opportunities. The problem with this is that in those urban areas, local authorities would already be struggling to meet the needs and demands of the people already. In the end, instead of going back to where they came from, and suffer the ignominy of having failed to make it in the big city, these people settle in informal settlements which do not have the necessary services. Even when services are provided, this is at a huge cost to the local authorities because the people cannot afford to pay for them.

According to Yankson, the housing challenges that are faced in most countries in the urban areas, are made worse by increasing poverty levels. With more and more people living below the poverty datum line, and barely making ends meet, securing accommodation is becoming more and more challenging. But the poverty is also a result of poorly performing macro-economic fundamentals. The economic challenges also affect the production of housing units.

There is an interesting phenomenon that is put forward by Yankson. Under normal circumstances, property values should decrease the further they are from the urban centre. This is because in most cases, the centre is more associated with commercial and ancillary uses. However, as the pressure on the centre increases, and there are issues of over-crowding and inadequate parking, most commercial services are finding themselves moving more and more away from the centre. The proliferation of shopping malls in Lusaka, Zambia and Durban, South Africa, further away from the centre is testimony to this. Whilst it reduces the pressure, it is also true that it inverts the issue of property values. With the decay in the centre, the periphery becomes more attractive, hence increasing the values of the properties. This brings to loggerheads the informality of the settlements which are supposed to house the poor people and the commercialisation of those areas. This challenge is further exacerbated by the fact that in most countries, land use regulations are generally outdated, and the processes to change them are tedious and would never move at the same pace as the developments themselves. In Zimbabwe, the President tried to address this by instructing all the local authorities to prepare master plans in accordance with the provisions of the Regional, Town and Country Planning Act (RTCPA), through the proclamation of the Presidential Call to Action. With the process now concluded, the implementation of the provisions has already seen some challenges with some rural local authorities objecting to plans by urban local authorities to annex and reserve land for future expansion. If this situation persists, then the problems such as those raised by Yankson will never be resolved.

Yankson concludes his chapter by advocating for an approach that works on understanding the different factors that affect housing and trying to match them in order to address the housing production and delivery challenges. It is important to note that this approach has to take into account the unique features and characteristics of each country, and in some instances, even within countries, the uniqueness of each local authority.

Nyadzani and Chipungu focus their chapter on social housing. Social development of any form seeks to lessen the blow of inadequacy of services. But it can also be understood from the point of view that in any country, there are those people who, by nature, will not be able to afford the services that are offered in their locality. Most social development programmes, especially in housing, are meant to address the inequalities of the past, and in the case of most African countries, the colonial past. Nyadzani and Chipungu focus on the social housing programmes implemented by the Government since the end of apartheid in 1994. The bedrock for social housing development in South Africa is the Social Housing Programme (DHS) which was viewed as a housing delivery mechanism capable of addressing the provision of low-cost rental stock, performing a catalytic role in stabilising areas experiencing urban decay and prompting urban regeneration and urban restructuring through economic, social and spatial integration.

What is important to note about social housing, or any social programme, is that it is not necessarily social especially on the deliverer. There is a cost to the programme, and this cost may end up being so prohibitive that it will not make sense to continue. Without the necessary checks and balances, it is also possible that a lot of people who should not benefit from these programmes end up doing so, at the expense of the deserving ones. Politics and corruption may end up rewarding the people who may already have properties on these programmes.

A more complex problem is actually that social development programmes can distort the market. This is an argument that has been put forth especially in social protection programmes, for example, food assistance programmes, where the food that is distributed comes in and distorts the food prices. In some instances, commercial providers of the food items have actually closed their businesses because they cannot compete. The situation is the same where the food prices are subsidised. This is the same situation with social housing programmes. Everything, from the cost of land to servicing, to even building materials can cause serious distortions on the market.

Social housing programmes are also mainly viewed from the communist, Marxist and socialist paradigms. Experiences from all over the world have shown that the houses, and the infrastructure around them are not well-taken care of because the people who stay in these housing programmes, do not have an economic attachment to them. In addition to this, the housing programmes are not necessarily viewed as productive assets, but the local authorities still have to ensure that they are properly serviced. Nyadzani and Chipungu, whilst agreeing to the important of social housing development programmes, proffer the view that ways be found to convert them into economic assets. That way, it is even possible to have a revolving fund that can be used to produce even more housing units.

Whilst Nyadzani and Chipungu focus on social housing, Chatiza tackles the issues of pro-poor housing from a social movement perspective. The basic argument by Chatiza is that unless there is proper representation of the poor in the housing debate, then very little progress will be made in the fight to address their issues. He gives the examples of social movements in Zimbabwe and Tanzania, that have done a lot of work in terms of pushing the housing agenda. The main challenge with this approach is that most of the movements have been supported by external donors. These donors fund programmes based on sectoral issues, and they also have time-specific projects. Whilst this is not a problem, it means that the housing agenda is only pushed depending on the availability of funds. Once these funds run out, it means that there will be a break in the fight. Nowhere is this highlighted than in the aftermath of the re-election of President Donal Trump in the United States. As soon as he was inaugurated, one of his first assignments was to sign executive orders, and the most impactful, at least, relative to the issues being discussed in this book being the order to suspend all foreign development and humanitarian assistance provided by the United States Agency for International Development (USAID).

The biggest challenge that the social movements have faced the world over, but especially in Africa and the developing world is that they have been labelled as fighting against the Government. Any policy and social pronouncements that they make are quickly (mis)interpreted as being in opposition to the Government. This has curtailed the progress and impact of their interventions.

So, there is a bigger question here about how urban communities could self-organise so that they can speak with one voice when they are pushing the housing agenda. It is important to understand that even in the most developed countries, lack of housing and poverty does not respect political boundaries and affiliations. There are a lot of examples that show that people who are affected by one problem can come together and be able to solve that problem, without any political or other undertones.

Perhaps another interesting strand of the chapter by Chatiza is the realisation that working on building the capacity of social movements to be better able to articulate the issues of housing production and delivery on its own does not necessarily help the situation too much, as it only addresses the demand side of the equation. More often than not, the local authorities (the supply side) that are supposed to work together with the social movements are not as prepared. This scenario can cause a lot of tension between the two institutions as one is not able to meet the legitimate demands of the other. From the work that has been done by Chatiza and others, there is an assumption that local authorities have the financial, organisational and technical capacity to respond to social movements. This is not always the case.

Ogunsanya brings into focus a conceptual, theoretical and technical approach to housing provision for the poor. The chapter is grounded in systems thinking. Systems thinking is something that is talked about a lot but is not easily implementable. This is because it cannot take place in a vacuum. But more importantly, it can also not take place in one sector of an organisation and not be present in the other sectors. As has already been argued by Yankson, housing issues are quite related to each other, but what Ogunsanya is arguing is that they should also be related to the rest of what goes on within an institution that is providing the interventions. Another argument by Ogunsanya is how technology can assist in the provision of housing. Technological advancements have made it possible to reduce the amount of time required to produce building materials, improve the quality of the building materials and ultimately reduce the amount of time required to produce a single housing unit. What has been a challenge so far is that the initial technology investments are quite high, and could be beyond the reach of most individuals, communities and local authorities.

A quite interesting angle to one reason why the adoption of technology has not been at a pace that could transform the housing industry is that in most cases, it comes with the reduction in the number of people employed. With the manual production of building materials, the number of people required is quite high. With technology, the number of people is potentially low, but the skill sets required are not only different but high. This goes against employment creation and provision of livelihoods to those that need this the most. As Ogunsanya argues, there is a tight rope to be manoeuvred here.

Both systems thinking and technology acceptance are supposed to lead to the production of 'smart' homes. In the context of the book, smart homes should contain most, if not all of the attributes such as energy management and efficiency, automation, occupant comfort and security, multi-functionality and adaptability and interactivity. What this narrow down to, is that a house is not only a place where a family has got shelter, but is increasingly also becoming an economic space. This became very apparent during the COVID-19 pandemic, when people had to work from home. But what is also important and shall be discussed further in this chapter is the fact that it is almost impossible to have smart homes without talking about smart cities.

But there is also a broader conversation to have in terms of whether a smart home, as envisaged by Ogunsanya can also be an affordable home. In most instances, this has not been the case. Therefore, it may be important to go back to the almost forgotten concept of appropriate technology.

Whereas Nyadzani and Chipungu focus on social housing, Darazo and Kawu dedicate their chapter to an almost similar issue, that of public housing. The main difference between the two is that in the case of the former, the housing is meant for people who cannot afford to provide their own and may not even be able to pay rentals and other service charges, whilst in the case of the latter, the beneficiaries will be able to at least pay for the rent. In most cases, public housing was a result of the need by the colonial masters to provide cheap accommodation for labour in urban areas and was never meant to be a permanent residence for the occupants. The idea was that once the beneficiary had enough money to build their own house, they would vacate the public house and make it available for the next person.

The biggest challenge with public housing, just as much as social housing, is that its availability is completely dependent on the ability and interest of the government to put up such kind of housing. With many governments in Africa struggling to provide even the most basic of services, public housing may not necessarily be a priority. In any case, with the liberalisation of most economies in the Global South at the behest of the World Bank and the International Monetary Fund, any form of subsidies is completely discouraged. The two institutions prefer a situation where issues such as housing, and other ancillary services such as education and health are left purely to market forces.

As Karazo and Aliwu also contend, whilst some governments are able to provide public housing, in most cases, there is no financial provision for the maintenance and upkeep of the buildings. This is the one reason why most of the public houses, together with the attendant infrastructure are almost always in a very bad state of disrepair. This situation is akin to the challenges faced by the provision of housing at mines. Once the minerals can no longer be mined, or the prices have gone down to the point where mining becomes uneconomic, and the staff is laid off, the houses are abandoned. It is for this reason that in Zimbabwe for example, one of the biggest mining companies in the country, Zimplats, decided to construct houses for its employees in the nearby towns of Norton and Chegutu. That way, the houses still retain some value even if the mine were to shut down.

Perhaps one of the most important conclusions made by the authors is that public housing should be provided as part of a public–private partnership (PPP). This approach reduces the financial and institutional burden on the government and allows a more market-based provision of accommodation. This may also mean that more deserving persons benefit from the initiatives as opposed to the issues discussed by Nyadzani and Chipungu where politics and corruption become entrenched in the programmes.

Le Roux write about the preponderance of backyard structures as a way of alleviating the housing problem. Backyard structures are an epitome of a housing crisis. Embedded in the crisis is also sometimes the need for more income by the main house owner. There are two main issues to contend with when it comes to backyard structures. The first is that in most cases, the structures actually overload the stand. These structures are quite popular in high density areas where the stand sizes are quite small. What then happens is that the land that is reserved for water, sewer and power servitudes is then used to erect the structures. This provides a lot of challenges in terms of routine maintenance of these services. Second, and in addition to this, the structures lead to overcrowding. When layout plans are designed, stand sizes correspond to a certain number of people staying in each house. The backyard structure obviously brings in more people than what the stand is originally planned for. This scenario brings with it other challenges such as the spread of diseases due to many people sharing a very small space. There are also chances that in the unfortunate case that a fire breaks out, it could be very difficult to get the fire tenders on site. Moreover, most of these structures are built without the approval, authorisation and supervision of the local authorities, and the quality of the structures cannot be guaranteed.

There is also an interesting phenomenon about the backyard structures. This is that the structures are not only necessarily used for accommodation. They could also be used for informal livelihood activities such as rearing of chickens for sale. This normally compounds the health problems that have been discussed above, because of the effects on the neighbours. Lehobo, Denoon-Stevens and Bhanye focus on the concept of mixed-use development. This can be looked at from two viewpoints. The first is that in terms of the zoning regulations, a house would not only be viewed as a provider or accommodation but could also be an economic, institutional and even recreational building. This mixed-use approach means that there is flexibility even in the design of the house to take this into account. The Government of Zimbabwe Statutory Instrument 216 of 1994 was an attempt to allow some hitherto unallowed uses within residential areas and stands to be permitted. This opened up avenues for the informal economy to thrive as small shops, commonly referred to as 'tuckshops' sprouted up at every corner. The biggest advantage of this was that most services became even more locally available than was the situation in the past, as people no longer has to move long distances to get to the shops. However, there was also an increase in health concerns, as there was not corresponding attention paid especially on foodstuffs by the health authorities.

The other way of looking at mixed-use development is when the concept of zoning becomes more flexible. In most countries, there are three housing zones, high, medium and low density, with very clear distinction between them mainly through the stand sizes. This has been seen as a colonial attempt to segregate communities and mainly according to race. In the suburb of Ivene in Gweru, Zimbabwe, the local authority trialled a mixed-use residential area where the three zones were allowed. Whilst it was not too popular, it provided an insight into what could be possible with more innovative thinking.

The key challenge with this approach is how it disrupts the property values. With the three distinct zones, it is very easy to maintain property values. But when there is a high-density house next to a low density one, the situation becomes difficult. It also requires a new way of preparing layout plans because the traditional way is that of ensuring that stand sizes are largely the same. This approach calls for a different perspective to planning, something that most local authorities are not amenable to.

Whilst in a previous chapter, Chipungu and Magidimisha-Chipungu focus on financing housing provision in contested lands, they move on to talk about another increasingly contentious area, that of land tenure. The authors' experiences borrow heavily from the land reform programme in Zimbabwe in 2000. What that programme did was to effectively do away with title deeds and declared all land as state land. Whilst this served the purpose of the acquisition of the land (without compensation, as was thought at that time), it also upsets the property market. Title deeds were the most acceptable type of collateral with the banks and the financial institutions. With the title deeds, the banks were assured that in the event of any default, they could still reclaim their money from the sale of the asset. In place of the title deeds, the Government put in place 99-year leases which were supposed to be redeemable with the banking system. However, these lease agreements did not offer the same type of security as the title deeds and the financial institutions were not too keen to accept them. Even the so-called pro-government banks were very hesitant to accept them.

As the authors rightly put it, the cancellation of the title deeds meant that whoever was on the land, could be deemed a 'squatter' or an illegal occupant and could be removed by force or otherwise. These people would have no recourse to the courts. The only upturn of this whole situation was that the Government was able to allocate land to those in need much quicker than they would have done if they had followed due process. However, as intimated earlier, without the security of title deeds, those people who built their houses on such land could never use it as collateral and were also open to be evicted in the same way that the previous owners had been.

Whilst the majority of the book contains chapters which focus on the policy, political, technical and financial aspects of housing provision, the work by Adamu is very specific to housing and its implications on gender, specifically women. This is based on the understanding that women, who bear the brunt of poverty and other inequalities, are perhaps the most deserving in terms of their access to social services especially housing. Whilst most African countries are patriarchal in nature, the responsibility to look after the family rests with women. However, in the event that the husband dies, and even if the house is jointly owned by both, there has been an emerging trend where the husband's relatives actually take custody of the house. In many countries now, legal provisions have been put in place so that the estate of the deceased is taken over by the surviving spouse.

But the problems are not only related to the actual structure. They are found at every step of the housing delivery process, from the acquisition of the land to the granting of title, to accessing housing finance, and the construction process itself. Most of the work done by gender advocates in housing provision has centred on creating a conducive atmosphere for women to have the same rights as men when it comes to access to housing. However, there are many lessons to be derived from the chapter by Adamu in terms of making housing available to other marginalised groups of society such as those living with disability. This will go a long way in making the mantra of 'housing for all' to be realised.

But as in any book project, the more questions that are answered, the newer and more important and difficult questions arise. This is the same with this book. The book was written in the aftermath of the devastating COVID-19 pandemic. How much has been learnt from the many and various experiences that each country, each community and each individual went through, and how can these lessons benefit housing provision? In the same manner, how prepared is the world for another pandemic of a similar, or worse nature.

It is not possible, nor should it be attempted, that one book covers all the issues affecting any thematic area, let alone, housing provision. This book was written within a certain context, respecting certain assumptions. However, there are more issues that have emerged and gained prominence since the project started.

Artificial intelligence is taking over the world. The world does not have a choice to adopt it or not, but only to understand how to be a factor within it. It begs the question of how housing provision will look like going forward. It is the same with climate change which has ravaged the world in many different ways, the flooding in Europe, the wildfires in America, the droughts in Africa, the reemergence of superpowerdom, with the battles being fought far away from the countries that initiate the chaos, a world economic order that is teetering on recession every few years, increasing levels of poverty and urbanisation. All these issues have to be taken into account as specialists and experts continue to grapple with the real issues of housing.

There is a growing discourse on smart cities. Many countries have tried to progress this concept as part of redefining urban planning and design. In the context of housing the term "smart" has a number of components, with the most important being that the urban space should be scientific in terms of design, it should be able to use and adapt technology, it must be market driven and should provide access to finance for all its inhabitants. Further, and as Yankson and others have argued, the development of housing should be systemic and should focus on integration of the various uses that make up the living space. In this way, it will be possible to have housing provision that thrives on order, convenience, safety and aesthetics.

This book was never meant to provide all the answers, but to start a conversation that is wide and broad enough to include all the stakeholders. That means that there is even more work ahead. The fact that there are successful models as espoused in the different chapters in this book, should be the fuel that continues to drive all of us towards the elusive goal of housing for all, especially the poor.

Joseph Kamuzhanje has over 30 years of experience in the development sector. He worked for 15 years in the Department of Physical Planning in the Ministry of Local Government, where he rose to be the Chief Planning Officer for Matebeleland South Province between 2002 and 2007. He holds a BSc in Rural and Urban Planning, and MSc in Regional Development Planning and Management and he is a Professor of Practice in Regional and Urban Planning. He has worked in Government, civil society (both humanitarian and development), and in the private sector. He has contributed to several books and papers on rural, urban, and regional development. He is currently working with the University of Venda (South Africa), University of KwaZulu Natal (South Africa) and the Federal University of Technology, Minna (Nigeria), on various publications for books and journals.

MIX
Papier aus verantwortungsvollen Quellen
Paper from responsible sources
FSC® C105338

If you have any concerns about our products,
you can contact us on
ProductSafety@springernature.com

In case Publisher is established outside the EU,
the EU authorized representative is:
Springer Nature Customer Service Center GmbH
Europaplatz 3, 69115 Heidelberg, Germany

Printed by Libri Plureos GmbH
in Hamburg, Germany